Theory and Practice of the Triple Helix System in Developing Countries

RIOT! Routledge Studies in Innovation, Organization and Technology

1. Innovation in the U.S. Service Sector
Michael P. Gallaher, Albert N. Link and Jeffrey E. Petrusa

2. Information and Communications Technologies in Society
E-Living in a Digital Europe
Edited by Ben Anderson, Malcolm Brynin and Yoel Raban

3. The Innovative Bureaucracy
Bureaucracy in an Age of Fluidity
Alexander Styhre

4. Innovations and Institutions
An Institutional Perspective on the Innovative Efforts of Banks and Insurance Companies
Patrick Vermeulen and Jorg Raab

5. Knowledge and Innovation in Business and Industry
The Importance of Using Others
Edited by Håkan Håkansson & Alexandra Waluszewski

6. Knowledge and Innovation
A Comparative Study of the USA, the UK and Japan
Helen Brown

7. Industrial Innovation in Japan
Edited by Takuji Hara, Norio Kambayashi and Noboru Matsushima

8. Managing and Marketing Radical Innovations
Marketing New Technology
Birgitta Sandberg

9. Mobility and Technology in the Workplace
Edited by Donald Hislop

10. Energizing Management Through Innovation and Entrepreneurship
European Research and Practice
Edited by Milé Terziovski

11. Innovating for Sustainability
Green Entrepreneurship in Personal Mobility
Luca Berchicci

12. Organizational Capital
Modelling, Measuring and Contextualising
Edited by Ahmed Bounfour

13. User-Innovation
Barriers to Democratization and IP Licensing
Victor R. G. Braun and Cornelius Herstatt

14. Working on Innovation
Edited by Christophe Midler, Guy Minguet and Monique Vervaeke

15. Organization in Open Source Communities
At the Crossroads of the Gift and Market Economies
Evangelia Berdou

16. Theory and Practice of the Triple Helix System in Developing Countries
Issues and Challenges
Edited by Mohammed Saad and Girma Zawdie

Theory and Practice of the Triple Helix System in Developing Countries

Issues and Challenges

Edited by
Mohammed Saad
and Girma Zawdie

Routledge
Taylor & Francis Group
New York London

First published 2011
by Routledge
270 Madison Avenue, New York, NY 10016

Simultaneously published in the UK
by Routledge
2 Park Square, Milton Park, Abingdon, Oxon OX14 4RN

Routledge is an imprint of the Taylor & Francis Group, an informa business

Typeset in Sabon by IBT Global.
Printed and bound in the United States of America on acid-free paper by IBT Global.

Library of Congress Cataloging-in-Publication Data

Theory and practice of the triple helix system in developing countries : issues and challenges / edited by Mohammed Saad and Girma Zawdie.
p. cm. — (Routledge studies in innovation, organization and technology ; 16)
Includes bibliographical references and index.
1. Technology transfer—Developing countries. 2. Technological innovations—Developing countries. 3. Academic-industrial collaboration—Developing countries. 4. Public-private sector cooperation—Developing countries. 5. Research and development partnership—Developing countries. I. Saad, Mohammed, 1950– II. Zawdie, Girma, 1948–
HC59.72.T4T54 2011
338'.064091724—dc22
2010021426

ISBN13: 978-0-415-47516-7 (hbk)
ISBN13: 978-0-203-83821-1 (ebk)

Contents

List of Figures ix
List of Tables and Boxes xi
Acknowledgements xv

Introduction 1
MOHAMMED SAAD AND GIRMA ZAWDIE

PART I
Knowledge Production, Sharing and Use through Triple Helix

1 **Triple Helix Knowledge Clusters: Accelerating Innovation and Creating Transformative Networks** 9
LEIGH W. JEROME

2 **Innovation Systems Interactions and Technology Transfer and Assimilation for Industrial Development: The Cases of South Korea and Mexico** 25
MARÍA ISABEL RIVERA VARGAS

3 **A New Regulatory Framework for the Relationship between University and Industry: The Brazilian Experience** 52
ANNE-MARIE MACULAN AND JOSÉ MANOEL CARVALHO DE MELLO

4 **Inter-Organizational Learning through South–South Cooperation: A Case Study Investigation** 67
NAJEH BOURAOUI, AMIRA BOUHAMED, JAMIL CHAABOUNI AND MOHAMMED SAAD

5 **Guarding the Gatekeeper: University Research Managers in a Triple Helix Environment** 87
JOHN KIRKLAND AND JULIE STACKHOUSE

6 Revisiting Agricultural Science for Development: From *Mode 1* to *Mode 2* 98
NORMAN CLARK

PART II
Operationalising the Triple Helix System and Bridging the Gaps

7 The Triple Helix Paradigm in Korea and Taiwan: A Test for New Forms of Capital 119
MATTHEW A. SHAPIRO

8 Mobilizing for Development: Putting the Triple Helix into Action in Ghana 146
JAMES DZISAH

9 Bridging the Gaps in the Triple Helix: A Case Study Based on the Challenge of the Indonesian Experience 161
DESSY IRAWATI

10 The Incubator Movement: A Comparative Analysis of Brazil and Mexico 176
MARIZA ALMEIDA, BRANCA TERRA AND MARIA DEL PILAR MONTSERRAT PEREZ HERNANDEZ

11 The Challenges Arising in the Evolution of the Triple Helix Institutional System: The Case of Malaysia 191
AZLEY ABD RAZAK AND MOHAMMED SAAD

12 Innovation in a Developing Country Context: Developing an Integrated Framework through India's Experience 207
SURJA DATTA AND MOHAMMED SAAD

13 Science Parks and University–Industry Collaboration in Malaysia: The Malaysian Experience 230
CHANDRA MALAIRAJA AND GIRMA ZAWDIE

PART III
Shift towards a Triple Helix Policy Paradigm and Implications for Developing Countries

14 Creative Reconstruction: A Triple Helix–Based Innovation Strategy in Central and Eastern Europe Countries 249
MARINA RANGA AND HENRY ETZKOWITZ

15 Not Yet Triple Helix III? Japanese MOT Policies and the Problem of Technology Exploitation 283
INGYU OH

Contributors 305
Index 311

Figures

3.1	Number of patents requested 1990–2004.	60
4.1	South–South cooperative agreements concluded by ALPHA.	76
4.2	Evolution of the financial turnover of STEFM.	78
4.3	Interdependence between three types of learning.	83
5.1	Presentation, negotiation and numeracy skills.	92
7.1	Tracing the effects of personal ties and repartnering.	126
7.2	Average number of patents through cross-sector R&D collaboration: By subsector: Korea.	127
7.3	Average number of patents through cross-sector R&D collaboration: By subsector: Taiwan.	128
7.4	Average number of total patents excluding cross-sector R&D collaboration: By subsector, Korea.	129
7.5	Average number of total patents excluding cross-sector R&D collaboration: By subsector, Taiwan.	129
10.1	Growth of incubators in Brazil.	184
10.2	Growth of incubators in Mexico.	184
11.1	Evolution of the triple helix system.	194
11.2	Ownership structure of innovating firms (1997–1999).	196
11.3	Ownership structure of innovating firms (2000–2001).	196
11.4	The process of data analysis.	199
14.1	Triple helix configurations.	253

14.2 Innovation performance indicators in CEE countries (% relative to EU25), 2005. 257

14.3 Percentage of researchers in the higher education sector by fields of science, 2002. 259

14.4 Enterprises with innovation activity in CEE countries, 2002–2004. 263

15.1 Evolution of the triple helix. 291

15.2 Triple Helix II toward venture creation. 297

Tables and Boxes

2.1	Research and Development Investment at Hyundai	34
4.1	Repartition of Operating Employees	77
7.1	Collaborative Tendencies: Aggregate Level, Korea	123
7.2	Collaborative Tendencies: Aggregate Level, Taiwan	124
7.3	Collaborative Tendencies: Subgroup Level, Korea	124
7.4	Collaborative Tendencies: Subgroup Level, Taiwan	125
7.5	Average Number of Patents through Cross-Sector R&D Collaboration: By Subsector: Korea	127
7.6	Average Number of Patents through Cross-Sector R&D Collaboration: By Subsector: Taiwan	128
7.7	Average Number of Total Patents Excluding Cross-Sector R&D Collaboration: By Subsector, Korea	128
7.8	Average Number of Total Patents Excluding Cross-Sector R&D Collaboration: By Subsector, Taiwan	129
7.9	OLS Results for New and Pre-Existing Capital's Effects upon Collaborative Output	131
7.10	OLS Results for New and Pre-Existing Capital's Effects upon Non-Collaborative Output	132
7.11	OLS Results for New and Pre-Existing Capital's Effects upon Collaborative and Non-Collaborative Output, by Sector Subgroups	133
7.12	OLS Results for New and Pre-Existing Capital's Effects upon Collaborative and Non-Collaborative Output, by Sector Subgroups	133

7.13	Rankings of Weighted Reasons for Repartnering: Korea	134
7.14	Rankings of Weighted Reasons for Repartnering: Taiwan	135
7.15	Rankings of Weighted Source of Personal Ties: Korea	136
7.16	Rankings of Weighted Source of Personal Ties: Taiwan	136
8.1	Snapshots of Various Development Plans in Ghana	153
11.1	Current Perceptions of Malaysian Universities	197
12.1	Examples of Relationships between Innovations and Types of Capital	212
13.1	SME Features versus the Knowledge Economy	232
13.2	Science Parks versus Industrial Parks	232
13.3	Science Park as a Resource Network	233
13.4	Tertiary Educational Institutions in Malaysia (2000 and 2005)	236
13.5	Science Parks and Technology Incubators in Malaysia	237
13.6	Number and Types of University Links between Science Park and Off-Park Firms	240
13.7	Types of University Links between Science Park and Off-Park Firms (Means and Standard Deviations)	240
14.1	Expansion of University Mission	254
14.2	Graduation from Tertiary Education and Annual Average Growth Rates (AAGR) in Total and Selected Fields of Study	257
14.3	FDI Intensity in CEE Countries, 1997–2004	261
14.4	Innovation and Entrepreneurship Indicators for CEE Countries (% of EU25 Average)	264
14.5	Innovation Activity and Cooperation during 2002–2004	267
14.6	Policy Measures Supporting Science–Industry Links in CEE Countries	268
15.1	U.S. and Japanese Triple Helix History	286

15.2 Performance of Triple Helix Systems in Japan, the U.S. and the UK 288

15.3 IOCR and Traditional Research Collaboration 293

15.4 Situations of University Venture Firms (March 2007) 295

15.5 Top Ten Universities with Most Venture Start-Ups (March 2007) 296

APPENDIX TABLES

7.1 Variable Notation 139

7.2 Robustness Checks 140

7.3 Robustness Checks (Industry Controls Included) 140

7.4 Robustness Checks (Industry, Respondent Controls Included) 141

7.5 Robustness Checks 141

7.6 Robustness Checks (Industry Controls Included) 142

7.7 Robustness Checks (Industry, Respondent Controls Included) 142

BOXES

14.1 Country Experiences in Cluster Development 271

Acknowledgements

It has been a privilege for us to work on this book with a number of inspiring colleagues from many different parts of the world. Special thanks are due to all contributors and reviewers who provided valuable feedback on our work. Our deepest thanks go to our respective dedicated families.

We are indebted to the following institutions for their support of the conference on 'The Triple Helix Paradigm for Development: Strategies for Cooperation and Exchange of Good Practice' which took place in September 2006 at the University of the West of England (UK). Their sponsoring and trust helped us to bring together research students, researchers and policymakers to the conference. These participants came from more than twenty different countries and essentially from the developing world.

- Bristol Business School (University of the West of England, UK)
- The Association of Commonwealth Universities and the Economic and Social Research Council (ESRC)
- The Cambridge-MIT Institute
- The Department of Technology and Society in the College of Engineering and Applied Sciences—Stony Brook (New York)
- *International Journal of Technology Management and Sustainable Development*
- Intellect Publisher
- Routledge Publisher

Our personal thanks go to Warwick Jones (Acting Dean), Judith Jordan (Associate Dean) and Rebecca England from the Business School for their continuous support throughout the different stages of this conference, from which this book was developed.

Introduction

Mohammed Saad and Girma Zawdie

This book brings together a collection of well-studied papers largely drawn from experiences in Africa, Latin America, Asia, Europe and North America, with the aim to explore from various perspectives the theoretical and practical aspects that underpin triple helix as a strategy for innovation and sustainable development. It also represents the first attempt to date to explore options for and difficulties in operationalising the triple helix system in the context of developing countries. The genesis of the book can be traced to a Bristol conference in September 2006 that sought to address the relevance of triple helix as a strategy for sustainable development in developing countries.

Triple helix is about the dynamics arising from the interactions between the three principal institutional spheres in an economy—namely, the university, industry and government—and the leading role the university assumes in this relationship to ensure that the knowledge it produces is useful enough to be widely shared and applied, ultimately translating into regional and, indeed, national development. As such, the triple helix relationship is more complex in character and more mission-oriented in purpose than what may be expected to be the case when the differentiated institutional spheres engage in relationships of 'arm's-length' type. In the triple helix system, the boundary between institutional spheres is necessarily blurred; universities assume entrepreneurial roles, shedding the traditional ivory tower image and focusing instead on the production of useful knowledge and its exchange across a wider network of key players in the socio-economic system; industry would be keen not only in knowledge application, but also in knowledge production and knowledge sharing as the way forward in the course of wealth creation. The role of governments would involve the task of facilitating university–industry interactions by setting out appropriate policy frameworks for defining, among other things, research and development priorities in the light of prevailing socio-economic circumstances and for allocating resources across these priorities.

In developing countries, where institutions are characterised by widespread fragmentation and rigidities, there is little or no scope for interlacing between institutions in terms of domains of responsibility; consequently, there is a low volume of interaction between the institutional players. Hence the low stock of social capital and the limited scope for innovation in these countries. Although the triple helix players are there, they are differentiated, and the system is essentially dysfunctional. Seen in this light, it can be argued that the triple helix culture is conspicuous for its absence in its proper form in developing countries. This is not, however, an argument against the relevance of the triple helix system to developing countries, but rather a challenge as to how the triple helix system can be operationalised in these countries to provide a sustainable basis for wealth creation and poverty reduction. A way forward in this respect would be for policy in developing countries to provide an overarching framework within which a knowledge-oriented triple helix culture would evolve in terms of network development and the emergence of active players as drivers of the system.

Questions may be asked as to whether developing countries in general are ripe at all in social, economic and technological terms for such a policy to take root. There is no doubt that making the triple helix system work in developing countries is a daunting policy challenge. It can be argued, however, that some countries that have made headway in terms of industrial development, like those in Latin America, the Indian subcontinent and South-East Asia, are better positioned for the implementation of the triple helix strategy than others, like those in sub-Saharan Africa, that are on the low end of the development rung. For all that, there is growing awareness in almost all developing countries about the need for a policy shift from the traditional North–South technology transfer practice to a policy position that is capable of providing a sustainable basis for technological progress and sustainable development.

As a policy handmaiden, this book explores the triple helix model of innovation and development in the light of experience that sheds light on the dynamics of the interactive relationships between government, industry and universities aimed at the creation, dissemination and sharing of knowledge in developing countries. A major point of inquiry in this regard is the extent to which the 'triple helix' system of relationships between university, industry and government would be expected to enhance the effectiveness of universities in developing countries as agents of innovation, industrialization and sustainable development. This is very much in line with the Schumpeterian thinking in which the distinction between the innovator and the entrepreneur is blurred.

The contents of this book would, needless to say, be of particular appeal to policymakers and researchers in developing countries. The multidisciplinary experiences and perspectives of the authors, including academics, policymakers and practitioners from fifteen different countries, make the focus of the book on triple helix in developing countries not merely an exercise in theory, but more significantly of practical policy import, calling as it does for a new and creative approach to development through the promotion of innovation, learning societies and exchange of best practice.

The book is divided into three main parts, each one addressing a key objective of the book. The first part, organized in six chapters, places a significant emphasis upon the role of universities in knowledge production, knowledge sharing and knowledge use, which are often viewed conceptually to be crucial for the successful implementation of the triple helix model. This part probes the implications of the 'triple helix paradigm' for the organization and management of the university system, and also for the modalities of technology transfer practices in developing countries. Cases from India, South Korea, Mexico, Brazil and Tunisia are featured to make the point that the triple helix system has taken root at least in some developing countries where policy appears to have been effective in mitigating the constraints on network and 'relational capital' development. Such experiences, however scattered, appear to suggest that a major challenge for policy—and indeed for the success of triple helix—in developing countries in general would be the provision of institutions and organizations that can help bring forth a thriving culture of trust, transparency, collaboration and effective governance. The development of such a culture would facilitate the emergence of strong synergies between academia, industry and government.

The second part, organized in seven chapters, focuses on the issues and challenges associated with the operationalisation of the triple helix system in developing countries. It investigates the nature of the relationships between the three institutional spheres of the triple helix structure—university-industry-government. In this part, too, triple helix initiatives are discussed, drawing on the experiences of a wide spectrum of developing countries ranging from Ghana at the lower end of the spectrum to Mexico, Brazil, India, Indonesia and Malaysia on the middle range, and South Korea and Taiwan at the upper end of the spectrum. In most developing countries, the institutional and organizational elements that contribute to the triple helix relationship exist, albeit in 'differentiated' and not in 'hybridized' and integrated form. This means that there is little or no scope for interaction between the three helices based on partnership and cooperation.

However, as the chapters in this part show, not all developing countries are in the same boat with respect to the development of the triple helix culture, although it can be maintained that in all cases, gaps have to be identified and bridged for the triple helix model to be increasingly effective. The Korean and Taiwanese cases show relative success in triple helix development, which is reflected in industrial and export success. In Malaysia, Indonesia, India and Brazil, there is strong government effort to provide institutional and organizational capacities that would make triple helix effective as a model for innovation and sustainable development. In

Ghana, on the other hand, triple helix is a different ball game, as it is in other sub-Saharan countries. In these countries, the lack of interaction among development actors explains the failure of development despite the massive government investment and huge donor funding efforts. Indeed, the triple helix system is dysfunctional here. Government research institutes are often far removed from research activities of universities; universities are hardly seen as a critical actor in the development process; and industry would rather source technology and consultancy services from foreign rather than local sources. The way forward for policy in these countries would be to embark on strategies for removing blockages that inhibit the integration and hybridization of functions across the institutional spheres that constitute the triple helix system. This would involve, *inter alia*, policy aimed at promoting the emergence of an open civil society that would pave the way for triple helix actors to organize and mitigate the factors that militate against the transformation of knowledge into innovation.

The third part, set in two chapters, is devoted to the discussion of best practices and lessons to be learned from developed countries that can help developing countries gain a better understanding of both the theoretical underpinning and operationalisation of the triple helix system of innovation.

In this part, two cases of best practice experience are highlighted. The first case is from the countries of Eastern Europe (CEE), where significant progress can be expected in the medium- to long-run period because the basic elements of a triple helix regime that are already active in these countries have been strengthened in recent years through various policy measures promoted by national governments and the European Union. A key factor in the advancement of a triple helix innovation strategy is related to an increased policy support for the emergence of entrepreneurial universities, which can foster regional development, innovative firm formation, clusters of high-tech firms from competitive research fields and the infusion of advanced technology into older firms.

The second case from Japan highlights the success and failure of current Japanese local and regional triple helix experiments. This case provides lessons in triple helix development that can be learned by developing countries.

The fifteen independent chapters, authored by different individuals with different backgrounds, aim to bring out the case for triple helix in developing countries. The underlying assumption in all cases is that where traditional development strategies, underpinned by North–South technology transfer, have failed to bear fruit in developing countries, there may be hope in policy initiatives that seek to promote capacity building to provide a sustainable basis for the development of a knowledge-based triple helix culture. The book does not by any means purport to have the last word on the role of triple helix in developing countries. Indeed, whereas the model may appear attractive, there is a sting in the tail in its operationalisation;

and it is mainly in drawing attention to the knowledge and institutional gaps that militate against the effective application of the triple helix system in developing countries in general that the book can duly claim significance for readership among academics, policymakers, practitioners and research students.

Part I

Knowledge Production, Sharing and Use through Triple Helix

1 Triple Helix Knowledge Clusters

Accelerating Innovation and Creating Transformative Networks

Leigh W. Jerome

INTRODUCTION

In these early years of the twenty-first century we are a global community plagued by intractable problems including disease, terrorism, poverty, illiteracy and other grand challenges that demand truly innovative and collaborative solutions (Seline 2008). Many urgent domestic and global social problems in education, community safety, the environment, job creation, affordable housing and healthcare call for flexibility and the savvy blending of expertise that will require us to cross traditional organizational and sector boundaries (Snyder and Briggs 2003).

The adoption of the Millennium Development Goals at the United Nations Millennium Summit in 2000 provided a global standard of reference for measuring and tracking improvements in the human condition, especially important in developing countries. Meeting the goals will require a substantial reorientation of policies, integration of new and established scientific and technological knowledge and cross-sector collaboration (Calestous and Yee-Cheong 2005). Our conventional public and private bureaucracies have been designed to solve circumscribed problems for established vertical constituencies through centrally managed programs and policies. These structures are not sufficient to address the interrelated problems we face today (Snyder and Briggs 2003). The complexity of today's challenges demands the blending of specialized knowledge from discipline and sector silos in order to create holistic and coordinated solutions. Public–private cooperation, rooted in and guided by science, is essential to the achievement of our social objectives (Wessner 2002). The potential offered by triple helix collaboration is significant. New patterns of cross-cutting partnering and commercialization have arisen in industry consortia, university–industry arrangements and hybrid public–private organizations that reflect prototypes for blending resources. Mounting evidence attests to the fact that controlled collaboration of government, academia and industry facilitates innovation and creative development while providing a good balance between highly specialized knowledge, social mission and profit motivations (Asheim and Coenan 2004; Leydesdorff and Fritsch 2005). Triple helix arrangements are the key to improving the conditions for innovation in our knowledge-based society (Etzkowitz 2002). Cross-cutting partnerships fortify organizational capabilities and harness complementary expertise so that new synergies between academia, industry and government can be realized.

The engineering of triple helix partnerships has coincided with the emergence of technological and scientific discovery that together promote a tighter coupling of information flows across customary boundaries (Dawes and Eglene 2004). Trilateral partners are able to leverage intellectual and material resources to create commercial products that take local, national and global missions into account. Moreover, triple helix arrangements bring the benefits of new products, new processes and new knowledge to market that are both profitable and tied to societal need and therefore more able to contribute to economic growth and human welfare.

Innovation is the process of transforming new knowledge and technologies into profitable products and services for national and global markets. Economic and social improvements result from the application of knowledge for productive activities (Calestous and Yee-Cheong 2005). Thus, improving the knowledge base of developed and developing countries and making use of cross-sector arrangements to apply that knowledge will impact economic growth. Thus, innovation contributes to greater intellectual capital, market creation, economic growth, job creation, wealth and a higher standard of living (Milbergs 2004).

We live in an era of global opportunity. The emergence of distributed collaborative networks is generating new opportunities for cross-sector partnering and the exchange of scientific and technological knowledge. This chapter will describe triple helix partnerships via collaborative networks that are transformative as a means for global knowledge exchange and accelerating innovation. Special emphasis will be put on the potential of distributed triple helix networks for enhancing nation development.

THE CHANGING INNOVATION LANDSCAPE

Successful innovation yields new products and services, gives rise to new markets and generates enterprise growth, thereby contributing to higher productivity, lower costs, increased profits and employment. Research has well established the importance and impact of innovation. Innovation is the main economic driver that allows organizations and nations to prosper, grow and sustain profitability (Drucker 1988; Thomke 2001). A competitive knowledge-based economy is powered by continuous technological innovation (Tyrrel 2007). The speed of technological change and market evolution makes the challenge to innovate urgent and continuous. Innovative firms generate higher global market share, higher growth rates, higher profitability and higher market valuations. Innovation also generates spillovers with cascading effects (Milbergs 2004). A nation's ability to solve problems and initiate and sustain economic growth depends on its capabilities in science, technology and innovation. Scientific and technical capabilities determine the ability of communities to provide clean water, good healthcare, adequate infrastructure and safe food (Calestous and Yee-Cheong 2005).

Whereas the value of innovation continues to expand, the traditional process of accomplishing innovation, whereby an organization houses and funds a centralized research and development (R&D) department, is being superseded (Tyrrel 2007). The common conception of innovation is as a linear progression from research to invention, and from invention to commercialization. But, modern innovation is rarely a linear process. Knowledge does not flow in a straight line, from basic research to applied R&D and then on to commercialization. If innovation was accomplished in such a direct manner, then public policy and corporate strategy could continue to focus on increasing R&D inputs and we would be done with it (Milbergs 2004). Innovation, however, has become more complex than a summation of knowledge inputs. Innovation is a dynamic process where feedback patterns interactively occur between activities, disciplines and sectors including actors from science, technology, learning, production, policy and demand (Edquist 1997). Thus, many innovative companies are disaggregating their R&D departments and distributing the innovation process across a network of external partners and remote sites. Distributed networks enable the allocation of activities according to strengths. A focus on strengths creates more effective and efficient R&D processes (Tyrrel 2007).

Convergences across disparate areas of science, engineering and technology require cross-cutting flows of information. Increasingly, firms, universities and research institutions are implementing cross-sectoral R&D, where research efforts are organized by problems rather than by disciplines (Calestous and Yee-Cheong 2005). Research questions are shifting in a parallel manner. Investigators are no longer asking *why* innovation is important; rather, the focus lies on *how* to innovate, how to accelerate innovation and how innovation processes can best be managed (Fredberg, Elmquis and Ollila 2008). Evidence suggests that as firms are able to collaborate with and draw ideas from many different perspectives, improvements and opportunities will multiply (UK, Department of Trade and Industry 2003).

Global connectivity and the ascendancy of knowledge and technological capabilities have brought about organizational structures and work models that contrast sharply with the brick-and-mortar structures and hierarchical arrangements of the industrial era (Thompson 2006). Increasingly, organizations are comprised of flexible systems whose resources, both technological and personnel, are more organically assembled and disassembled according to the shifting needs of specific projects (Hartog, Keegan and Verburg 2007). Further, work demands and the teams required to accomplish them are often distributed rather than co-located, representing global value chains. Finally, the ability to absorb new ideas and turn them into action has become critical to high-innovation performance (UK, Department of Trade and Industry 2003).

These insights about the innovation process and the changing economic demands of a global market have led to the development of an open innovation model. In open innovation, ideas are allowed to flow outside the originating organization, to wherever the ideas can be most efficiently handled at each stage of the R&D process (Tyrrel 2007). The boundaries between organizations and their surrounding environment are becoming more porous to enable innovation to move more fluidly between them (Chesbrough 2003). Whereas it is clear that diverse sources of knowledge represent an important trigger for stimulating innovation, there is little systematic research that lays out models and methodology for establishing effective cross-cutting collaborations and knowledge exchange (Bathelt, Malmberg and Maskell 2004). To more effectively accomplish complex projects with trilateral partnerships, codified standards are needed where competencies are established for effective knowledge exchange across firms and sectors (Ngo-Mai and Raybaut 2007).

TRIPLE HELIX INNOVATION

The recent growth in science and technological knowledge has compelled scientists to begin addressing complex scientific and humanitarian problems from multiple perspectives. Interdisciplinary partnerships that reach across universities, industry and government sectors are increasing in scale and diversifying in scope. Regulatory and statutory legislation have concurrently evolved in support of these relationships.

A shift to more open and collaborative R&D is underway; however, there remains a paucity of rigorous and scholarly groundwork from which to construct evidence-based models of cross-disciplinary, cross-sector engagement (Campbell et al. 2004). Our evolution into a knowledge-based society makes requisite the emergence of a research platform where knowledge can be pursued under the rigorous conditions of basic science while operating within a flexible distributed structure, able to take advantage of imagination and diversity (Branscomb 1999).

The easy availability of large amounts of digital information and the widespread use of computer technologies promotes interdisciplinarity. Virtual spaces allow people to meet synchronously or asynchronously, exchange information, share ideas and documents and otherwise co-create (Dubé, Bourhis and Jacob 2006). As collaboration becomes more accessible, it is also becoming more essential to the advancements in basic science. Contemporary research questions tend to span multiple scientific disciplines (Metzger and Zare 1999). In fact, the creative fusion of disciplinary perspectives is a core ingredient in the success of many scholarly pursuits such as bioinformatics and structural biology (Sherman and Miyataki 2004). Further, the synergy between multiple disciplines has led to the advent of new fields of study, including bioengineering, cognitive neuropsychology, genomics and photonics.

Whereas the value of interdisciplinarity is widely accepted, there are several noteworthy barriers that systematically inhibit the promotion of interdisciplinary collaborations (Metzger and Zare 1999). For the true value of interdisciplinarity to be realized in basic science research, an enormous shift must take place in the working culture of collaboration (Sherman and Miyataki 2004). Fundamentally, our institutional systems stifle collaborative endeavours by fostering the isolation of disciplines (Dauphinée and Martin 2000). Further, practical concerns often impede interdisciplinary communication and comprehension. For example, scientists of different disciplines do not share the same jargon or statistical language. This becomes problematic in achieving consensus on common problems (Robertson, Martin and Singer 2003; Zeleznik, Agard-Henriques and Schnebel 2003). There also exist issues of data recording, language, coding and analysis (Treloar and Graham 2003); difficulties monitoring quality assurance (Zeleznik, Agard-Henriques and Schnebel 2003); and publication implications, such as authorship (Treloar and Graham 2003). These interdisciplinary challenges highlight the need for emergent models that can guide the essential processes of collaborative research (Legris, John and Pierre 2003). Cross-disciplinary advances hold tremendously positive prospects for humanity. For example, early disease detection, prevention and intervention have had direct consequences in terms of human welfare and healthcare costs (Wessner 2002). Whereas scientific discovery continues to accelerate, the impressive advances realized in the last century have been unevenly distributed across the world's population. To transfer new scientific knowledge equitably, we must validate new models for collaborative engagement that increase global access to information and facilitate the efficient transfer of emergent findings (U.S. Department of Health and Human Services 2005). The quick transfer and practical application of academic discovery has demonstrable benefits for developing economies. Collaborative research endeavours offer hope for solving critical problems that are fundamental to our maturing global community. Carefully evaluated, evidence-based partnerships can offer society a means of enhancing both the welfare of its citizens and the security of the nation (Salicrup, Harris and Rohrbaugh 2005). Trilateral collaborations target public missions through the execution of virtuoso science (Guston 2000). Examples of triple helix partnerships aimed at alleviating poverty are discussed in detail in Chapters 8 and 10 of this volume.

TECHNOLOGY TRANSFER AND COLLABORATIVE INNOVATION

During the 1990s, government agencies consulted with academia and the private sector when they needed to acquire specific expertise. Over the last decade, new patterns of collaboration and commercialization have emerged that more directly couple industry, university and government agencies. By combining expertise, assets and resources, triple helix partnerships facilitate technological advancement and efficient knowledge transfer (Schacht 2008). Collaborative value chains are able to transfer knowledge and promote serendipitous spillovers, which elite specialization precludes. Flexible partnering also offers the potential for exploiting technological leads, thinking in integrated ways about requirements and providing new sources of supply for products and services.

In our rapidly changing social and technological environment, the speed at which a product, process or service is brought to the marketplace is a crucial factor in competitiveness. Economic benefits accrue when a technology or technique is efficiently brought to the marketplace where it can be sold to generate income and/or applied to increase productivity (Schacht 2008). Collaborative efficiency and capability are central competitive discriminators for achieving technology transfer as they supply the necessary exchanges of specialization, experience and ideas about common problems to engender an environment conducive to success.

There is a strong case to be made for enhancing technology transfer to developing countries, for the acceleration of humanitarian solutions as well as to the benefit of the overall global community. One of the problems hindering the achievement of the UN Millennium Goals is the absence of adequate infrastructure for research and knowledge transfer (Calestous and Yee-Cheong 2005). Greater opportunity for technology transfer will allow developing countries to advance technologies that are most consistent with their own regional needs and resources, thus enabling the development of sustainable solutions for local and regional priorities (OECD 2002; Varmus et al. 2003; Saha, Satyanarayana and Gardner 2004). Facilitating knowledge and technology transfer in developing countries also increases the overall development of technologically capable partners. Technology transfer can thereby create a more robust global network for the application of scientific discoveries, methods, procedures, techniques and equipment for promoting health and social development. Technologies that are leveraged globally extend scientific knowledge, adding value and social returns to existing inventions (Saha, Satyanarayana and Gardner 2004).

Collaborative innovation occurs when actors are able to successfully bridge traditional silos; release their respective expertise; integrate corresponding information and perspectives; and, in consequence, accelerate the development of synthesized ideas, products and processes. Engagement in cooperative innovation projects widens the horizon of organizations and nations by increasing sensitivity to new developments and projects. Within collaborative R&D networks, mutual needs promote development rather than a focus on the defeat of the opponent (Agapitova 2003).

The survival and performance of collaborations often rest on factors associated with cross-sector, interdisciplinary relationship building (including leadership, trust, transparency, motivation, competence and fairness) to create an enabling environment for risk management, conflict resolution and the generation of integrated solutions (Dawes and Eglene 2004). The ready availability of data and evidence-based approaches for collaboration is critical to understanding the strengths, weaknesses, costs and outcomes of the nation's technology transformation process (Campbell et al. 2004). Innovative network dynamics are important for domestic, regional, national and international contexts as they promote cross-sector and cross-cultural interactions that are essential for the successful functioning of complex innovation systems.

GLOBAL NETWORKS FOR INNOVATION—BUILDING LINKAGES

Building cross-sector, interdisciplinary bridges between governments, communities, the private sector and non-governmental organizations (NGOs) will provide benefits for multiple social issues. The development of effective global networks can help intervene into the social and political discord that results from diseases and disasters (Blumenthal and Schlissel 2007). Global networks can, moreover, facilitate communications that increase trust and confidence toward improved international relations (Blumenthal and Schlissel 2007).

Targeting the welfare of developing countries is not only a humanitarian goal, it is also intricately intertwined with the security of all countries, making global network development a critical venture (Calestous and Yee-Cheong 2005). The burden of poverty and disease on the social fabric of our global community is a serious economic and security issue. Countries ravaged by poverty and disease are more likely to be unstable politically; more likely to require inflated foreign assistance; and they are less likely to develop economically into strong international trading partners (Folkers and Fauci 2001; Institute of Medicine 1997; Salicrup, Harris and Rohrbaugh 2005). Many countries now classify human development challenges, such as HIV/AIDS, as national security issues, recognizing the emergence of a single global community that requires collective action to deal with issues that were once considered strictly national (United Nations 2003).

Science, technology and innovation underpin every one of the UN Millennium Development Goals. It is inconceivable that gains can be made in health and environmental concerns without focused science, technology and innovation policy and coordination (Calestous and Yee-Cheong 2005). In the 2000 United Nations 'Brahimi Report' specific actions were outlined for improving information sharing between civilian and military groups during complex emergencies. In particular, the report recommended the design of new systems for collaboration, information management and planning capabilities involving military, humanitarian, human rights and other entities engaged in complex emergency situations (Solomon and Brown 2005). In order to effectively promote their respective missions, the military and the humanitarian assistance organizations need both their own spheres for operation and information integrity, as well as collaborative spaces for working toward common goals, prioritizing needs and establishing coordinated objectives (Solomon and Brown 2005). Achieving the goals requires a cross-disciplinary and holistic approach. The creation and transfer of scientific, logistic and technological knowledge is not achievable by a single office or laboratory. Knowledge blending is required, whereby a dynamic process is in place for the sharing of insights, problems, issues, experimental approaches and outcomes across disciplinary boundaries (Calestous and Yee-Cheong 2005). Practical examples of adoption and operationalisation of triple helix approaches both in developed and developing countries are described and evaluated in Part II ('Operationalising the Triple Helix Model and Bridging the Gaps') and Part III of this book ('Shift towards a Triple Helix Policy Paradigm and Implications for Developing Countries').

Interoperability is required for boundary spanning and to establish successful global trilateral networks. Interoperability, in this context, refers to the conditions that enable separate organizations to share information toward a common end. Interoperability includes technical and political compatibility as well as the will and the means to communicate, to cooperate and to share a common culture of communication. When systems are not interoperable, information becomes stove-piped within a single organization and systems cannot easily collaborate or transfer knowledge (Solomon and Brown 2005).

Tactically, all aspects of humanitarian ventures should become flexible components of an integrated effort. There is currently no systemic approach that has been designed to match essential needs with the resources that are actually available. Further, few projects have built-in methods for assessing their efficacy or sustainability. Fewer still have ever scaled up beyond initial pilot stages. And nearly all have been designed, managed and executed by residents of the developed world with virtually no provisions to allow the community residents to say what they want, contribute to deciding which projects serve their needs or that adopt local innovations. Finally, nearly all programs lack exit strategies or safeguards against the dependency of local governments (Garrett 2007).

Triple helix networks can play an instrumental role in accelerating the development of a more integrated, systemic approach to global issues; however, these partnerships need to be properly constructed and ably led with clear goals, shared contributions and regular evaluations. Current research on trilateral partnerships and distributed networks is limited and fails to address many critical issues (Daniel, Schwier and McCalla

2003). Collaborative models and methodologies are needed for implementing and evaluating these varied efforts. As a global community, we have established the Millennium Goals that outline our most pressing and complex issues. To achieve these goals, we must distil an evidence-based methodology for expediting collaborative engagement and innovation including analytic tools for evaluating the collaborative environment, essential requirements and platforms for ensuring a rich and integrated system and the dynamic application of strategies that facilitate successful distributed partnerships.

TRIPLE HELIX KNOWLEDGE CLUSTERS

Whereas there is an abundance of scientific data, there is often a pronounced deficit in our ability to apply it meaningfully (United Nations 2003). Innovation requires more than the emergence of a good idea or a promising prototype. The efficacy of new developments must be empirically substantiated and then pushed out as a product or as codified knowledge, within a societal context. Innovation favours boundaries and intersections where ideas are able to collide and blend. Partnerships that leverage resources and create opportunities for knowledge exchange and spillovers are more likely to generate novel solutions through the transfer of scientific knowledge into tangible product development. Removing barriers to collaboration, supporting distributed structures and facilitating an exchange of science and technology resources influence the orientation of research efforts to become more applied and thus more likely to address societal needs. Specific issues related to the implementation of innovation in developing countries are addressed in Chapters 9, 11, 12 and 13 of this volume.

Innovation clusters, such as in Silicon Valley, represent a key mechanism for removing barriers between institutions, cultures and private firms (Granovetter 1973, 1985). Clusters, or agglomerations, of trilateral resources that are geographically co-located offer partners the opportunity to work in close proximity and exchange resources and ideas. Innovation clusters predispose individuals and firms toward innovative action (Agapitova 2003).

The Internet and information and communications technologies (ICT) have been developed in order to permit non-co-located actors to come together in distributed innovation networks. The availability of global connectivity has created R&D infrastructures without walls (Seline 2008). Cross-cutting networks are emerging where varied participants are collaborating at a distance to learn from each other and build on each other's contributions, under the guidance of a network organizer (Hagel and Brown 2005).

'Knowledge clusters' are an example of a distributed innovation network. Knowledge clusters represent an evolution of traditional, co-located innovation clusters, utilizing distributed environments to share research; blend knowledge sets; and generate new ideas, partnerships and funding opportunities

(Agapitova 2003). Where Communities of Practice (CoP) are *self-organizing* networks that arise spontaneously, knowledge clusters are *intentionally* established networks. Knowledge clusters bring together multidisciplinary scientists and cross-sectoral innovators, funders and specialists in expertly facilitated sessions. The distributed sessions create previously unrecognized opportunities to bridge geographic boundaries, exchange explicit and tacit information, blend knowledge sets, catalyze new knowledge, promote swift trust and foster serendipity (Snyder and Briggs 2003; Jerome 2009). Knowledge clusters represent a new model for collaboration. They validate accelerants of innovation that are made possible by collaborative technologies and the rise of the Web as a participatory platform.

Like co-located teams, knowledge clusters require a mobilization of resources that will allow the emergence of a cohesive team with clear goals. Research has identified a set of predictable challenges associated with collaborative partnering. Differences in organizational missions, structures and processes, human resources policies, reward systems and cultures create problems that make communication, coordination, cooperation and, ultimately, collaboration less effective (Jarvenpaa and Leidner 1998; Zolin 2003). Further, distributed teams must contend with additional challenges such as sub-optimal communicative channels, absent interactive cues and difficulties related to time, space and culture. Existing models of collaboration are immature and do not adequately address persistent barriers such as bureaucratic inertia, fear of change, turf-minded participants and the truth that many change efforts are much too dependent on charismatic champions whose exits spell the demise of promising innovation (Snyder and Briggs 2003). To date, the literature about distributed teams has been technology-focused, often neglecting key aspects of collaboration, such as blending diverse knowledge and analytics, accommodating variance in perspectives, combining expertise, dividing tasks, establishing a common lexicon, managing expectations and problem solving. Knowledge clusters are process-focused, rather than technology-focused, targeting the important aspects of collaboration, such as resolving diversities of knowledge and analytics, accommodating variance in perspectives, combining expertise, dividing tasks, establishing a common lexicon, managing expectations and problem solving.

There are at least three essential requirements that distributed networks must have in place to foster successful collaboration: (a) an enabling environment for innovation, (b) robust ICT and (c) an evidence-based system for collaborating in a distributed environment. Greater examination of the nature of distributed knowledge clusters will lead to new insights about how co-located and distributed networks work. Research is needed that explores distributed networks of various sizes, representing participants from different sectors and in a variety of socio-cultural contexts. From these investigations, we will be able to extract a classification of types of knowledge clusters and validate applied models for social problems requiring global, interrelated solutions (Roberts 2006).

TRANSFORMATIVE NETWORKS

The proliferation of triple helix models for collaboration and the emergence of distributed capabilities provide a mechanism that can bridge formal organizational boundaries and increase the collective knowledge, skills and professional trust and reciprocity of individuals who serve in these organizations (Snyder and Briggs 2003). Innovation is a dynamic process involving interactions among a wide range of actors who form a system of mutually reinforcing learning activities. Innovation is dependent on infrastructure; adaptation to new technologies; educational systems; policies and policies on science, technology and innovation; the nature and composition of the private sector; and the capabilities inherent in the public sphere. Innovation also depends on the flows of knowledge, resources and people between public and private domains of knowledge and the structures, tools and strategies by which information on specific innovations is shared, developed, commercialized and diffused (Calestous and Yee-Cheong 2005).

The transformational capacity of trilateral networks is very real but we must adapt for this potential to be realized. The nature of technological innovation requires a shift from specialized silos of knowledge generation to more open, interactive and collaborative platforms that broaden access to science globally (Calestous and Yee-Cheong 2005). When knowledge that is trapped in vertical silos can be released into networks of cross-sector collaboration, actors will develop new methods for co-creating in value chains that reach across distance, expertise and organizations. International linkages have the power to recharge local regions and provide opportunities for international knowledge sourcing, knowledge flow and knowledge spillover (Saxenian and Hsu 2001). Industry, academia and government, joined in distributed networks, will discover more integrated and holistic opportunities for effecting social and economic transformation. The emergence of cross-cutting networks and technologies puts pressure on the outdated sectoral systems and their actors to readapt their structures, institutions and strategies to embrace new opportunities and demands—not just once and in a disruptive fashion, but continuously (Dolata 2008).

New skills development and novel partnering models will require training in global technology management, multidisciplinary issues, cross-sector partnering and the cultivation of collaborative knowledge clusters. Training will significantly improve opportunities for new product development and technology transfer (Eiss and Balakrishnan 2008).

As we realize the potential of global transformative networks, dynamic knowledge exchange and interactive learning between creators, producers and consumers of information are increasingly driving innovation (Houghton and Sheehan 2000). Ongoing technological dynamics create novel spaces and first-mover advantages, put established actors under demands to renew their strategies and organizational fit, alter existing patterns of competition and collaboration, require the continuous adjustment of legal and regulative frameworks and evoke changes in societal problem perceptions and patterns of economic development (Dolata 2008). A knowledge economy is, in effect, a network of networks, driven by the accelerated rate of change and learning, where the opportunity and capability to get access to and join knowledge clusters determines the socio-economic position of individuals and firms (Houghton and Sheehan 2000).

CONCLUSION

Many of the improvements in human welfare over the past century are attributable to technological innovation in public health, nutrition and agriculture. These improvements have greatly reduced mortality rates and increased life expectancy (Calestous and Yee-Cheong 2005). Technology is making it possible to build the platforms and linkages for networks to become truly global.

Collaborative innovation is on the rise, facilitated by the robustness of the Internet as a participatory platform. The collision of cross-sector, multidisciplinary knowledge sets through the creation of distributed collaborative networks holds great promise for boosting economies, improving health and human welfare and moreover for the emergence of a global community with well-grounded hope for all people to take their rightful place in a progressive and innovative world. A developing country's induction into privileged circles of technology transfer, trade negotiation, economic treaties and preferential status depends in part on its technological capability. There is no substitute for scientific and technological bases, which undergird everything from agricultural self-sufficiency to public health to indigenous technology advances (Calestous and Yee-Cheong 2005).

Because wealthy nations depend on global development, which in turn relies on stability and increased prosperity, it is in their direct interest to help improve the state of the world. A commitment to sustainable development to address paramount global challenges, such as the provision of public healthcare, will increasingly create positive economic impact (Schwab 2008).

The development of transformative networks across developed and developing countries will encounter challenges such as sectoral inertia, concerns that nation power will be diminished by the sharing information and fear of change. But innovation diffusion and the development of collaborative knowledge networks follow a ripple effect (Gloor et al. 2007). As the old structures struggle to adapt, new collaborations will continue to form with remarkable speed, creating new opportunities, priorities and national interests that will bring people, societies and cultures together in real and virtual combinations (Solomon and Brown 2005). The innovation landscape is in flux. For the foreseeable future, the landscape will continue changing on many levels where innovation and triple helix collaboration are likely to intersect. These dynamic scenarios will yield countless opportunities for creating novel solutions to intractable social problems.

REFERENCES

Agapitova, N. 2003. 'The role of social networks on innovation and industrial development: Social dimensions of industrial dynamics in Russia'. Paper presented at DRUID Summer Conference on 'Creating, Sharing and Transferring Knowledge: The Role of Geographical Configurations, Institutional Settings and Organizational Contexts', Copenhagen.

Asheim, B. T., and L. Coenen. 2004. 'Knowledge bases and regional innovation systems: Comparing Nordic clusters'. *Research Policy* 34:1173–1190.

Bathelt, H., A. Malmberg and P. Maskell. 2004. 'Clusters and knowledge: Local buzz, global pipelines and the process of knowledge creation'. *Progress in Human Geography* 28:31–56.

Blumenthal, S. J., and E. Schlissel. 2007. 'Health diplomacy: A prescription for peace'. Available HTTP: http://www.huffingtonpost.com/susan-blumenthal/health-diplomacy-a-prescr_b_73267.html (accessed 26 December 2009).

Branscomb, L. M. 1999. 'The false dichotomy: Scientific creativity and utility'. *Issues in Science and Technology* (Fall). http://www.nap.edu/issues/16.1/branscomb.htm (accessed 29 August 2003).

Calestous, J., and L. Yee-Cheong. 2005. 'Innovation: Applying knowledge in development, UN Millennium Project Task Force on Science, Technology and Innovation'. http://www.unmillenniumproject.org/documents/Science-complete.pdf (accessed 26 December 2009).

Campbell, E. G., J. B. Powers, D. Blumenthal and B. Biles. 2004. 'Inside the triple helix: Technology transfer and commercialization in the life sciences'. *Health Affairs* 23:64–76.

Chesbrough, H. W. 2003. *Open innovation: The new imperative for creating and profiting from technology.* Boston, MA: Harvard Business School Press.

Daniel, B., R. A. Schwier and G. McCalla. 2003. 'Social capital in virtual learning communities and distributed communities of practice'. *Canadian Journal of Learning and Technology* 29:3.

Dauphinee D., and J. B. martin. 2000. Breaking down the walls: Thoughts on the scholarship of integration: Acad. Medicine 75(9): 881–886.

Dawes, S. S., and O. Eglene. 2004. 'New models of collaboration for delivering government services: A dynamic model drawn from multi-national research'. Center for Technology in Government. http://dgrc.org/dgo2004/disc/presentations/collaboration/dawes.pdf (accessed 30 December 2008).

Dolata, U. 2008. 'The transformative capacity of new technologies—how innovations affect sectoral change: Conceptual considerations'. MPIfG Discussion Paper, Max Planck Institute for the Study of Societies, Cologne.

Drucker, P. F. 1988. 'The coming of the new organization'. *Harvard Business Review* 66 (1): 45–53.

Dubé, L., A. Bourhis and R. Jacob. 2006. *Interdisciplinary Journal of Information, Knowledge, and Management* 1. http://ijikm.org/Volume1/IJIKMv1p069–093Dube.pdf (accessed 26 December 2009).

Edquist, C. 1997. 'Sustems of innovation approaches - their emergence and characteristics.' In (1997) Systems of Organisations, ed. C. Edquist, C., Pinter / Cassell, London.

Eiss, R. B., and U. R. Balakrishnan. 2008. 'Academic licensing to global health product development partnerships'. *MIHR*. http://www.tmgh.org/assets/MIHR-TMGH%20Sequel%20Booklet.pdf (accessed 26 December 2009).

———. 2002. 'Networks of innovation: Science, technology and development in the triple helix era'. *International Journal of Technology Management and Sustainable Development* 1:7–20.

Folkers, G. K., and A. S. Fauci. 2001. 'The AIDS research model: Implications for other infectious diseases of global health importance'. *JAMA* 286 (23): 458–461.

Fredberg, T., M. Elmquis and S. Ollila. 2008. 'Managing open innovation—present findings and future directions'. Chalmers University of Technology. http://www.vinnova.se/upload/EPiStorePDF/vr-08–02.pdf (accessed 26 December 2009).

Garrett, L. 2007. 'The challenge of global health'. *Foreign Affairs*. http://www.foreignaffairs.com/articles/62268/laurie-garrett/the-challenge-of-global-health (accessed 26 December 2009).

Gloor, P., M. Paasivaara, D. Schoder and P. Williams. 2007. 'Finding collaborative innovation networks through correlating performance with social network structure'. *International Journal of Production Research* 23 (49): 1357–1371.

Granovetter, M. S. 1973. 'The strength of weak ties'. *American Journal of Sociology* 78:1360–1380.

———. 1985. 'Economic action and social structure: The problem of embeddedness'. *American Journal of Sociology* 91:481–510.

Guston, D. H. 2000. 'Technology transfer and the use of CRADAs at the National Institutes of Health'. In *Investing in innovation: Creating a research and innovation policy that works*, ed. L. M. Branscomb and J. H. Keller. Cambridge, MA: Harvard University Press, 85–111.

Hagel, J. and Brown, J. S. 2005. The Only Sustainable Edge: Why Business Strategy Depends on Productive Friction and Dynamic Specialization. Boston, MA: Harvard Business School Press.

Hartog, D., A. E. Keegan and R. M. Verburg. 2007. 'Limits to leadership in virtual contexts'. *Electronic Journal for Virtual Organizations and Networks* 9:54–62. http://www.ejov.org/projects/264/Issues/eJOV%20Special%20Issue%20.Limits%20of%20Virtual%20Work.%202007/eJOV_SPI4_Den%20Hartog_Leadership%20in%20Virtual%20Contexts.pdf (accessed 30 December 2008).

Houghton, J. W., and P. Sheehan. 2000. 'A primer on the knowledge economy, Center for Strategic Economic Studies'. Paper presented at the National Innovation Summit, Victoria University, Melbourne, February.

Institute of Medicine. 1997. 'America's vital interest in global health: Protecting our people, enhancing our economy and advancing our international interests'. *Board of International Health, National Academy Press, Washington, DC*. www.nap.edu (accessed 26 December 2009).

Jarvenpaa, S. L., and D. E. Leidner. 1998. 'Communication and trust in global virtual teams'. *JCMC* 3:4. http://hyperion.math.upatras.gr/commorg/jarvenpaa/ (accessed 30 December 2008).

Jerome, L. 2009. 'Building an institute for triple helix research innovation in the Pacific Region'. Report of congressional research grant, the Institute for Triple Helix Innovation, Honolulu, HI.

Legris, P., I. John and C. Pierre. 2003. 'Why do people use information technology? A critical review of the technology acceptance model'. *Information and Management* 40 (3): 191–204.

Leydesdorff, L., and M. Fritsch. 2005. 'Measuring the knowledge base of Regional Innovation Systems in Germany in terms of a triple helix dynamics'. Paper presented at the 5th Triple Helix Conference, Turin, Italy, May.

Metzger N., and R. N. Zare. 1999. 'Interdisciplinary research: From belief to reality'. *Science* 283:642–643.

Milbergs, E. 2004. 'Measuring innovation for national prosperity'. *National Innovation Initiative—Innovation Framework Report, 3.0, IBM Corporation.* www.innovationecosystems.com (accessed 6 June 2009).

Ngo-Mai, S., and A. Raybaut. 2007. 'Swift trust and virtual team dynamics'. http://gredeg.idefi.cnrs.fr/web/Colloques/Dime/Papers/Raybaut%20NgoMai%20-%20dime%20nice%2011052007.pdf (accessed 30 December 2008).

OECD. 2002. 'Conference on biotechnology for infectious diseases: Addressing the global needs'. Paper presented at Rapporteurs' Report, Lisbon, 7–9 October.

Roberts, J. 2006. 'Limits to Communities of Practice'. *Journal of Management Studies* 43:3.

Robertson, D. W., D. K. Martin and P. A. Singer. 2003. 'Interdisciplinary research: Putting the methods under the microscope'. *BMC Medical Research Methodology* 3:20.

Saha, R., K. Satyanarayana and C. A. Gardner. 2004. 'Building a "cottage industry" for health (and wealth). The new framework for IP Management in India'. *IP Strategy Today* 10:23–58.

Salicrup, L, R. Harris and M. Rohrbaugh. 2005. 'Partnerships in technology transfer: An innovative program to move biomedical and health technologies from the laboratory to worldwide application'. *IP Strategy Today* 12:1–15.

Saxenian, A., and J. Y. Hsu. 2001. 'The Silicon Valley-Hsinchu connection: Technical communities and industrial upgrading'. *Industrial and Corporate Change* 10:4.

Schacht, W. H. 2008. 'Cooperative R&D: Federal efforts to promote industrial competitiveness'. Congressional Research Service Brief, Library of Congress. http://digital.library.unt.edu/govdocs/crs/permalink/meta-crs-9714:1 (accessed 7 July 2009).

Schwab, K. 2008. 'Global corporate citizenship. Working with governments and global society'. *Foreign Affairs.* http://www.weforum.org/pdf/CGC/Schwab_Article.pdf (accessed 17 July 2009).

Seline, R. and Miller, S. 2009. 'The Federal Role in Catalyzing Innovation: Beyond the Beltway and Through the Networked Economy. Science Progress, Fall/Winter. http://www.scienceprogress.org/wp-content/uploads/2009/01/issue2/seline_miller.pdf (accessed 15 August 2010).

Sherman, R., and G. K. Miyataki. 2004. 'Report to the John A. Burns School of Medicine Regents' Task Force'. University of Hawaii. http://www.hawaii.edu/uhmfs/documents/jabsom_final_report_10–19–04.html (accessed 30 December 2008).

Solomon, R., and S. J. Brown. 2005. 'Creating a common communications culture: Interoperability in crisis management'. United States Institute of Peace, VDS, No. 17.

Snyder, W. M., and X. de S. Briggs. 2003. 'Communities of Practice: A new tool for managers'. Washington, DC, IBM Center for the Business of Government. http://www.businessofgovernment.org (accessed 26 December 2009).

Thomke, S. 2001. 'Enlightened experimentation—the new imperative for innovation'. *Harvard Business Review* 79 (2): 66–75.

Thompson, K. 2006. 'Social network analysis: An introduction'. *Social Networking* 16. http://www.bioteams.com/2006/03/28/social_network_analysis.html (accessed 30 December 2008).

Treloar, C., and I. D. Graham. 2003. 'Multidisciplinary cross-national studies: A commentary on issues of collaboration, methodology, analysis, and publication'. *Qualitative Health Research* 13:924–932.

Tyrrel, P. 2007. 'Sharing the idea: The emergence of global innovation networks'. *The Economist Intelligence Unit, The Economist.* http://www.gencat.cat/diue/doc/doc_41824369_1.pdf (accessed 26 December 2009).

UK, Department of Trade and Industry. 2003. 'Competing in the global economy: The innovation challenge'. http://www.berr.gov.uk/files/file12093.pdf (accessed 13 May 2009).
United Nations. 2003. 'Expanding public space for the development of the knowledge society'. Report of the ad hoc expert group meeting on knowledge systems for development, Department of Economic and Social Affairs, New York.
U.S. Department of Health and Human Services. 2005. 'Moving medical innovations forward—new initiatives from HHS'. Report of the Medical Innovation Task Force. http://www.hhs.gov/reference/medicalinnovations.shtml (accessed 26 December 2009).
Varmus, H., R. Klausner, E. Zerhouni, T. Acharya, A. S. Daar and P. A. Singer. 2003. 'Grand challenges in global health'. *Science* 302 (5644): 398–399.
Wessner, C. W., ed. 2002. *Government–industry partnerships for development of new technologies, National Research Council, Board on Science, Technology and Economic Policy*. Washington, DC: National Academies Press.
Zeleznik, J., B. Agard-Henriques and B. Schnebel. 2003. 'Terminology used by different health care providers to document skin ulcers: The blind men and the elephant'. *Journal of Wound Ostomy Continence Nursing* 30:324–333.
Zolin, R. 2003. 'Swift trust in hastily formed networks'. Paper presented at the Cebrowski Institute, Navel Postgraduate School. http://www.nps.edu/Cebrowski/Docs/swifttrust100302.pdf (accessed 30 December 2008).

2 Innovation Systems Interactions and Technology Transfer and Assimilation for Industrial Development

The Cases of South Korea and Mexico

María Isabel Rivera Vargas

INTRODUCTION

The interactions of national and international innovation systems both with nether and upper regions, as conceived by Freeman (1995), particularly having transnational corporations as key actors, have convincingly demonstrated their importance for catching up, particularly in developing countries. Consequently, Porter (1990) could not be more accurate in his argument that the intensification of global competition has made the role of the home nation more important, not less. And it is within this context that the triple helix paradigm would need to be considered with all its potential to promote the interactions between university, industry and government for industrial development. Principally, the role of government in the design and implementation of national policies for catching up in technology is essential in the case of developing countries. This has been demonstrated with the assertive participation of government and *chaebols*[1] in the successful industrialization process of Korea, and, for omissions and cultural differences, in the not so favourable outcome obtained in Mexico.

The aim of this chapter is to investigate the important role that government could play within a triple helix paradigm promoting national and international innovation systems interactions for technological catching up and industrial development. The first section of this chapter describes the analytical framework which is used as a basis for analysing and interpreting the research findings. The second section depicts and analyses the national and international innovation systems interactions leading to the catching up process in Korea; identifying within this process the triple helix model and the outstanding government systemic planning of it, as well as the organizational expeditious learning strategies in *chaebols*. Following this, the third section examines how, in the case of Mexico, despite some declared intentions and provisions of policy design, there were some specific mechanisms and conditions to enforce these policies that were missing, so that the triple helix model did not produce the intended result, as in Korea. To conclude, the results in Korea and Mexico are analysed and contrasted using the typology of national and international innovation systems interactions advanced by Perez (1998).

THE ANALYTICAL FRAMEWORK

Recent studies seek to analyse the interrelation among the various aspects of the internalization process and between these and the development of national productive and innovative systems. Several of these studies seem to agree with the neo-Schumpeterian framework. According to this, the competitiveness of a country is determined by the degree of development of its innovative system influencing the country's trade performance and the internationalization of its firms. The studies of Narula (1996) and Narula and Wakelin (1995, in Perez 1998) as well as Keesing (1966) support this theoretical framework. The results of the econometric analysis in Narula and Wakelin confirm the existence of significant relations between 'country characteristics', on the one hand, and exports and foreign investment entering or leaving the country on the other. The findings of Keesing (1966), in reference to international trade and the amount of skilled people in a country, arrived at the same conclusion almost thirty years before.

Also analysing the role of innovative systems in the international transfer of technology, Mowery and Oxley (1995) find that economies that benefited most from the innovation process have opted to strengthen their national absorptive capacity through the adoption of national innovation systems. The role played by policies in South-East Asia was considered very illustrative and of particular importance. These authors claim that government policies can delay or accelerate the absorption of the advanced countries' technologies, helping or hindering the development of domestic industry as a result. Other studies have found how similar policy experiments in different countries provide different outcomes, revealing the wide variety of factors at play and the complexity of their interaction (see Lall 1985; Westphal et al. 1985; Lynn 1985; Ozawa 1982, 1985). Cantwell (1989) and Perez (1998) explained how the growing international movements of goods and production factors could give rise to two antithetic processes. On the one hand, one of the processes could lead to the genesis or acceleration of cumulative processes of growth. On the other hand, another process could lead to the decline of national industries. This decline is explained in relation to the system's ability to exploit the opportunities presented by the greater international mobility of goods or production factors, or inversely to their inability to mobilize the resources necessary to combat competition by foreign firms. Perez (1998) illustrates the first case with a situation in which the encounter takes place among highly developed industrial systems in the same product area. This is the case where all the countries involved benefit from the stimuli created by stronger technological competition for the control of world markets among firms from different countries (Cantwell 1995; Perez 1998).

An example of the second case, or the vicious circle possibility, is a diametrically opposite situation that takes place when internationalization processes displace the industry of the country receiving the commercial flows or foreign direct investments (FDIs). In this case, as indicated by Cantwell (1995) and Perez (1998), the encounter between the innovative and productive systems of different countries severely penalizes the one subject to the passive internationalization of its industrial system. Furthermore, the competing country's firms appropriate a proportion of the latter's national productive and innovative resources without this becoming a stimulus to the growth of the country's national industry. In addition, Rodríguez-Clare (1996) explores how multinational corporations (MNCs) affect underdeveloped regions through the generation of positive or negative linkages. The linkage effect of multinationals on the host country is more likely to be favourable when the goods produced by MNCs use intermediate goods intensively, when there are large costs of communication between the headquarters and the production plant and when the home and host countries are not too different in terms of the variety of intermediate goods produced. When these conditions are reversed, MNCs could even hurt the developing economy by creating enclave economies within them.

Perez (1998) analysed the links among FDIs, government policies and industrial development in the internalization process of Japanese and American multinational firms in Europe in two different points in time: 1980 and the early 1990s. Within the logic of the internalization of production, he recognizes the internalization of the economic systems through active or passive processes of internalization, where the MNCs are major protagonists. Perez defines the foreign expansion of resident firms as an *active internationalization process*, whereas the *process of passive internationalization* is defined as the involvement of foreign firms in a countrty's economic system. These interrelated processes give a place to the establishment of increasingly closer links among the productive and innovative systems of different countries. The process of internationalization of economies involves winners and losers rather than producing unequivocally positive or negative effects for the industries of the countries concerned. Perez identified multiple factors interrelated within a complex dynamic system that determines the prevalence of one or another of these effects and therefore the consequences of the passive internationalization of national productive and innovative systems. He shows the most salient factors in this dynamic to include the following: (a) the initial technological gap between domestic and foreign firms, (b) the level and pace of the expansion of the foreign presence in the country, (c) the strength of the market's selective mechanism and (d) the existence of government policies designed to encourage the technological development of local firms and to favour technological exchange between the two groups of firms (Perez 1998).

In the theoretical construction about the forms of interactions between national innovative and productive systems, Perez (1998) concurs with Cantwell (1995). He divided the consequences of MNCs' direct investments on the industrial systems of the recipient countries into two distinct components: (a) positive externalities or spillover effects; and (b) the displacement effect on domestic industry. Perez also went further and from his results was able to propose a typology that corresponds to the encounter between different innovative and productive systems. He proposes three different forms of encounter that could apply at the level of individual industry or country:

> Type I is complementary between national innovative and productive systems. In this case the internationalization of individual national economies enables the merger of scientific and technological knowledge which, although of differing origin, is closely complementary. This process gives rise to a third innovative and productive system of international scope which produces positive externalities for all the countries involved. Displacements effects are reduced to the minimum and they correspond at most to the reallocation of resources from industrial sectors in decline to those benefiting from the internalization of the economy . . . Type II involves a relation of substitutability between competing innovative and productive systems. The internationalization of country A's firms into country B leads to the substitution of local producers by foreign firms. There is maximum displacement of domestic firms and minimal technological spillovers for country B . . . Type III comprises completely (or almost) internationalized innovative and productive systems . . . largely identifiable with the industrial systems of small-sized countries. In many of these cases, indeed, it is inappropriate to talk of national innovative and productive systems, so great is the role played by the international component in developing the competitive advantages of resident firms . . . These, in fact, are industrial systems which have for long made a virtue out of necessity by compensating for the small size of their domestic markets through the massive internationalization of their productive systems. (1998, 169–170)

Perez concludes by proposing future research with closer analysis that takes into account the interactions among the four interrelated factors mentioned earlier and the abandonment of the linear relations between foreign presence and technological development used in most of the analysis. He also proposes the expansion of the theoretical approach adopted in the discussion of the national innovative system by explicitly acknowledging the existence of relations or points of intersection among national systems engendered by the processes of industrial internationalization. These have been rarely analysed up to now. The type of interactions between the productive and innovative system of modern economies represented by the MNCs and the national innovation systems in Korea and Mexico, as well as the different outcomes, are the subject of analysis in the study reported here.

The interactions of national and international innovation systems in Korea and Mexico take into consideration definitions of national innovation systems suggested by Cimoli and Dosi (1994), Freeman (1987) and List (1841). Cimoli and Dosi propose integration of the following three elements in the definition of a national innovation system: (a) the firms; (b) the networks through which cooperative links between the firms and the research institutions, universities and public or private organizations that promote science and technology are established; and (c) the institutional structure as a body that stimulates or inhibits innovation in the productive system.

Freeman's and List's definitions agree for the most part with that of Cimoli and Dosi, but also include within the innovation systems concept features that now are found at the heart of any contemporary study, such as user-producer interactive learning, knowledge accumulation, adapting imported technology, promotion of strategic industries and the role of the state in coordinating and carrying through long-term policies for industry and the economy.

INNOVATION SYSTEMS INTERACTIONS IN KOREA

Perez (1998, 1) identified Korea as one of those countries that has been able to exploit the opportunities afforded by the globalization of national productive and innovation systems. This section describes and analyses the interaction of the Korean National Innovation System with nether and upper regions, particularly with transnational corporations through its catching up process, which starts with the duplicative imitation stage. It then moves to the creative imitation, until innovation is achieved.

The first phase of the Korean catching up process could be identified with the national innovation systems' interactions, particularly with MNCs through which Korean firms assimilated assembly operations. This is the case of Hyundai, Samsung and more than two hundred firms in different industries—not just electronics, automobiles (Kim 1980, 1998) and semiconductors (Kim 1997), but also shipbuilding (Amsdem 1989; Kim 1985), iron (Amsdem 1989; Amsdem and Kim 1985) and machinery (Amsdem and Kim 1986). Firms in all these different industries followed a similar expeditious learning pattern for catching up. Using the case of Hyundai, the following section analyses the different phases of the Korean technological learning in the catch up process.

THE CASE OF HYUNDAI

Phase One: Assimilating Assembly Technology

Duplicative Imitation: Learning by Doing and by Using

Without previous knowledge in automobile production, Hyundai formed a task force in 1967 with selected personnel who had strong project management and engineering backgrounds from its construction division. It also recruited some other engineers with production experience from other auto producers (Kim 1998). The next step was an overseas assembler agreement with Ford, whereby Hyundai agreed to assemble Ford compact cars on a semi-knock-down (SKD) basis. Ford consented to transfer 'packaged' technology to Hyundai with a set of explicit knowledge, such as blueprints, technical specifications and production manuals, and to provide the training of Hyundai engineers at Ford sites. Additionally, Ford also agreed to send to Hyundai ten of its engineers to help translate the transferred explicit knowledge into tacit knowledge and to transfer Ford tacit knowledge on procurement planning, procurement coordination, production engineering, process engineering, production management, welding, painting, after service and marketing. Hyundai also received assistance from its suppliers' engineers, who set up equipment and trained its technicians. Thereafter, the most competent engineers trained by Ford were assigned to production engineering departments (Kim 1998).

Within this learning process, the Korean government played a very important role designing new policies and formulating promotion plans for strategic industries. In the case of the automobile industry, the government imposed stringent demands by making a radical policy change. The pressure on the automobile industry was to shift from assembly production of foreign cars to the development of locally designed Korean cars (Kim 1998). In addition, in 1973, the government formulated its long-term plan for the promotion of the automobile industry and required four automobile companies to submit detailed plans to develop Korean cars. The government's plan included very specific conditions such as 95 per cent local content, that the production cost should be less than U.S.$2,000 and that the product should be introduced to the market by 1975. These demands and deadlines to meet them set a major challenge to the automobile industry. Such a challenge can offer an appropriate opportunity for strategic and technological transformation. Several authors claim that constructing and then resolving organizational crises can be an effective means of opportunistic learning (Nonaka 1988; Pitt 1990; Schon 1967; Weick 1988). To foster growth within this industry, the government established six principles, namely, protection of the local market from new entrants and from new foreign knock-down imports, a significant tax reduction, promotion of vertical integration leading to new business opportunities, preferential financing, tax concessions and administrative decree to guarantee a large market share for the indigenous model. Furthermore, the government coordinated the educational and scientific policies to support this learning effort by providing scholarships for studies abroad in strategic areas.

Phase Two: Initial Design Technology

Assimilating Licensed Technology: Learning by Doing and by Using

The requirement imposed by the Korean government on the automobile industry to develop locally designed Korean cars (Kim 1998) was the stepping-stone for the second major jump of the Hyundai technological learning process.

Hyundai approached twenty-six firms from five countries to acquire and assimilate foreign technology. For car design technologies Hyundai approached ten firms from Japan and Italy. For stamping shop equipment technologies, four firms from the United States and Japan were contacted. For casting and forging plants technologies, Hyundai sought assistance from five companies from the United Kingdom and Germany. The engine technology was transferred from two firms from Japan and the United Kingdom. In addition, after acquiring unpackaged technology, Hyundai engineers obtained from the suppliers further knowledge such as observation tours, not only of the suppliers' own plants, but also of the leading automobile manufacturing factories that were using the suppliers' technology. This enabled Hyundai engineers to convert explicit knowledge to tacit knowledge and to gain significant insight into large-scale, modern automobile manufacturing systems.

Within the automobile industry, Hyundai acquired more foreign technologies through licensing agreements from a greater variety of sources than its competitors in Korea. But the assimilation of foreign license technology was not enough. Hyundai also hired for a three-year period[2] British managerial and technical expertise for the successful development of its first indigenous car. These highly qualified expatriates played crucial roles in helping Hyundai engineers convert explicit knowledge supplied by licensors into tacit knowledge and integrate specific tacit knowledge into an effective production system. Hyundai also hired moonlighting Japanese engineers to troubleshoot problems (Kim 1998).

Phase Three: Deepening Design Technology

Reaching Creative Imitation: Assimilating Limited License Technology

Foreseeing a drop in car sales as a result of rising gasoline prices, Hyundai decided to develop the next generation FF[3] car for the North American market. Thus, Hyundai approached several major car producers such as Volkswagen, Ford and Renault to obtain FF technology. In 1981 Mitsubishi agreed to license engine, transaxle, chassis and emission control technology to Hyundai, demanding in return 10 per cent equity share and management participation. However, Hyundai agreed only with the 10 per cent equity share and arranged to keep total management and the right to import parts and technology from Mitsubishi's competitors and to compete directly in Mitsubishi's own markets. In addition, Hyundai sourced body styling from Italdesign and constant velocity joint technology from British and Japanese providers.

It appeared that Hyundai's knowledge base, built while developing and manufacturing the first indigenous Korean car, was sufficient to assimilate FF car design and manufacturing technologies without any outside engineering support. However, the need for technological learning in the following three areas started to surface.

- car manufacturing to meet the most stringent safety and environmental requirements
- adoption of CAD/CAM systems (computer-aided design and computer-aided manufacturing systems) and ALC system (assembly line control system), as well as the development of transfer machines, which would lead to full computerization
- construction of a test-ground[4] necessary to conduct preparatory work

During this stage, specifically trying to learn about CAD/CAM systems, Hyundai relied on the literature review, internalizing and converting explicit knowledge into tacit knowledge. The acquisition of an advanced computer system and a graphic plotter was a second step, therefore acquiring packaged foreign technology. This step was followed by an in-depth study of Mitsubishi's operational CAD/CAM system. As a fourth step of this specific learning process, a Hyundai-designated team undertook an in-depth study of available alternative software packages. As a result, Hyundai selected the French Dasso Aerospace package called 'Catia' and conducted preparatory work for almost three years. Finally, Hyundai completed the FF car model for exportation in 1985. Branded Excel, the car passed both emission and safety tests before being exported to U.S. market in 1986. According to Kim (1998), in 1987 Excel became the best-selling imported car of the year in the United States, overtaking Nissan Centra, Honda Civic, Subaru DL/GL and Toyota Corolla.

Phase Four: Own Design Technology

Learning by Research

Although Hyundai was successful in manufacturing the subcompact FF car, Excel, relying on limited license technology, during the next phase of its technological learning process the company faced a technological barrier. Mitsubishi and other foreign suppliers were unwilling to share their latest technology. This led to Hyundai lacking the appropriate level of absorptive capacity to keep upgrading quality to compete in the North American market. This crucial need for technology prompted Hyundai to further develop an extensive research and development (R&D) network, not only in Korea, but also abroad, in order to increase its absorptive capacity and be able to create its own design technology.

Although Hyundai's R&D efforts were initiated in 1978 to develop its own research capability (Kim 1998), these efforts started to take shape only in 1984 with the establishment of the Advanced Engineering and Research Institute, created to develop its own engines and transmissions. This automaker also founded the Passenger Vehicle R&D Centre and the Commercial Vehicle R&D Centre. In addition, Hyundai established joint R&D laboratories with local universities and established its international network for R&D by opening the Hyundai American Technical Centre, Inc., in Ann Arbor, Michigan, and the Hyundai Styling Studio in Los Angeles to conduct R&D on cars for the U.S. market in 1986. As part of this international network, Hyundai set up a technical centre in Frankfurt to monitor technological developments in Europe and to design and engineer new cars for the European market. In 1997, it also launched R&D activities in Japan in order to penetrate its market. Setting up this international R&D network required a colossal effort that included recruitment of highly qualified engineering personnel and training overseas. More than three hundred R&D personnel had received training overseas before the development of its 'alpha' engine project was officially launched in 1984.

As in the former stages, the process to achieve the goal for phase four began with the mastery of all available English and Japanese literature on engines and transmissions. Afterwards, Hyundai entered into an agreement for technical training in engine design with British Ricardo Engineering Company. The next step was to hire two Korean experts with doctoral degrees from American universities and with experience in engine development at General Motors and Chrysler. According to Hyun (1995), Hyundai also hired an engineer from British Ricardo for a period of three years. Finally, after fourteen months of 'trial and error', the first prototype was made; and after 288 changes in the engine design, Hyundai improved its functioning and achieved perfection in 1992 (Hyun 1995). It outperformed comparable Japanese engines. The successful development of the 'alpha' engine led to the new generation, 'beta' engines, making Hyundai completely independent of foreign licenses in engines for mid-size, compact and subcompact cars (Hyun and Lee 1989). For Kim (1998) this approach led to the reduction to zero in royalty payments for compact and subcompact cars.

Many Korean-American engineers from U.S. universities, some with experience from General Motors and Chrysler, were recruited. Most of the thirty-five senior research engineers at Hyundai's Advanced Engineering and Research Institute were U.S. trained at the PhD level. A great emphasis was also placed upon the training of its engineers, which included short training, observation and long-term graduate programs. The number of R&D scientists and engineers sent abroad for training increased from

Table 2.1 Research and Development Investment at Hyundai

(billiions wons and percentrages)									
	1975	*1978*	*1982*	*1984*	*1986*	*1988*	*1990*	*1992*	*1994*
Sales	30	216	430	669	1906	3411	4656	6079	9052
R&D	1.1	5.4	7.9	22.7	79.5	116	190.4	248.8	400
R&D % of	3.5	2.5	1.8	3.4	4.2	3.4	4.1	4.1	4.4
Number	197	381	725	1298	2247	2459	3418	3192	3890

Source: Kim, L.

seventy-four in 1982 to 351 in 1986. As illustrated in Table 2.1, from 197 researchers in 1975 Hyundai R&D personnel reached 3,890 in 1994, thus accounting for nearly 10 per cent of the company's total employment. The investment in research also increased sharply from 1.1 in 1975 to 400 billion Korean won in 1994. Hyundai invested in R&D in higher proportions than its domestic competitors such as Daewoo and Kia.

As a result of continuous R&D efforts, Hyundai achieved the largest domestic market share in 1995. It also became an important technology exporter to Thailand, Egypt, Zimbabwe, the Philippines, Malaysia and other countries.

Synthesis and Analysis

Hyundai's technological learning process shows how it was nourished constantly and permanently through its interactions with international innovation systems, specifically MNCs at all stages of development achieved. For example, during the first stage of development, Hyundai was without any experience in automobile production and received valuable migratory knowledge related to auto assembly through an assembly agreement with Ford and, also, through the acquisition of packaged technology. Hyundai received from Ford a set of explicit knowledge, such as blueprints, technical specifications and production manuals. Furthermore, Hyundai engineers were trained by the multinational Ford company, both at the Ford and Hyundai sites. In addition, Hyundai recruited engineers from other foreign auto producers with previous experience in car production. These hired foreign engineers brought in migratory knowledge that raised Hyundai's level of tacit knowledge related to automobile production. In addition, the interaction of this company with international innovation systems paid off when hiring expatriates or when poaching experienced personnel from foreign corporations.

During the second phase of the process, Hyundai acquired unpackaged technology transfer through technology license agreements. The third phase of the Hyundai learning process could also be characterised by direct and indirect interactions of its national and international innovation systems because Hyundai relied on international literature reviews, observation tours in foreign MNC plants and packaged as well as unpackaged foreign technology transfer. Throughout the fourth phase, Hyundai's interactions with international innovation systems transpired through direct and indirect relationships with higher education institutions, among others. In addition, as in all the previous phases, the access to specialized English and Japanese literature represented an indirect link to international information sources.

The adoption of these expeditious learning processes can be linked to the leadership of the Hyundai group, recognized by many as far-sighted and bold in terms of risk-taking (Kirk 1994). There are also important managerial and organizational learning strategies involved in the course of technological learning. Among the most important organizational learning strategies that could be observed in all phases of the technological learning process is the *proactively constructed crisis*, which leads to an increase in *intensity of effort*. The *systemic planning* efforts involved in this technological learning process can be observed in the policies of the government through, *inter alia*, investment in education and human resource development, R&D, the provision of incentives in the form of protection barriers and promotion plans for strategic industries, all of which were at the heart of Korea's industrialization experience.

The Triple Helix Paradigm Presence

The process described in the preceding reflects the triple helix paradigm functioning in Korea, characterised by the mutually supportive interacting actions of government, universities, research institutions, *chaebols* (or industrial groups) and MNCs. The Korean government played an active role in the management of the economy, designing and implementing adequate public policies, providing the infrastructural R&D and human resource bases for industrialization.

The universities and scientific research institutes were able to collaborate and coordinate adequately with the industrial sector and the industrial groups, *chaebols*.

INNOVATION SYSTEMS INTERACTIONS IN MEXICO

Within the logic of the internalization of production, the analysis of the industrialization process of Mexico could easily illustrate the consequences of the passive internationalization of its national productive and innovative systems advanced by Perez (1998). Mexico has demonstrated its inability to exploit the opportunities presented by the greater international mobility of goods or production factors, or to mobilize the resources necessary to combat competition by foreign firms.

This section discusses the interaction of the Mexican National Innovation System with nether and upper regions, particularly with MNCs representing an element of the international innovation system. The discussion highlights the passive internationalization process illustrated through the specific case of the electronics and information technology industry[5] in Jalisco, which offers an example of a passive internationalization process leading to the predominance of FDI in this sector. By 1997, the FDI in this industry represented U.S.$451 million. In addition, according to Whiting (1992), by 1988, 10 per cent of IBM worldwide production of PC components was reported to be produced in Mexico, and from the total of computers made in Mexico, 60 per cent were built in Guadalajara.

The Origin of the Electronics Cluster

The foundation of the electronics industry in Jalisco dates back to the 1950s and 1960s, when large MNCs, such as IBM, Siemens, Kodak, Motorola and Burroughs[6] established operations in Guadalajara. The OECD registers 1975 as the year of growth of this industry. In fact, it was only after the launch of the Program for the Promotion of the Computer Industry (PMSEC) by the Mexican government, and after IBM began operations in Guadalajara as a sole owner in 1986, that the cluster began to take shape. Thus, basically, it was mainly during the decade of the 1980s and essentially during the 1990s that this industry registered impressive dynamic growth. The growth during the 1980s was stimulated by the new industrialization policy promoted by the Mexican government, which allowed total ownership to foreign capital. The rapid growth of the electronics cluster in Guadalajara during the 1990s took place within the contextual framework of the opening of the Mexican market to imports and the North American Free Trade Agreement (NAFTA). An important number of the most important corporations recognized as world-class innovators and designers established operations in the metropolitan area of Guadalajara, particularly in the industrial corridors of 'El Salto'. Among the noteworthy are: IBM, Kodak, NEC, Motorola (later ON Semiconductors), Hewlett-Packard, Siemens, Lucent Technologies (later Phillips Consumer Communication), Compac, Intel, Telmex, etc. This cluster of electronics firms, identified as original equipment manufacturers (OEMs), has been able to attract international and mixed capital suppliers, contract equipment manufacturers (CEMs) and specialized suppliers (SSs).

Contextual Framework

The change in the industrialization policy issued by the Mexican government by the mid-1980s, relaxing the 51 per cent limit for foreign ownership of capital in joint ventures to allow 100 per cent foreign ownership, was aimed to attract FDI in electronics, which was considered a strategic industry for the development of the nation. Mexico targeted FDI in electronics by designing more friendly policies and incentives such as offering tax exemption and other incentives with the expectation of obtaining the transfer of technical assets, managerial skills and markets. The Mexican government, which was successful in attracting FDI in electronics, demanded from international investors the fulfilment of the following four main conditions in exchange: to increase exports; to acquire technology through transfer arrangements; to establish relationships with Mexican academic institutions; and to train indigenous suppliers (OECD 1990; Whiting 1992).

Consequently, by the end of the 1980s and early 1990s, among the thirty firms registered in the cluster, there were seven local firms and only nine joint ventures, in which ownership structure was 51 per cent Mexican capital and 49 per cent foreign capital. All the rest were totally foreign owned.

Along with the change in the industrialization policy to promote FDI, the scientific and technological policy of the nation was modified after the mid-1980s. The adjustments registered in the National Development Plan, 1989–1994, addressed the following:

- The private industrial sector was to become a major protagonist of the generation and assimilation of technology.
- A modification in the industrial property systems was introduced in order to promote the deregulation and protection of the external technological flows in order to increase its intensity.
- A greater emphasis was placed upon the formation of scientific and technological capacities in order to ensure the efficient and effective assimilation of foreign technology by the national productive system.

In addition, by the mid-1980s the Ministry of Education (SEP) was stating its demands for research and technological development more emphatically, as well as calling for closer cooperation between the industrial sector and the universities. Consequently, SEP in conjunction with the National Association of Universities and Higher Education Institutions (ANUIES) initiated programs which included the National Program for Higher Education (PRONAES) and the Integral Program for the Development of Higher Education (PROIEDES) in 1984 and 1985, respectively. Within the framework of these programs, the National Program for Educational Development 1995–2000[7] aimed to achieve the following objectives[8]: to increase the response of the educational sector to the demands of the community, especially the productive sector; to consolidate and to strengthen the partnership and links between higher education and the productive sector; and to promote joint research projects with industry. To what extent were these objectives achieved through interactions of the national and international innovation systems and through the transfer and assimilation of technology?

Innovation Systems Interactions and Technology Transfer in Mexico

Traditionally, technology transfer has been conceptualized as the selling or contracting of technology through any of the commonly utilized formal mechanisms or legal instruments, such as licensing, patents, packages, trademarks, turnkey plants and consulting technical services. Given the nature of the electronics cluster formation in Guadalajara, which represents a passive internalization process, the focus here is on the informal mechanisms of technology transfer. This study excludes,[9] therefore, all those commercial transactions or formal mechanisms mentioned earlier. Technology transfer is defined as the transfer or spillovers which, according to the literature, occur as a result of the MNCs operations in a foreign country and have major impact on developing countries. These informal mechanisms include: *training and turnover of employees*; *reverse engineering*; *personal contacts*; *study groups between academia and the industrial sector*; *involuntary leakage of technical information*; and *training of local suppliers* (Bonin 1972; Robinson 1988; Stewart and Nihei 1987; United Nations 1977; Kim 1980; Larsen, Wigand and Rogers 1987; Wigand 1990; Williams and Gibson 1990).

The Findings

Of the thirty-two activities[10] more commonly undertaken internationally between the academic, industrial and governmental sectors, thirty-one were found in the region.[11] However, only seventeen cooperative activities were identified between some of the foreign high-technology electronics corporations included in the study and some of the national firms and research and higher education institutions in the sample. Of these activities, fourteen are considered direct links and three are considered indirect links. In the following paragraph, we will address only the most strategic direct activities leading to technology transfer.

One of these strategic activities relates to *applied research linkages*, like in the case of applied research project links between the research institute CINVESTAV and three corporations recognized as world-class innovators, namely, IBM, Hewlett-Packard and Lucent Technologies. The director of this Mexican research institute indicated that the outcome of the joint research project included several technological developments, including, among others, an impact printer, a motherboard based on a bluelighting, multi-layer printed circuit cards and more than one hundred printed circuit boards (personal communication, 18 November 1997). However, by 1997, the majority of the electronics corporations requiring applied research projects were located outside the country. Neither the higher education institutions nor the corporations interviewed reported being engaged in basic research linkage projects. By 1999, the ongoing applied research projects were not in electronics or related areas.

Consequently, within the academic–industry relationship examined, no co-publications and no joint efforts for securing patents were reported. This is despite the fact that some patents resulted from the joint applied research activities between the MNCs and CINVESTAV. The reason given by the director is that usually the contracts between the industry and academia stipulated that the patent resulting from the research project belongs to the firm funding the project.

The absence of cooperative projects in applied research might be explained by the fact that the foreign corporations chose the option of building their own research facilities. Six of the eight global innovators incorporated in the sample of thirteen corporations examined in this study have installed their own research facilities in their Guadalajara plants and about three hundred Mexican engineers were working as researchers at these MNCs by 1997. The number of corporations with research and design functions in Guadalajara grew to eight in 2000 when Intel started operating there, focusing only on the function of electronics design. To start its operations, this MNC bought TDCOM, a Mexican firm specializing in electronics design, and ASIC, which was created in 1998 by CINVESTAV. Some of these corporations with R&D laboratories have obtained a good number of internationally recognized patents and one of them obtained product responsibility in printer design.

The Triple Helix Presence

The triple helix paradigm as conceived by Etzkowitz (2003) is still very incipient in Mexico. It began to take shape in the country around the second half of the 1980s with changes in the educational and scientific-technological policy of the country, registered in the National Development Plan of 1989–1994. Although it is just indicative planning, during more than twenty years some progress was observed, albeit without the generalization of the paradigm. On the contrary, triple helix is still a very selective process. The selection in the participation seems to be related to institutional plan consistency, consolidation of functions and experience in cooperation. Thus, MNCs play a very important role not only in university–industry cooperative relations, but also in the electronics cluster formation of Jalisco. That is, without the establishment of some of the most important corporations recognized as world-class innovators and designers in the metropolitan area of Guadalajara, the electronics cluster in Jalisco could not have achieved its current level of integration.

The process of development of this electronics cluster reflects an example of a triple helix paradigm in which the role and the influence of the government and MNCs are highlighted. This is evident through the cooperative interactions and relationships between academic institutions and indigenous as well as multinational firms promoted by the scientific-technological, educational and industrial policies designed by the government, as well as the funding provided by the government and in some cases by corporations. The government provided the funding of more than 75 per cent of the scientific, technological and educational activities at all levels. It funded, for example, the creation in Guadalajara of a graduate and research institution in agreement with one of these corporations, IBM, which also participated with 50 per cent of the funding. In consequence, CINVESTAV was created in Guadalajara in 1988 as a transplanted unit of the National Polytechnic Institute (IPN) founded in Mexico City in the 1930s. Its purpose since its foundation in Guadalajara was to respond to the electronics industry's needs. CINVESTAV Guadalajara specifically created the Semiconductors Research Centre (CIS) to provide technological service and work jointly with the electronics industry in the region. This is the first and the most representative and transcendent action, which fits into the triple helix model. As it was demonstrated in the description of innovation systems interactions, this research centre has become the 'angular stone' in the growth and development of the electronics cluster in Jalisco, even in ulterior stages, as it will be demonstrated.

Shaking the Three Helices

A certain event had to occur, and as a result, the three helixes had to make the decision to achieve a certain industrial restructuring. This unexpected event was the crisis in the electronics cluster of Guadalajara that reached its peak in 2001 when five large MNCs left for China and Malaysia with a subsequent decline in FDI, employment and exports in the region. This crisis highlighted the crucial need for an alternative strategy that can be associated with the triple helix model. This alternative strategy was related to an industrial restructuring aimed at promoting the shift from assembly and manufacturing to electronics design.

The triple helix paradigm can be perceived as linked to the cluster's strategic industrial restructuring, which was conceived, planned and promoted by representatives of the three sectors commonly identified within the triple helix paradigm. The roles played by the government, MNCs and academic institutions were crucial in the development of world-class electronics designers and entrepreneurial capacity.

The interactions between these three institutional actors, which led to the industrial restructuring, included new course development and funding. The government, represented by the State Council of Science and Technology (COECYTJAL) convinced CINVESTAV to design new courses and be in charge of the training required to form world-class electronics designers and entrepreneurial capacity. This indigenous research institution represented the academic sector within the model, whereas the industrial sector is represented by the multinational Intel. CINVESTAV has worked with Intel in the design and creation of new training programs, which in turn have been financed by COECYTJAL and have been equipped with cutting-edge equipment provided by Intel.

One of the programs created is the Advanced Program in Electronics Design and Software (PADTS), whose main objective was to train five hundred highly qualified and world-class electronics designers in five years.

Besides its technical collaboration with CINVESTAV in the design of the program, Intel invested about U.S.$50,000 in equipment, and COECYTJAL, a government agency, in turn provided the funds to sustain the educational program and the scholarships for the students.

The other scheme is the Program for the Creation of Technological Entrepreneurship, which aims to produce at least thirty entrepreneurs in two years. This program provided the origin of a cooperative interaction within the triple helix context. The participation of each helix was significant and exhibited higher levels of integration that accounted for the electronics cluster of Jalisco evolving from assembly and manufacturing to design.

As a result of these programs, the design activities at Intel registered a 50 per cent increase between 2002 and 2004. By the end of 2007, the program PADTS had achieved 50 per cent of its initial goal of training five hundred electronic designers. In general, in the region there was a growing investment in R&D and therefore in innovation. This created opportunity for synergy to the point that some of the MNCs that had left for China and Malaysia in the past decided to come back and install their functions of research and design in the region. One of these corporations is Motorola, later on transformed into ON Semiconductors. It created a design centre named Freescale, which is dedicated to applications and semiconductor design. But most important is the creation and growth of more than 150 small software firms as a result of the 250 world-class electronics designers who graduated from PADTS and the technological entrepreneurial group formed by the recently created program. In the electronics cluster of Jalisco, there is currently a value creation based on technology design (Medina 2004). As a result of this process, the electronics cluster of Jalisco has modified its composition qualitatively. In 2004, it comprised thirteen OEMs; seventeen CEMs; more than four hundred SSs; twenty-seven design houses, nineteen of which were small companies; and 151 small firms dedicated to software development (Medina 2004).

As it can be seen, the impact of the triple helix paradigm was of major significance and relevance for the development of the electronics cluster in Jalisco as it has helped to promote the leap to the stage of design function.

The Planning of a Technological Learning Process

The interaction between national and international innovation systems is significantly reflected in technology transfer to the region, which has been taking shape since 2005. This is promoted by the 'Group of Technology Transfer', composed of a National Industrial Chamber, several MNCs, small indigenous firms, government agencies and some universities. The work of this group could also be identified as part of the triple helix paradigm. Representing the industrial sector are the most important MNCs with operations in Jalisco, such as Intel, IBM, Hewlett-Packard, Siemens, Freescale, Jabil, etc. The industrial group also included some of the small indigenous software companies (recently created through the two special programs just reported) and the National Chamber of Electronics and Telecommunications Industry (CANIETI). The government sector is represented by COECYTJAL and the Productive Chain of Electronics Industry Suppliers (CADELEC). The academic helix consisted of two higher education institutions and the United States-Mexico Foundation for Science (FUMEC). The international innovation system within this group was represented by the MNCs mentioned earlier and FUMEC. The national innovation system included COECYTJAL and CADELEC, as well as the two higher education institutions and the software small firms. The program's main objective was to transfer technology to the small indigenous software firms aiming to develop core capacities, so that key competencies could be acquired and consolidated in the cluster.

This group of technology transfer has designed strategies for the short and long run. The short-run strategy has as a main objective sharing best practices with the small indigenous software firms. The long-run strategy aims to stir up the planning of a technological learning process among the small indigenous software firms with the objective of building the key competencies on the basis of the strategic capacities previously identified in the region by a study of FUMEC.

The short-run strategy was implemented in August 2005 with a program consisting of courses offered by the participants in the 'Technology Transfer Group' to the small indigenous software companies. The courses offered included: Structured Solution to Problems, Strategic Marketing, Project Management, Methodology of Inventive Solution to Problems, Software Testing, Hardware Testing and the Planning of the Technological Learning Process in Korea: Lessons for Latin America, etc.

The long-run strategy was instigated in order to motivate and involve the small indigenous software companies in a self-evaluating process. This self-evaluating process has the objective of identifying the core competencies of these companies and helping them to define their main goals for development in the long run before they can start planning their technological learning process and developing the key competencies in the region. Studies promoted by this group in the region have identified technical capacities in testing and embedded software. This has led the group of technology transfer planning to promote a technological learning process in the region in these two areas: testing and embedded software.

The triple helix paradigm is selective but with more participation from the foreign and indigenous organizations. However, there are still only two indigenous academic institutions participating in the triple helix relationships.

Technological Spillovers

Three informal mechanisms of technology transfer were found to be active within the interactions between the national and international innovation systems taking place in Guadalajara. The next sections identify the informal mechanisms activated by the MNCs operations and discusses their impacts.

Training and Turnover of Employees Impact

MNCs play an important role in the training of staff in host countries, giving rise to the occurrence of technological spillovers through the turnover of employees (Baranson 1978; Caves 1974; Lewis 1958; Mansfield and Romeo 1980; Robinson 1988). However, during the first decade of the cluster, this study[12] did not find any evidence of a major MNC impact in the region in spite of the training provided by eight of the thirteen corporations included in the sample. Training was provided but there was hardly any spillover gain to be had by indigenous firms upon turnover of personnel. As the MNCs reported, the turnover of employees was very low in this region (an average of 0.5 per cent in 1997). Although there was a general perception of turnover increasing after 1997, this turnover of personnel took place mainly among MNCs and not from MNCs to indigenous firms. If anything, indigenous firms and institutions were losing their best trained staff to MNCs who offered better-paid jobs.[13] In addition, 78 per cent of the training provided to corporate employees and 100 per cent of the training offered to professors and students was essentially concentrated at the operations level.

Reverse Engineering Impact

According to Pavitt (1985), the reverse engineering and the links between engineering and production departments are central to the process of the international transfer of technology. The literature has extensively addressed the role of reverse engineering in the assimilation of foreign mature technology in Japan (Tsurumi 1980; Ozawa 1985; Lynn 1985) and in the industrialization of the Asian NICs through the case of Korea (Rosenberg and Frischtak 1985; Schive 1990; Gereffi and Wyman 1989). However, in Guadalajara, the reverse engineering practiced was very scanty and elementary. It was limited to certain imitations and adaptations of electronic cards. Higher education and research institutions as well as the corporations and indigenous firms reported very limited efforts in reverse engineering (Rivera 2002). This could be explained, *inter alia*, by shortfalls in the technical and financial resources that were required for reverse engineering[14] and the law regarding intellectual and industrial property, which is more restrictive now than it was when Japan and the other newly industrialized South-East Asian countries were assimilating foreign technology.

The empirical research reported that only four higher education and research institutions from the nine included in the study[15] declared that they had performed adaptation to electronic cards and micro-controls donated by some of the corporations. These institutions confirmed that they analysed and assimilated the technology incorporated in those donated electronic cards and micro-controls, and they designed electronic cards according to their own needs. Thus they modified the functions, in some cases replacing the materials used with low-cost options, and mitigated the limitations and problems of the original cards, such as overheating, for example. In some cases, the cards were reproduced and modified with the purpose of being used in the universities' electronics laboratories for teaching practices. In other cases, students and professors at the higher education and research institutions modified and adapted these electronic cards to specific instruments or equipment utilized in small and micro indigenous industry. None of the indigenous firms or the corporations reported activities of reverse engineering by the end of 1999. However, another study[16] carried out during the period 2001–2002 reported the adaptation of flexible manufacturing and quality control courses to their own production systems by indigenous suppliers of the electronics corporations.

Training of Indigenous Suppliers Impact

Phillip Lasserre (1982) highlighted training of suppliers to be of crucial importance as a mechanism for technology transfer. Singer (1988) considers the training of suppliers as one of the mechanisms through which MNCs transfer technology across national borders. Several authors agree with Singer, but they also argue that the training MNCs offer to local suppliers is essentially limited to quality control methodologies (Bonin cited in Paquet 1972; Schive 1990; Stewart 1988; Stewart and Nihei 1987). In the case of Guadalajara, from 1997 to 1999 five of the thirteen corporations included in the sample reported they had trained a total of twelve indigenous suppliers, mainly in quality control. The training provided in manufacturing skills and efficiency in the production process was diversified, with nine courses offered to eight of the twelve indigenous suppliers by two corporations. Rivera and Regino's study (2004) also reported that five indigenous suppliers received some training in quality control and two of them also received some training in flexible manufacturing.

Thus, during the first stage of the cluster, most of the twelve indigenous firms acting as suppliers of the electronics industry in Guadalajara received some training, mainly in quality control. Therefore, 100 per cent of the training received by the indigenous suppliers is concentrated in operative capacity. But most of these suppliers were in the packaging industries; the rest were from the plastic industry (two) and the metal mechanic industry (one). The absence of the indigenous electronics firms amongst these suppliers is both striking and intriguing.

However, by the end of 1989, Guadalajara had seven indigenous firms in electronics. Some of these were producing integrated circuit boards and buffers. Other local firms were capable of producing and designing their own printers and computers (Wilson 1996). There was thus evidence of the existence of a certain level of capacity in the indigenous electronics and metal mechanics firms.

The global production strategies of MNCs, among them their international network of suppliers, is designed, decided and contracted at their parent companies. Ortiz (2000) reported the CEMs claim that they are often unable to buy Mexican-made materials because their MNC clients, who are the OEMs, have purchase commitments elsewhere. Thus, the barriers created by the highly concentrated global electronics industry and the 'domination of the United States and Asia in electronics makes it very difficult for newcomers to enter this field' (Amsden 2004). Thus, even if the training provided by the corporations could have been consistent and formal, and even if some technical and financial support could have been seriously and systematically provided by the government to the indigenous firms, it was very unlikely to have brought forth their effective integration as suppliers in the electronics industry productive chain.

The weakness of the extent of integration of the electronics industry into the Mexican economy is reflected in the low contribution of the industry to local value added. In Jalisco, the value added[17] by the electronics industry to the export sector was reported to have jumped from 2 per cent in 1994 to approximately 15 per cent in 1998 and 16 per cent in 2000 (CANIETI, personal interviews, 1998, 2000).

This situation also explains why by the end of 1980s, 71 per cent of the small indigenous electronics firms have been pushed out of the market and 40 per cent of those still operating during the first half of the 1990s were displaced. This virtual wipe-out of Guadalajara's fledgling indigenous firms is a powerful indicator of the limited investment capacity created in the region during the first two decades of MNC operation in Mexico considered in this chapter.

After the electronics cluster crisis of 2001, the results of national and international innovation systems interactions for the training of world-class electronics designers and technological entrepreneurs could be perceived to have been productive with the formation of more than 150 small software firms and the improvement and strengthening of investment, design and innovation capacities. Moreover, the technology transferred by the Group of Technology Transfer since 2005 could be seen as leading to the initiation of a technological learning planning process, which offers the possibility of further strengthening the created key design and innovation capacities.

CONCLUSION

Analysis of the Korean technological learning process, through the case of Hyundai,[18] revealed how Korea pursued a strategy of independence by developing absorptive capacity and therefore technological capability by preparing the groundwork for the acquisition, assimilation and development of foreign technology. Although this process is characterised by a significant reliance on international innovation systems, specifically MNCs, it is important to acknowledge the important role played by the leadership of the Hyundai group as risk-takers in investment ventures in Korea (Kirk 1994). This process was also considerably influenced by several idiosyncratic features of organizational learning within the catching up process. These are important managerial and organizational learning strategies such as creative crisis, intensity of effort and systemic planning involved in this colossal technological learning process.

To the specific characteristics of the unique Korean learning process just pointed out, the availability of a well-trained human resource (United Nations Development Program 1994) and the Korean culture of hard work should be added (Kim 1997, 1998).

However, the findings observed in Guadalajara, after almost two decades of MNC operations with tax exemption and incentives, can be seen as examples of practices and procedures which can give rise to the 'vicious circle' phenomenon, which Cantwell (1989, 1995) and Perez (1998) identified as resulting from the passive internationalization process. The 'vicious circle' is expressed in terms of the minimum technological spillover and the displacement of the indigenous electronics and computer firms from the region. It is caused by the encounter between the innovative and productive systems of these countries, where there is an initial technological gap, as in the case of Mexico. This suggests that the policies and strategies set up by the Mexican government to promote FDI in the electronics and computer industry in order to obtain technology, capital and markets did not achieve their objectives. This is largely accounted by the limited role played by the government in the provision of structural reforms and proper mechanisms to promote interaction and coordination between MNCs and indigenous firms and institutions to achieve the declared objectives (Perez 1998; Cantwell 1995).

Policies and strategies were set to promote technology transfer and to take advantage of the opportunities offered by the international mobility of goods and production factors. They were also aimed at boosting exports by training local suppliers and establishing relationships and links with the academic and scientific institutions of the region. However, this did not happen. In the case of Jalisco, the economic promotion law[19] issued by the state government in 1997 confirms the existence of strategies with that purpose. However, these policies and strategies were effectively nothing more than good intentions.

For the most part, this state of affairs can be explained by the lack of coordination and coherence in policy design and implementation, duplication of programs, inadequate and inconsistent technical and financial support to the small and medium-size firms (Dussel 2001) and the lack of adequate orientation to enhance the technological learning process of indigenous suppliers (Rivera and Regino 2004). These policy shortfalls experienced in Mexico, together with other factors identified by several authors,[20] explain Mexico's failure and Korea's success in the catching up process.

Contrasting these dissimilar technological learning processes and outcomes found in these two distant regions based on the typology advanced by Perez (1998) about the different types of interactions of national and international innovation systems, it can be said that the technological learning experience in Korea can be categorized as Type III, which, according to this author, is characterised by a national innovative and productive system where the role played by the international component in developing the competitive advantages of resident firms is significant. In the case of Mexico, on the other hand, the experience in Guadalajara's electronics industry, which saw the displacement of indigenous firms and the generation of minimal technological spillovers during the first two decades of MNC operation, bears the characteristics of the Type II category of technological learning.

NOTES

1. Business groups in Korea.
2. During the period 1974–1977, according to Kim (1998).
3. Front engine, front-wheel drive.
4. It is a special and conditioned space for testing.
5. To simplify, from now on it will be identified as 'electronics industry'.
6. According to Palacios (1994) Burroughs arrived in Guadalajara in 1968 but later changed its name to Unisys and finally closed down in 1991.
7. Cited in OECD (1997).
8. The national policies for science and technology as well as for higher education mention the social, political and economic aspirations and expectations from higher education institutions. These expectations have been translated into objectives and strategies for research, university–industry links, and assimilation, adaptation, development, creation and diffusion of technology in the programs created to support and achieve those purposes.
9. Nevertheless, if a case of a formal mechanism of relevance to this study was found, it will be reported.
10. See Cummings (1995, 1998); Jones and Jenkins (1999); Stewart and Gibson (1990); Klingstrom (1987); Powers et al. (1988); Rosmalen (1998); Stauffer (1980); Wigand (1990); William and Gibson (1990). The thirty-two activities described in Appendix A (Rivera 2002): 179–182.
11. Rivera (2002).
12. Ibid.
13. See Castillo (1992) and Rivera (2002).
14. Reverse engineering and adaptation require a technological enterprise and as great an amount of resources as those required for the original research and design effort (Pavitt 1985).
15. Rivera (2002).
16. Rivera and Regino (2004).
17. The value added was estimated by CANIETI using differences in export and import values without including labour.

18. The analysis of Hyundai is generalized to Korea given that it has been demonstrated by a series of studies covering more than two hundred firms that a similar process has occurred in other industries, see p. 4.
19. Secretary of Economic Promotion (1997).
20. See Castells and Laserna (1989); Cummings (1994); Gereffi and Wyman (1989); and Hanson (2008).

REFERENCES

Amsdem, A. 1989. *Asia's next giant: South Korea and late industrialization*. New York: Oxford University Press.

———. 2004. 'Import substitution in high-tech industries: Prebisch lives in Asia!' *CEPAL, Review* 82 (April): 77–91.

Amsdem, A., and L. Kim. 1985. 'The acquisition of technological capability in Korean industries'. World Bank, mimeograph.

———. 1986. 'Technological perspective on the general machinery industry in the Republic of Korea'. In *Machinery and economic development*, ed. Martín Fransman, 93–123. London: Macmillan Press.

Baranson, J. 1978. *Technology and the multinationals: Corporate strategies in a changing world economy*. Lexington, MA: Lexington Books.

Bonin, B. 1972. 'The multinational firm as a vehicle for the international transmission of technology'. In *The multinational firm and the nation state*, ed. G. Paquet, 111–126. Don Mills, Canada: Collier-Macmillan.

Cantwell, J. 1989. *Technological innovation and multinational corporation*. Oxford: Basil Blackwell Ltd.

———. 1995. 'Multinational corporations and innovatory activities: Towards a new, evolutionary approach'. In *Technological innovation, multinational corporations and new international competitiveness: The case of intermediate countries*, ed. José Molero, 21–58. New York: Harwood Academic Publishers.

Castells, M. and Laerna, R. (1989). The new dependency: Technological change and socioenconomic restructuring in Latin America. *Sociological Forum*, 4(4).

Castillo, P. D. 1992. 'Technology transfer via multinational high technology firms: The Guadalajara, Mexico experience'. Master's thesis, University of Texas at Austin.

Caves, R. 1974. 'Effects of international technology transfer on the US economy'. Washington, D.C. *National Science Foundation*: 36–48.

Cimoli, M., and G. Dosi. 1994. 'De los paradigmas tecnológicos a los sistemas nacionales de producción e innovación'. *Comercio Exterior* 44 (8): 669–684.

Cummings, W. (1995). The Service University Mimeo.

———. (1998). The Service University in comparative perspective. Higher Education 35(1), pp. 1–8.

Dussel, E. 1999. 2001. 'Condiciones y retos de las pequeñas y medianas empresas en México, Estudio de casos de vinculación de empresas exitosas y propuestas de política'. In *Claroscuros: Integración exitosa de las pequeñas y medianas empresas en México*, ed. Enrique Dussel, 17–62. Mexico City: Editorial Jus, S. A. de C. V.

Etzkowitz, H. 2003. 'Innovation in innovation: The triple helix of university-industry-government relations'. *Social Science Information* 42 (3): 293–338.

Freeman, C. (1987) *Technology policy and economic performance: Lessons from Japan*. London: Frances Pinter.

———. 1995. 'The "National System of Innovation" in historical perspective'. *Cambridge Journal of Economics* 19:5–24.

Gereffi, G., and D. Wyman. 1989. 'Determinants of development strategies in Latin America and East Asia'. In *Pacific dynamics: The international politics of industrial change*, ed. Stephan Haggard, and Chung-In Moon, 23–52. Coulder, CO: Westview Press.

Hanson, M. 2008. *Economic development, education and transnational corporations*. New York: Routledge.

Hyun, Y. 1995. 'The road to self-reliance: New product development of Hyundai Motor Company'. Paper presented at International Motor Vehicle Program Annual Sponsors Meeting, Toronto, 4–7 June.

Hyun, Y., and J. Lee. 1989. 'Can Hyundai go it alone?' *Long Range Planning* 22(2): 63–69.

Keesing, D. B. 1966. 'Labor skills and comparative advantage'. *American Economic Review* 56 (2): 249–258.

Kim, L. 1980. 'Stages of development of industrial technology in a developing country: A model'. *Research Policy* 9 (July): 254–277.

———. 1985. 'Innovation in a newly industrializing country: A multiple discriminate analysis'. *Management Science* 31 (3): 312–322.

———. 1997. 'The dynamics of Samsung's technological learning in semiconductors'. *California Management Review* 39 (3): 86–100.

———. 1998. 'Crisis construction and organizational learning: Capability building in catching-up at Hyundai Motor'. *Organization Science* 9 (4): 506–521.

Kirk, D. 1994. *Korean dynasty: Hyundai and Chung Ju Yung*. Armonk, NY: M. E. Sharpe.

Klingstrom, A. (1987). Cooperation between higher education and industry. Proceedings from the seminar in Uppsala. Uppsala, Sweden: Uppsala University Textgnuppen.

Lall, S. 1985. 'Trade in technology by a slowly industrializing country: India'. In *International technology transfer: Concepts, measures, and comparisons*, ed. N. Rosenberg and C. Frischtak, 45–76. New York: Praege.

Larsen, J. K., R. T. Wigand and E. M. Rogers. 1987. 'Industry–university technology transfer in microelectronics'. Report submitted to the National Science Foundation.

Lasserre, P. 1982. 'Training: Key to technological transfer'. *Long Range Planning* 15:51–60

Lewis, A. 1958. 'Economic development with unlimited supplies of labor'. In *The economics of underdevelopment*, ed. A. Agarwala and S. Singh, 400–449. London: Oxford University Press.

List, F. 1841/1904. *The national system of political economy*. London: Longman.

Lynn, L. 1985. 'Technology transfer to Japan: What we know, what we need to know, and what we know that may not be so'. In *International technology transfer: Concepts, measures, and comparisons*, ed. N. Rosenberg and C. Frischtak, 255–276. New York: Praeger.

Mansfield, E., and A. Romeo. 1980. 'Technology transfer to overseas subsidiaries by U.S.-based firms'. *Quarterly Journal of Economics* 95:737–750.

Medina, F. 2004. 'The case of Jalisco and the Program of the State Council of Science and Technology'. Paper presented at the Challenges and Perspectives on Technological Development in Mexico Seminar, University of Guadalajara CUCEA, Guadalajara, 10 December.

Mowery, D., and J. E. Oxley. 1995. 'Inward technology transfer and competitiveness: The role of national innovation systems'. *Cambridge Journal of Economics* 19:67–93.

Narula, R. 1996. *Multinational investment and economic structure*. London: Routledge.

Narula, R., and K. Wakelin. 1995. 'Technological competitiveness, trade and foreign direct investment'. In *New challenges for European and International business, Vol. 2, Proceedings of the 21st Annual Conference of the European International Business Academy, Urbino, Italy, 10–12 December*, ed. R. Schiattarella. Rome: Cofindustria.

Nonaka, I. 1988. 'Creating organizational order out of chaos: Self renewal in Japanese firms'. *California Management Review* 30 (3): 57–73.

OECD. 1990. *Foreign direct investment and industrial development in Mexico*. Paris: CEDEX OECD.

———. 1997. *Exámenes de las políticas nacionales de educación: México Educación Superior* [*Review of national policies for education: Mexico higher education*]. Mexico City and Paris: OECD-CEDEX.

Ortiz, F. 2000. *Help wanted in Guadalajara hi-tech factories*. Mexico: Reuters.

Ozawa, T. 1982. 'A newer type of foreign investment in third world resource development'. *Rivista Internazionale di Scienze Economiche e Commerciali* 29 (12). Also reprinted in Readings in Transnational corporations and the Natural resource Industries. 1993. ed. Brue McKern, vol. 10 UNLTNC. London: Routledge.

———. 1985. 'Macroeconomic factors affecting Japan's technology inflows and outflows: The postwar experience'. In *International technology transfer: Concepts, measures, and comparisons*, ed. N. Rosenberg and C. Frischtak. pp. 222–254. New York: Praeger.

Palacios, J. J. 1994. *Foreign direct investment and technology transfer in the Pacific Rim: The case of the electronics industry in two proto-silicon valleys*. Vancouver, BC: Institute of Asian Research at the University of British Columbia.

Paquet, G., ed. 1972. *The multinational firm and the nation-state*. Toronto: Collier-Macmillian.

Pavitt, K. 1985. 'Transfer among industrial advanced countries'. In *International technology transfer, concepts, measures, and comparisons*, ed. N. Rosenberg and C. Frischtak, 1–23. New York: Praeger Publishers.

Perez, T. 1998. *Multinational enterprises and technological spillovers*. Amsterdam, The Netherlands. Harwood Academic Publishers.

Pitt, M. 1990. 'Crisis modes of strategic transformation: A new metaphor for managing technological innovation'. In *The strategic management of technological innovation*, ed. R. Loveridge and M. Pitts, 253–272. Chichester, UK: Willey.

Porter M. 1990. *The comparative advantage of nations*. New York: Free Press.

Powers D., Powers, M., Betz, F., an Aslanian, C. (1988). Higher education partnership with industry. San Francisco, CA: Jossey Bass Publishers.

Rivera, V. 2002. *Technology transfer via university–industry relationship: The case of the foreign high technology electronics industry in Mexico's Silicon Valley*. New York and London: Routledge Farland.

Rivera V., and J. Regino. 2004. 'Entrenamiento y Aprendizaje Tecnológico en la Subcontratación: El caso de los proveedores endógenos en la industria electrónica' ['Training and technological learning in subcontracting: The case of the indigenous suppliers in the electronics industry']. *Comercio Exterior* 3 (54): 196–206.

Robinson, R. 1988. *The international transfer of technology: Theory, issues and practice*. Cambridge, MA: Ballinger Publishing Company.

Rodríguez-Clare, A. 1996. 'Multinationals, linkages, and economic development'. *American Economic Review* 86 (4): 852–873.

Rosenberg, N., and C. Frischtak, eds. 1985. *International technology transfer, concepts, measures, and comparison*. New York: Praeger Publishers.

Rossmalen, K. (1998). Strategic alliances from contract to organisation. *Industry and Higher Education* 12(3): 139–146.

Schive, C. 1990. *The foreign factor: The multinational corporation's contribution to the economic modernization of the Republic of China*. Stanford, CA: Hoover Institution Press.

Schon, D. 1967. *Technology and change: The new Heraclitus*. Oxford: Pergamon.
Singer, H. W. 1988. *Technology transfer by multinationals. Part I and II*. New Delhi: Ashish Publishing House.
Stewart, C., and Y. Nihei. 1987. *Technology transfer and human factor*. New York: Arno Press.
Stauffer, T. (1980). Agenda for business and higher education. Washington, D.C.: American Council on Education.
Stewart, F. 1988. 'Technology major issues for policy in the 1980s'. In *Technology transfer by multinationals Part I*, ed. H. W. Singer, N. Hatti and R. Tandon, 17–45. New Delhi: Ashish Publishing House.
Stewart, H., and D. Gibson. 1990. 'University and industry linkages: The Austin Texas, Study'. In *Technology transfer: A communication perspective*, ed. F. William and D. Gibson, 183–205. Newbury Park, CA: Sage Publications.
Tsurumi, Y. 1980. *Technology transfer and foreign trade: The case of Japan, 1956–1966*. New York: Arno Press.
United Nations. 1977. *Transnationals in world development—an examination*. New York: UN.
———. 1988. *Transnational corporations in world development. Trends and prospects*. New York: United Nations Organization, Center on Transnational Corporations.
United Nations Development Program. 1994. *Human resource development report, 1994*. New York: Oxford University Press.
Weick, K. 1988. 'Enacted sensemaking in crisis situations'. *Journal of Management Studies* 25 (4): 305–317.
Westphal, L., L. Kim and C. Dahlman. 1985. 'Reflections on the Republic of Korea's acquisition of technological capability'. In *International technology transfer, concepts, measures, and comparisons*, ed. N. Rosenberg and C. Frischtak, 167–221. New York: Praeger Publishers.
Whiting, V. R. 1992. *The political economy of foreign investment in Mexico*. Baltimore, MD: The Johns Hopkins University Press.
Wigand, R. 1990. 'University and microelectronics industry: The Phoenix, Arizona, Study'. In *Technology transfer: A communication perspective*, ed. F. Williams and D. Gibson, 132–152. Newbury Park, CA: Sage Publications.
Williams, F., and D. Gibson, eds. 1990. *Technology transfer: A communication perspective*. Newbury Park, CA: Sage Publications.
Wilson, P. 1996. *Las nuevas maquiladoras de Mexico: Exportaciones y desarrollo local* [*Mexico's new maquiladoras: Exports and local development*]. Austin: University of Texas at Austin.

3 A New Regulatory Framework for the Relationship between University and Industry

The Brazilian Experience

Anne-Marie Maculan and
José Manoel Carvalho de Mello

INTRODUCTION

Brazil today is attempting to structure a proper system of innovation capable of integrating the efforts of universities and research institutions and the business sector in a lasting and consistent manner. The three actors—government, academic and research institutions and the business sector—are changing their strategies with regard to technological training, innovation and the recognition of the value of knowledge.

Over the last three decades the research system in Brazil has shown notable progress. But the capacity of the industry to innovate has not kept pace with this progress and continues to be quite modest. This situation is the result of the particular characteristics of the process of delayed industrialization.

Over the last decade, the government has adopted a set of measures with the objective of encouraging universities to intensify the transfer of knowledge to companies, by establishing a regulatory framework favourable to innovation.

Universities were given the mission of playing a more direct role in the development of the local economy, to be active instruments in the technological training of companies and to promote technological entrepreneurship. Some of them, acting on their own, had already taken such initiatives during the preceding years. The new regulatory environment ratified and opened new possibilities.

This chapter examines the impact of the new regulatory environment and focuses on the changes introduced to the interactions between the system of research and the system of production to intensify and spread the transfer of knowledge to companies.

During the 1990s, the policy of the Brazilian government with respect to support for research and innovation included measures to create opportunities for the relationships between universities and companies to become stronger and break with the long-standing lethargy in innovation. The encouragement of interaction between the two institutional spheres of the triple helix—university and industry—evolved into an essential goal as a determinant of gains in competitive capability of industry and consequently economic development. A new regulatory environment was configured that included many features of the policies implemented in various countries around the world with the same objectives. New opportunities appeared that allowed the principal actors in the system of innovation to cooperate more effectively and more intensively in this effort.

Brazil's policy of science, technology and innovation has incorporated a fundamental concept of the triple helix. The emphasis was placed on the importance of recognizing the value of knowledge generated in the universities. The commercialization of knowledge through the licensing of patents represents one of the most important modes for the transfer of technology and a new mark of the relationship between universities and companies. The normative dimensions of the triple helix influenced many of the measures adopted. The 'Innovation Law' adopted at the end of 2004 is the principal tool of the new set of regulations that is discussed in detail in this chapter.

The capacity for innovation of companies is now incorporated as one of the objectives of Science, Technology and Innovation (ST&I) policy. But the achievement of this goal requires a reorganization of the institutions of research as a whole. From universities it is expected not only that they carry out research following the overall guidelines of ST&I policy, but that they also assume the mission of commercializing the knowledge generated and of transferring the technologies developed to companies or to social organizations. This new responsibility meets the social and economic objectives defined at the beginning of 2003, which at that time were not part of the role given to universities.

In fact, both the system of higher education and research and the system of manufacturing and production had been structured separately to allow each of them to follow their own path. The need for links between the two systems was not taken into consideration. Hence, the process of delayed industrialization as it took place in Brazil did not lead to the formation of a proper national system of innovation with the characteristics identified by Lundvall (1988), Nelson (1993) and Freeman (2002). The science and technology (S&T) policies of the 1970s were designed to develop endogenous technology capable of replacing external technology that was acquired by Brazilian industry in the form of equipment or in know-how contracts.[1]

By the 1990s, the economic environment had changed substantially. The continued process of opening the economy to foreign investment in sectors and activities that had formerly been protected and the privatization of state-owned public service companies led to the emergence of a competitive environment that heretofore had been unknown. The desire to be more competitive caused companies to adopt more efficient management technologies, and the capacity to innovate became an essential goal.

Greater interaction between higher education, research and industry was considered to be an important means to encourage innovation and strengthen the ability of local companies to compete. Researchers and universities were given the task of becoming direct agents of the economic development, instruments for the technological training of companies and disseminators of entrepreneurship.

The reforms in the organization and financing of public research created a new model of economic and technological development. The goal was now to encourage Brazilian companies to become actively involved in the economics of knowledge. But it should not be forgotten that the reforms in the system of higher education that gave universities and research institutions these new missions sometimes provoked strong resistance and tensions on the part of the academic community.

This chapter presents the changes in the regulatory environment introduced to allow Brazilian universities to commercialize the results of their research activities and afford greater interaction with companies. The new regulatory environment is relatively recent and the impact is still hard to evaluate. Closing the distance between the system of research and the system of production requires the adoption of measures that must be accepted, both by the universities and the companies. The process currently under way is being led in a very dynamic way by the government through its various decision-making levels and seconded by initiatives developed by the other helixes.

This chapter is divided into three main sections. The first section describes the process of structuring and consolidation of teaching and research, as well as the persistence of limited innovative capacity over the period 1950–1980, against the background of a process of delayed industrialization. The second section examines the new roles that began to be assumed by the universities in the context of globalization and the opening of the economy. The last section examines the recent context of the new regulatory environment and the impact on the relationship between the universities and the companies.

TEACHING, RESEARCH AND DELAYED INDUSTRIALIZATION

Consolidation of Teaching and Research

The gap between universities and the business sector and between research and innovation is rooted in the historical characteristics of the late-developing process of industrialization. It was not until the decade of the 1950s that Brazil began a process of extensive industrialization, based on the substitution of imports. Prior to that time, universities and some isolated institutions of higher education were institutions involved only in the training of human capital. But the demand by the manufacturing sector for qualified human resources was fairly limited, reflecting the low level of industrial development that had been achieved.

During the period of military rule (1964–1984), the government intensified efforts to achieve industrial and technological self-sufficiency through the marked presence of public companies. These companies had a dual role which included the production of basic industrial inputs (steel, minerals, energy) and the establishment of centres for research and development (R&D) for the development and diffusion of technological advances in strategic sectors (nuclear, aviation, data processing and computing).

Universities were given the responsibility for the training of human resources to manage the process of industrialization. This led to a rapid growth of the system of higher education. During the 1970s there were attempts to promote institutional research activities and the training of highly qualified professionals to develop endogenous technology for the business sector. These activities were based on a linear concept of innovation and the belief that there would be a natural and automatic spillover of research activities to companies. But the utopia of autonomous technological development never happened. Domestic companies continued to be extremely passive in terms of innovation. They were engaged in a slow process of learning and acquisition of knowledge that merely allowed a limited grasp of production technologies similar to the processes described by Lall (1992) or Bell and Pavitt (1993).

The persistent difficulty in achieving regular and systematic interaction between the universities and companies derives in large part from the problematic relationship between the process of industrialization and the creation of a system of higher education and research. In 1968, the federal government began a reorganization of the university system, sanctioned under the act called the University Reform Law,[2] to meet the training needs of human resources and the generation of scientific and technical knowledge required by the process of industrialization. Postgraduate and research activities were introduced to complement the teaching mission at the undergraduate level. A complex institutional system was implemented around federal government agencies that were given the task of promoting postgraduate studies, grant scholarships for master's and doctoral students and evaluating the quality of the courses offered.

The S&T policy during this time stressed the adoption of programs designed to create an appropriate infrastructure for research and provided a significant level of human resources. The principal instrument for this policy was the National Scientific and Technological Development Fund, created in 1969.[3] The availability of public funding was, on one hand, linked to the policy of economic development through import substitution and, on the other hand, on a policy of support for higher education that included the creation of graduate degree programs, which increased from 125 in 1969 to 974 in 1979.

The plans for scientific and technological development created between 1968 and 1979[4] were designed to reduce the dependency on a startled technology by encouraging the development of a so-called endogenous technology with the assumption that the technology developed by the research institutions would be used by local companies. The military government attempted to promote the 'nationalization' of technology in strategic production sectors such as petroleum, electric power, steel, telecommunications and aviation. These extremely ambitious objectives, however, were not completely achieved. However, this policy did promote the formation of qualified human resources that were needed to carry on the process of industrialization. Furthermore, it allowed the state-owned companies to start accumulating high levels of technological competencies.

One portion of public funding was made available to the manufacturing sector for innovation; but the impact was minimal because of the lack of demand on the part of companies that, rather than investing in internal research, gave preference to the importation of machinery or the transfer of technology from abroad. The internal funding allocated by companies for R&D did not exceed 3 per cent of the total of government funding and was concentrated in a few state-owned enterprises.

Some of the more dynamic research institutions, especially the engineering and agricultural technology institutes (Maculan and Soares 2000), were able to establish more effective relationships and achieve significant cooperation with some industrial sectors and state-owned companies. The means that were used were mostly confined to the provision of technological services—testing, analysis, verification—and, in some cases, research projects to expand the knowledge of complex technologies contained in the contracts the state companies had with foreign suppliers of know-how or equipment.

Taken as a whole, industrial companies were characterised by their passive technological and innovative behaviour and by the lack of investment in research. The import substitution industrialization model dispensed with research activities and only attempted to absorb and internalize the technology incorporated in the production capacity of foreign subsidiaries that were being set up in Brazil.

However, the companies—with the exception of state-owned companies—did not even show that they understood the importance of incorporating the human resources trained locally through research.[5] For the Fordist model of industrial mass production that was current at the time, all that was needed was an abundant supply of cheap labour, for which the level of education was not an important criteria. The requirements for technological assistance were met from offshore. This made the process of apprenticeship and on-the-job training difficult. But, as Viotti (2002) has noted, it is this process that makes possible the spread of technology and incremental innovation, and it plays an important role in late-developing economies. University-trained researchers and scientists, instead of shifting to industrial production and innovation activities, remained in their own system of teaching and fundamental research. This has, however, led to an increase in the availability of, and access to, higher education.

In summary, during the period of the import substitution industrialization (1950–1980), the interactions between the universities and the companies took on a more traditional framework of training human resources. The system of production and the system of higher education and research developed rapidly, but in an autonomous manner in relation to each other, although both were under the guidance of the federal government. The specific characteristics of this period left behind a structure that isolated the universities and companies from one another. The helix of government led the process of development and guided the trajectory of the other two helixes, but was unable to reduce the gap between them.

Limited Capacity for Innovation

The S&T policy that prevailed between 1970 and 2000 led to the expansion and consolidation of a system of postgraduate research, which is considered to have the best and most mature structure in South America (Schwartzman 2002). Preparation of highly qualified human resources—at the master's and doctoral levels—as well as the production of scientific research began to expand at very rapid rates. The number of graduates with master's and doctorates increased from 868 doctorates and 3,647 master's degrees in 1987, to 8,094 doctorates and 27,630 master's degrees in 2003. It reached 9,366 doctorates and 32,500 master's degrees in 2006. It is important to note that in 2006, there were 118,000 postgraduate students in Brazil, of which forty-four thousand were in PhD programs. These students were enrolled in nineteen hundred master's degree programs and approximately one thousand doctoral programs taught by approximately thirty-three thousand professors with doctorates. According to Thorn and Soo (2006), with more than eight thousand doctoral graduates per year, Brazil is the best example of a successful attempt to strengthen graduate education in the region. There was also a significantly large increase in the number of scientific publications associated with this expansion in the level of postgraduate studies. The number of articles produced in Brazil and published in international scientific journals indexed in the Institute for Scientific Information (ISI) increased from 1,891 in 1981 to 9,591 in 2000, and reached 19,428 in 2007. The production of 'science' in Brazil is ranked fifteenth in the world; representing 2.02 per cent of the total scientific output worldwide (CAPES 2008).

Whereas the results achieved in universities and research institutions in terms of the number of advanced degrees awarded and the generation of knowledge were impressive, the same cannot be said of companies in terms of technological development and innovation. The low level of technological dynamism and limited capacity for innovation in Brazilian companies were clearly revealed in the results of the three PINTEC innovation surveys.[6] In the most recent innovation survey, for example, covering the 2003–2005 period (IBGE 2007), a total of 91,055 industrial companies with ten or more employees were surveyed. It was found that approximately only one-third (33.4 per cent) of the companies surveyed innovated by developing products and/or implemented processes that were technologically new or substantially improved at least for the company. Moreover, the innovative degree was extremely reduced because only in 3.2 per cent of the cases the product developed was in fact new to the domestic market. The comparison with the behaviour of European companies is quite unfavourable. The survey results for the Community Innovation Survey (CIS) for the three-year period 1988–2000 show that this percentage was 21 per cent for Belgium, 12 per cent for Portugal and Spain, 23 per cent for Germany, 17 per cent for France, 25 per cent for Denmark and 22 per cent for Italy (Viotti and Baessa 2007).

The PINTEC data (IBGE 2002, 2005, 2007) shows that the predominant innovative activity in these industrial companies was the acquisition of machines and equipment (representing 1.34 per cent of expenses over net income from sales), and that a smaller emphasis was given to internal activities of R&D (representing 0.57 per cent of expenditures over net income from sales). The predominance of the acquisition of machines and equipment, in large part imported, confirms that the path of Brazilian industry continues to be strongly dependent on external sources for technology and that companies continue to be essentially receivers of technologies developed by others; in general, outside the domestic production system (Maculan 2005). Only 16.24 per cent of the companies of the total surveyed declared expenses relating to internal R&D activities. To conduct their internal R&D activities, these companies hired approximately forty-three hundred postgraduate researchers on a full-time basis during the period under consideration (2003–2005). To give an idea of the modest size of this number, it is worth highlighting that in 2003 approximately eight thousand PhDs and 27,600 master's degrees were awarded.

In summary, the asymmetry between the research and postgraduate system and the industrial system persisted. On the one hand, we have universities and research institutions performing well, producing knowledge and graduating students with master's and doctorate degrees. On the other hand, we have companies with poor performance in innovation, low levels of internal R&D activities and a limited number of researchers with master's and doctorate degrees.

NEW ROLES FOR THE UNIVERSITY

Initiatives from the Universities

From the beginning of the 1990s, the Brazilian economy has been subjected to profound changes that reprogrammed and rearranged the path of growth. The new pathways were the result of several different factors, which included the deregulation of public services (telecommunications, water, electric power distribution), the privatization of large state companies in mining, the opening of various sectors to foreign investment (energy, petroleum, finance), the opening of the economy, the reduction or removal of tariff and non-tariff barriers to international trade, the attraction of direct foreign investment and changes at the industrial property laws. New models for management and administration began to be put into general use that favoured a wave of organizational innovation in Brazilian companies as a means to initiate new standards of performance and facilitate their integration into the globalized economy.

During this period, universities involved in R&D activities started introducing a set of initiatives to prepare themselves for a more active role in the system of innovation.[7] These universities try to adopt a more entrepreneurial role with the creation of Business Incubators, Technology Parks and Offices for the Transfer of Technology, which showed a dramatic increase over the past decade (Etzkowitz, Mello and Almeida 2005).

According to estimates by the National Association of Organizations for the Promotion of Entrepreneurships Innovators (ANPROTEC 2005), the number of incubators in Brazil increased from merely two in 1988[8] to 150 in 2000, reaching 339 in 2005. In 2005 the activities of the incubators involved 5,618 companies, of which 2,327 were incubated, 1,613 were associated and 1,678 were graduated, generating a total of 28,449 direct employment jobs. During the same year forty-two projects for Technology Parks were registered of which fifteen were effectively operating (Vedovello, Judice and Maculan 2006). Although many of the incubators and parks encourage the creation of small companies in traditional sectors of the economy, a significant proportion of these firms dedicated themselves to projects that were actually innovative.

The universities and public research institutions, principally those which are viewed as leading in scientific research, became increasingly concerned with the protection of the knowledge generated during the process of their research activities, motivated in large part by the potential of patents as a mechanism for transfer of technology and the source of financial funding through licensing. The adoption of a new law for intellectual property[9] brought new possibilities for the protection of patents for medication, foodstuffs and chemical products. The creation of technology transfer offices (TTO) within the universities,[10] responsible for the management of the intellectual property rights as well as the establishment of financial incentives for the researchers that hold the patent rights, contributed to a growing interest of the issue of intellectual property in the scientific community.

The evolution of the applications for patents by the universities in the National Institute of Industrial Property (INPI) between 1990 and 2004 (Figure 3.1) shows, first of all, a positive inflection of the number of applications beginning in 1996, when the new law of intellectual

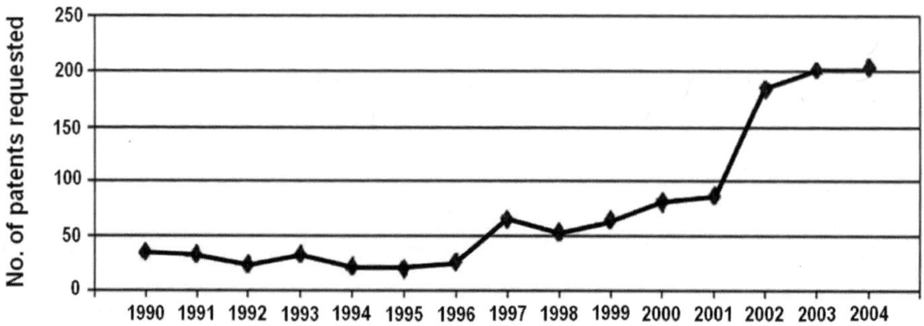

Figure 3.1 Number of patents requested 1990–2004. *Source*: INPI (2007).

property was introduced, followed by a change in the level over the three most recent years (2002–2004). This change appears to be related to the conjugation of an increase in interest by researchers and universities in applying for patents as well as the increase in the number of researchers and postgraduate research programs at the universities. It should be pointed out that since the year 2000 the INPI has had a specific program whose objective is to increase the share of universities in the intellectual property system, spreading a culture for the protection of intellectual property rights.

Moreover, looking at the patent applications at the INPI between 1999 and 2003, it follows that the institution with the greatest number of required patents was the University of Campinas (not a company!). Out of the twenty largest institutions that required patents, five were universities and two were development agencies (Carvalho 2006). In the U.S. patent office[11] 106 patents were granted to Brazilians in 2004, in other words 0.6 per cent of the patents registered[12] in that year (Rezende 2005), whereas Brazil's share in worldwide production of scientific articles was three times greater.

The universities, in addition to generating knowledge and training highly qualified human resources, began to assume the mission of incubating technological projects in their incubators and technology parks with impressive results. They also assumed the mission of protecting and commercializing the knowledge that they were producing. However, the results, achieved through all the initiatives carried out on the basis of activities initiated by the universities themselves, failed to significantly alter the low technological dynamism and innovation in the industrial sector.

The scenario described here suggests that the exercise of a more entrepreneurial role on the part of universities, with activities devoted to the incubation of companies, the creation of technological parks and the protection and commercialization of knowledge, was unable to produce a significant impact on the system of innovation. They were activities, in large measure, informal and voluntary, isolated and unrelated to the industrial policies of the 1990s (Viotti 2008; Sbragia, Moreira and de Cota 2008).

Companies continued to express limited interest in the initiatives offered by universities. The efforts and the investments by universities appear not to have been sufficient to open up a new wave of more systematic interactions with the companies that would lead them to have greater capacity for innovation. It is likely that these companies still have insufficient capacity internally to absorb and benefit from the knowledge generated at the universities.

A New Regulatory Environment

In an attempt to overcome the effects of this adverse environment, the government has been trying to implement a significant set of important and complex S&T policy measures since the middle of this decade in order to create an environment that will lead to a greater involvement by companies in the process of technological innovation.

Of immediate importance was the creation of an integrated industrial, technological and foreign trade policy, introduced at the end of 2003 and designed to increase the efficiency of the structure of production, the capacity for innovation and the exports of the Brazilian companies.[13] This policy established an ambitious set of guidelines and priorities that were difficult to implement. Despite the difficulties and drawbacks that have to be confronted to achieve its objectives, the most important feature of this policy lies in the explicit articulation between the industrial policy and the technology policy (Viotti 2008).

Two key pieces of legislation were enacted as part of this policy, the Innovation Law[14] sanctioned on December 2004, and Law 11.196, informally known as the 'Good Act', sanctioned on November 2005. The Innovation Act, inspired by the Bayh-Dole Act and by the French law on innovation, proposes a set of measures and incentives to encourage innovation and research by companies, by setting up the three following objectives:

1. creating an environment conducive to strategic partnerships between universities, technological institutions and companies
2. encouraging the involvement of scientific and technological institutions in the innovation process
3. fostering in-house innovation

The 'Good Act', in turn, consolidates and expands pre-existing tax incentives, it also establishes new and substantial incentives for companies that engage in R&D activities and technological innovation.[15]

The Innovation Law and the Good Act offer new ways to promote innovation in companies. The economic subvention (a mix of venture capital and tax incentives) deserves special attention. The government, through its agencies for promoting S&T, grants significant financial resources to companies that decide to develop in-house research activities. This measure allows the financing of the cost of R&D for products and innovative processes. In addition, the Good Act permits the funding of the remuneration of researchers, with master's or doctorate qualifications, who are engaged in innovative activities within companies located in Brazil. Since 2006, various bids have been published for the selection of projects submitted by the companies.

These two laws give new meaning to the entrepreneurial role that universities have been engaged in on an informal or voluntary basis, covering them under a broader and more extensive policy and giving them legal standing. With the Innovation Law, the creation of Technological Innovation Nucleus (NIT) was made obligatory in universities and public research institutions. The NITs are given the authority to deal with the various aspects of commercialization of knowledge, management of innovation and interactions with companies.

Universities and research institutions are creating the NITs, as a rule, based on the structures of the existing TTOs. The Ministry of Science and Technology will have to monitor the compliance with this clause in the Innovation Law using information provided by the universities and research institutions.

In 2007, seventy-two NITs were in operation, compared to forty-three in 2006—an increase of 67 per cent. The royalties obtained through technology transfer—due either to the transfer of expertise or through patent licensing—increased from less than R$1 million[16] in 2006 to almost R$5 million in 2007.[17] The NITs reported that 830 patent protection requests were filed in Brazil, of which 186 were co-titled and eighty-three were registered abroad. The data shows a significant number of requests for co-titled patents and a significant increase in royalties from the transfer of technology. This suggests an increase in the level of cooperation between the university and companies.

The activities in which the university had been engaged have now extended to include a new dimension in terms of entrepreneurship and the incubation of start-up companies. The new regulatory environment conveys a strong incentive for the universities to participate effectively in the creation of technology-based companies using the new instruments available under the Law of Innovation.

The agency for the promotion of innovation in the Ministry of Science and Technology recently introduced the First Company Program (PRIME).[18] Under this scheme, the intention is to recognize the experience acquired by the incubating companies and technological parks in the promotion of entrepreneurship and to feed it with the benefits of economic subvention and tax incentives. The goal is to make available R$1.3 billion[19] to start-up companies during the next four years. The government hopes to support eighteen hundred requests for new projects by 2011. Eighteen incubators have been selected to prepare the request for proposals and identify the candidates. The companies will be supported by the PRIME program and will be eligible for a double financing. It is likely that a good part of these eighteen hundred companies will be spin-offs from the universities, in other words, companies created to commercially exploit the knowledge generated by the research activities in the universities.

The effective impacts of the original provisions of the Innovation Law and the Good Act have not yet been subjected to a complete evaluation. But the initial results that have been identified appear to indicate that the adoption of these laws is in fact providing the incentive for forward movement in the integration of companies with universities. This is also creating an environment which is more favourable to the development of innovation and the appreciation of the value of knowledge by the universities.

CONCLUSIONS

Brazil has reached the twenty-first century with a diversified structure of production and a system of research of reasonable size, measurable either by the quality and the quantity of graduates or by the number of scientific publications in international academic journals. However, the two systems appear to be quite unrelated. The process of industrialization in Brazil favoured the passive absorption of technology by companies and did not take into consideration the importance of learning or technological training as paths for innovation. Because of this history, the systems of production and innovation have little synergy and the interactions between the actors are insufficient to generate innovation.

Over the years, very few of the results achieved in research have managed to flow to the productive sector, which has remained distant from scientific and technological developments. This has led to a serious lack of interaction between the structure of production and the structure of scientific research. The use of the research infrastructure that is installed mainly in the public universities to increase the level of technological competencies and innovation capability of the Brazilian companies has become the most important objective of recent S&T policy. The universities are being strongly encouraged to take economic advantage of the results of their research, the protection of their creation, the patenting of technologies that they develop, and they are encouraged to provide support to the process of entrepreneurship and the generation of technology-based companies.

On the other hand, the economic use of the results of research, and patenting and licensing policies of universities as expressed in the recent Innovation Law, continued to create apprehension on the part of a significant portion of the academic community. But the new regulatory environment that ratifies the previous initiatives taken by universities with better performance in research offers important opportunities. Interactions between universities and government are certainly more intense and broader and are permitting the definition of a legal framework more appropriate for the innovative experiences of universities. The responses so far from the industry 'helix' to the new context are quite tenuous but appear to be increasing.

ACKNOWLEDGEMENTS

One of the authors, Mello, gratefully acknowledges the support from the National Research Council, Brazil, through the research grant # 311684/2006–8. The authors also would like to thank the editors of this book for their helpful comments on this chapter.

NOTES

1. I and II Plano Brasileiro de Desenvolvimento Cientifico e Tecnológico (PBDCT), II Plano Nacional de Desenvolvimento (PND).
2. Law 5.540, 11 November 1968.
3. Decree Law 719, 31 July 1969.
4. Three 'Basic Plans for Scientific and Technological Development' were passed between 1972 and 1985.
5. The process of industrialization started in the 1950s and resulted in the installation of multinational corporations in the more dynamic sectors, in the creation of state companies in strategic and basic sectors and in the emergence of domestic companies that should have been prepared technologically to develop the domestic market, preferably in relationships with foreign companies.
6. PINTEC is a large industrial survey of innovation performed by the federal statistics agency -IBGE in 2000, 2003 and 2005.
7. For more details about these types of initiatives see Chapter 10 of this volume.
8. At the Federal University of São Carlos in the State of São Paulo and at the Federal University of Santa Catarina.
9. Law 9.279/96 of 1996.
10. In 2004 there were already thirty offices in operation in universities (Lahorgue, Ritter and Mello 2005).
11. United States Patent and Trademark Office.
12. The same proportion as in 1990.
13. Ministério do Desenvolvimento da Indústria e Comércio Exterior. Diretrizes de Política Industrial, Tecnológica e de Comércio Exterior, Brasília, 2003. See www.desenvolvimento.gov.br/sitio/
14. Law 10.973, 2 December 2004.
15. Law 11.196, 21 November 2005.
16. Approximately U.S.$500,000.
17. Approximately U.S.$2.5 million.

18. See http://www.finep.gov.br/imprensa/noticia.asp?cod_noticia=1466.
19. Approximatively U.S.$600 million.

REFERENCES

ANPROTEC. 2005. 'Panorama 2005'. www.anprotec.org.br/ArquivosDin/Panorama_2005_pdf_11.pdf (accessed 20 December 2008).

Bell, M., and K. Pavitt. 1993. 'Technological accumulation and industrial growth: Contrasts between developed and developing countries'. *Industrial and Corporate Change* 2 (2): 157–210.

CAPES. 2008. 'Produção científica brasileira é a 15ª em todo o mundo'. http://www.capes.gov.br/servicos/sala-de-imprensa/36-noticias/1990 (accessed 12 December 2008).

Carvalho, P. E. 2006. 'O papel da propriedade industrial no estímulo à inovação—a experiência da UNICAMP'. *Seminário Transferência e Inovação Tecnológica.* Curitiba: UFPR. http:// www.inova.unicamp.br/paginas/palestras.php (accessed 10 August 2007).

Etzkowitz, H., J. M. C. Mello and M. Almeida. 2005. 'Towards "Meta-innovation" in Brazil: The evolution of the incubator and the emergence of a triple helix'. *Research Policy* 34:411–424.

Freeman, C. 2002. 'Continental, national and sub-national innovation systems—complementarity and economic growth'. *Research Policy* 31:191–211.

IBGE. 2002. 'Pesquisa Industrial Inovação Tecnológica 2000'. Rio de Janeiro: IBGE. http://www.ibge.gov.br/home/estatistica/economia/industria/pintec/pintec2000.pdf (accessed 24 October 2002).

———. 2005. 'Pesquisa Industrial de Inovação Tecnologica 2003'. Rio de Janeiro: IBGE. http://www.ibge.gov.br/home/estatistica/economia/industria/pintec/2003/default.shtm (accessed 20 June 2005).

———. 2007. *Pesquisa de Inovação Tecnológica 2005.* Rio de Janeiro: IBGE. http://www.ibge.gov.br/home/estatistica/economia/industria/pintec/2005/pintec2005.pdf (accessed 28 September 2007).

INPI (2007) 'Universidades Brasileiras—Utilização do Sistema de Patentes de 2000 a 2004', Rio de Janeiro, available at: www.inpi.gov.br (accessed 1 July 2007).

Lahorgue, M. A., E. Ritter and J. M. C. Mello. 2005. 'Economic development mission in brazilian universities'. Paper presented at the Triple Helix 5 International Conference: The Capitalization of Knowledge—Cognitive, Economic, Social and Cultural Aspects, Turin, Italy (unpublished).

Lall, S. 1992. 'Technological capabilities and industrialization'. *World Development* 20 (2): 165–186.

Lundvall, B. A. 1988. 'Innovation as an interactive process: From user-producer interaction to the national system of innovation'. In *Technical change and economic theory*, ed. G. Dosi, C. Freeman, R. Nelson, G. Silverberg and L. Soete, 349–369. London: Pinter Publishers.

Maculan, A. M. 2005. *Capacitação tecnológica e inovação nas empresas brasileiras: balanço e perspectivas.* Rio de Janeiro: FGV Cadernos EBAPE.BR. Edição Especial sobre Gestão da Inovação Tecnológica no Contexto de Economias Emergentes: Experiências de Empresas no brasileiro. http://www.ebape.fgv.br/cadernosebape/asp/dsp_lst_artigos_edicao.asp.?code=29 (accessed 6 July 2006).

Maculan, A. M., and C. F. Soares. 2000. 'Os pesquisadores e a transferência de conhecimentos para a indústria'. Paper presented at the NPGCT FEA USP, *XXI* Simposio de Gestão da Inovação Tecnológica, São Paulo, 7 October.

National Institute of Industrial Property. 2007. 'Universidades Brasileiras—Utilização do Sistema de Patentes de 2000 a 2004'. http://www.inpi.gov.br (accessed 1 July 2007).
Nelson, R. R., ed. 1993. *National systems of innovation. A comparative analysis*. New York and Oxford: Oxford University Press.
Rezende, S. 2005. *Ciência, tecnologia e inovação para o desenvolvimento nacional: o papel do MCT*. Brasília: CGEE. http://www.cgee.org.br/cncti3/ (accessed 18 November 2005).
Sbragia, R., N. V. A. Moreira and M. F. de Cota. 2008. 'Technological innovation in Brazil: Evaluation of regulatory framework advancements'. Paper presented in the VI Globelics Conference, Mexico City. Available: http://globelics_conference2008.xoc.uam.mex/Con19.html (accessed 3 March 2009).
Schwartzman, S. 2002. *Higher education and the demands of the new economy in Latin America. Background paper for the 2003 LAC Flagship Report*. Washington, DC: World Bank. http://www.schwartzman.org.br/simon/pdf/flagship.pdf (accessed 3 June 2008).
Thorn, K., and M. Soo. 2006. 'Latin American universities and the third mission. Trends, challenges and policy options'. World Bank Policy Research Working Paper 4002, August. http://www.wds.worldbank.org/servlet/WDSContentServer/WDSP/IB/2006/08/30/000016406_20060830142439/Rendered/PDF/wps4002.pdf (accessed 23 November 2006).
Vedovello, C., V. Judice and A. M. Maculan. 2006. 'Strategic issues related to the technological parks in Brazil'. Paper presented at the 4th Globelics International Conference, Innovation Systems for Competitiveness and Shared Prosperity in Developing Countries, Trivandrum, Kerala, India. http://www.globelicsindia2006.org/sessions.php#_Session_II-4_Regional_Innovation_Sy (accessed 12 October 2006).
Viotti, E. B. 2002. 'National learning systems: A new approach on technological change in late industrializing economies and evidences from the cases of Brazil and South Korea'. *Technological Forecasting and Social Change* 69:653–680.
———. 2008. 'Brasil: de política de C&T para política de inovação? Evolução e desafios das políticas brasileiras de ciência, tecnologia e inovação'. In *Avaliação de políticas de ciência, tecnologia e inovação: diálogos entre experiências internacionais e brasileiras*, ed. CGEE. Brasília. http://www.cgee.org.br/sobre/finalidade.php (accessed 3 March 2009).
Viotti, E. B., and A. Baessa. 2007. 'Innovation in Brazilian, Argentine and European industries: A comparison of innovation surveys'. In *Technological innovation in Brazilian and Argentine firms*, ed. J. De Negri and L. Turchi, 223–244. Brasília: IPEA. http://www.ipea.gov.br/sites/000/2/livros/technological_innovation_ingles.pdf (accessed 3 April 2008).

4 Inter-Organizational Learning through South–South Cooperation

A Case Study Investigation

Najeh Bouraoui, Amira Bouhamed, Jamil Chaabouni and Mohammed Saad

INTRODUCTION

Innovation is recognized as the driver of economic growth. The triple helix model of university-industry-government is increasingly becoming the requisite basis for innovation and development in a knowledge-based society. This chapter focuses on the study of the industry sphere of the triple helix model. The main objective is to investigate inter-organizational learning through strategic alliances in the industrial area. The chapter proposes to explore three types of learning that can be generated in South–South cooperation: internalization of partners' capabilities, learning with alliance partners and/or learning to manage alliances.

This study is based upon longitudinal research undertaken between 2004 and 2007 of one case study relating to a Tunisian group of companies called ALPHA. This group has concluded two joint venture (JV) agreements. The first was set with an Egyptian partner and the second with a Libyan partner. Through the two JVs, the different partners have internalized each other's competencies and have developed new knowledge and know-how. ALPHA has also developed a competence in how to manage alliances, especially in the context of developing countries. The contribution of this empirical research is to show that there is interdependence between these three types of learning.

INNOVATION THROUGH TRIPLE HELIX RELATIONSHIPS

In a knowledge-based economy, organizations are looking for ways to gain or maintain a competitive advantage in the marketplace. Both knowledge and innovation are sources of competitive sustainable advantage (Matthews 2003). Innovation has become one of the main engines of long-term economic growth and is closely linked to knowledge, which is the major value-creating factor in modern society. Innovation is a complex process involving a significant number of actors and sources of learning, knowledge and skills (EU, A Background Paper 2008). A triple helix model of university-industry-government as relatively equal, interdependent and interacting institutional spheres is increasingly becoming the requisite basis for innovation and development in a knowledge-based society (Etzkowitz and Dzisah 2007). The 'triple helix' is a spiral model of innovation that captures multiple reciprocal relationships at different points in the process of knowledge capitalization. Industry, which is one of the three helices of the model, is the locus of production. The government helix is perceived as the source of regulation and contractual relations that guarantee stable interactions and exchange. The university, core helix of the model, is the source of new knowledge and technology (Etzkowitz 2003). The first dimension of the triple helix model is the internal transformation in each of the helices, such as the development of lateral ties among companies through strategic alliances. The second is the influence of one helix upon another. The third dimension is the creation of a new overlay of trilateral networks and organizations from the interaction among the three helices (Etzkowitz 2002). This chapter focuses on the first dimension of the triple helix model and more particularly on the study of strategic alliances in the industrial area.

As companies are becoming aware of international competitiveness, they are striving to achieve a sustainable competitive advantage through the initiation of strategic cooperation. According to Arino and De La Torré (1998), it is becoming more difficult for organizations to remain self-sufficient in an international business environment that demands both focus and flexibility. Firms use alliances for a variety of reasons to gain competitive advantage in the marketplace, to access or internalize new technologies and know-how beyond firm boundaries, to exploit economies of scale and scope or to share risk or uncertainty with their partners (Kale, Singh and Perlmutter 2000). Within the strategic cooperation literature, learning is traditionally seen as a key motive for the formation of alliances (Inkpen and Dinur 1998; Mowery, Oxley and Silverman 1996; Larsson et al. 1998; Kale, Singh and Perlmutter 2000). In fact, there is a growing amount of research dedicated to examining learning within organizations as well as learning between organizations. Recent studies have noted that inter-organizational learning is the more preferable route to effective and efficient learning capability development (Dyer and Singh 1998; Ratten and Suseno 2006).

This chapter examines inter-organizational learning through alliances, which is the most popular mode of new market penetration. In the context of alliances, much research has focused on learning phenomenon from several streams (Kale, Singh and Perlmutter 2000; Parise and Sasson 2002; Soekijad and Andriessen 2003). First, firms can learn about their alliance partners and internalize their critical knowledge, capabilities or skills (Hamel 1991; Mowery, Oxley and Silverman 1996; Inkpen and Dinur 1998; Kale, Singh and Perlmutter 2000). The second stream of literature refers to learning wherein the partners learn about management of collaboration process and working better with each other as their relationship evolves (Doz 1996; Arino and De La Torré 1998). Finally, a third part of the literature has found that organizations can learn about alliances (Gulati, Nohria and Zaheer 2000). In this stream, organizations may want to acquire knowledge that can be used for the management of alliances in general. This knowledge can also help them to improve the design of new future alliances (Anand and Khanna 2000; De Man 2005).

Many studies, to date, have focused on alliances that are made between firms which belong to developed countries (North–North), or between firms belonging at the same time to developed and developing countries (North–South). These North–North and North–South cooperations aim at the transfer of technology and access to new markets (Beamish 1987; Gherzouli 1997; Delande 2003). A marginal issue paid an attention to the alliances between firms belonging to developing countries (Moatassime 1983; Voduzek 1988; Ouédraogo 2003). Voduzek (1988) and Ouédraogo (2003) consider South–South cooperation as a means by which developing countries could reduce their economic delay through their capacity to acquire and develop the resources and competencies which are not available in a market. These arrangements can then be used as a lever of industrial modernization.

This chapter attempts to explore and provide evidence from the literature to address the important role learning has in these relationships and to inspire future research in this study area. Using data from a longitudinal case study of a Tunisian company (ALPHA) which has concluded two South–South JVs, this chapter intends to address the following research question. If the South-South cooperation offers opportunities of learning, what types are to be generated: internalization of partners' capabilities, learning how to build an effective relationship with a partner and/or learning to manage alliances?

The conceptual arguments are based on a review of relevant academic literature. This chapter is structured around four sections. First, it introduces the concept of organizational learning and establishes a connection between organizational learning and strategic alliances. Second, the chapter identifies and discusses three types of learning in strategic alliances. The third section is devoted to discussing the methodology, results and case study. Finally, areas for future research are proposed.

CONCEPTUAL FRAMEWORK

Organizational Learning

In recent years, researchers in organization theory (Levitt and March 1988), economics (Nelson and Winter 1982), as well as strategic management (Dussauge and Garette 1995) have identified organizational learning as one of the most important subjects for scholarly inquiry. The multidisciplinary nature of the subject has resulted in a plethora of conceptualizations of organizational learning (Karthik 2002). Learning has been defined in a number of different ways. From Argyris and Schön's (1978) perspective, organizational learning is a process by which a firm acquires information, knowledge, understanding, know-how, techniques and practices that lead to changes in its routines. According to Shrivastava (1981), organizational learning refers to the process by which the organizational knowledge base is developed and shaped. Fiol and Lyles (1985) describe organizational learning as the process of improving actions through better knowledge and understanding. Learning enables organizations to build an organizational understanding and interpretation of their environment and to begin assessing viable strategies. It results in associations, cognitive systems and memories that are developed and shared by members of the organization. According to Levitt and March (1988), organizational learning is presented as occurring through the encoding of inferences from past experiences into routines that include rules, procedures, strategies, technologies and cultures that guide firm behaviour. Routines reflect the organizational knowledge base and change in response to organizational experience history. Huber (1991) defines organizational learning as encompassing information acquisition, distribution and interpretation as well as organizational memory. More recent perspectives (Teece, Pisano and Shuen 1997) view learning as a capability that can be nurtured and developed by firms and that is better equipped to explicate the learning process in strategic alliances.

Although there is no unanimity of what organizational learning is, most definitions of this concept have in common that organizational learning involves the transfer, internalisation, transformation, creation and application of new knowledge and know-how by the organization. Two themes characterise the organizational learning literature. The first one focuses on intra-organizational learning processes. The second focal point is related to inter-organizational learning processes (Lane and Lubatkin 1998; Kale, Singh and Perlmutter 2000).

Intra-organizational learning refers to learning within organizations. During this process, knowledge and information are exchanged, shared, developed and stored in intra-organizational structures and routines (Huelsmann, Lohmann and Wycisk 2006). Moreover, organizational learning has been linked to the study of networks and the interaction between organizations, using the term *inter-organizational learning* (Larsson et al. 1998). By learning from and with other organizations, each element of the network hopes to benefit, compared to acting alone. Inter-organizational learning has been regarded as a way of developing the organizational learning literature by conceptualizing another unit of analysis, which is the strategic alliance (Lane and Lubatkin 1998; Kale, Singh and Perlmutter 2000).

Consequently, organizational learning can be understood as an umbrella term, encompassing both intra- and inter-organizational learning. Intra-organizational learning deals with the internal learning processes within a single organization. Inter-organizational learning, which is studied in this chapter, describes learning processes between and with other organizations (Holmqvist 2003).

Learning in Strategic Alliances

Alliances between organizations have become an increasingly important aspect of strategic management (Parise and Henderson 2001). They are relatively enduring inter-firm cooperative arrangements, involving flows and linkages that use resources and/or governance structures for the joint accomplishment of some strategic objectives in the short or long term (Parkhe 1993; Parise and Sasson 2002). Strategic alliances can take many forms: technical exchange, co-production agreements, sale and distribution ties, R&D coalitions, licensing agreements and JVs that result in the formation of separate legal entities (Yoshino and Rangan 1995). Notwithstanding this wide array of organizational forms, this chapter recognizes that they converge on certain salient characteristics. Thus, a strategic alliance is defined as possessing the following features:

- Two or more firms that unite to pursue a set of agreed goals yet remain independent subsequent to the formation of the alliance.
- The partner firms share the benefits of the alliance and control over the assigned tasks.
- The partner firms contribute on an ongoing basis to one or more strategic areas, e.g. technology, product development or marketing.

The study of alliances has basically emphasized the use of different perspectives: economic, strategic, and learning (Dussauge and Garette 1995). As to the first perspective and mainly from the transaction cost theory, hierarchical governance structures are better than market contracting when opportunistic behaviours are probable, and therefore, transaction costs are high (Williamson 1975).

Strategic perspective involves using cooperative strategy to enhance a firm competitive position through market power or efficiency (Dussauge and Garette 1995). From this perspective, we found the most important reason for the existence of cooperative strategy, especially from the theoretical perspective of the resource-based view of the firm (Peteraf 1993).

Learning explanations view alliance as a means to learn or absorb critical skills or capabilities from partners (Karthik 2002). Despite the different emphasis of these perspectives, existing literature has begun to focus increasingly on the phenomenon of learning in alliance situations. Research (Kale, Singh and Perlmutter 2000; Parise and Sasson 2002; Soekijad and Andriessen 2003) has revealed that different types of learning occur in alliances. The first type of learning involves acquiring the partner's knowledge and know-how. The second is to learn about cooperation process in order to gain a better understanding about key success and failure factors, and to be able to build an organizational memory of this knowledge. The third type is about learning to manage cooperation.

Internalization of Partner's Skills

The first stream involves the acquisition of the learning and the partner's knowledge and know-how (Kale, Singh and Perlmutter 2000). In some cases, firms do not have in-house all the necessary capabilities needed to implement a particular strategy. A strand in the alliance literature (Hamel 1991) has suggested that the solution in that case is to enter into alliances with partners who have the requested capabilities. This type of alliance helps organizations tap into and use the capabilities of their strategic partners in order to transfer them to their in-house operations (Hennart and Zeng 2005). So alliances are seen as a means of trading access to each other's complementary capabilities, and as a mechanism to fully acquire or internalize partner skills. In particular, relationships that involve close inter-partner collaboration help firms learn, absorb and internalize the tacit knowledge and skills possessed by their partners (Hamel 1991; Soekijad and Andriessen 2003). Some resource transfers are 'sticky' or inert, where the complexity, codifiability and tacitness of knowledge itself create barriers to knowledge transfer between firms (Kogut and Zander 1992). Ambiguous, difficult to communicate and idiosyncratic, tacit knowledge is essentially individual, experimental and assimilated through practice (Grant 1991). It is disseminated through social interaction of individuals, where shared mental models and communities of interaction allow individual know-how to be transformed into organizational knowledge (Kogut and Zander 1992; Nonaka 1994; Wang and Nicholas 2005). Learning can be about the acquisition of new technical skills or technological capabilities from the partner firm, financial expertise, experienced managerial personnel, ways to approach the product development process, access to market and distribution channels (Mowery, Oxley and Silverman 1996). In such a learning type, members gain an understanding of partners' skills by questioning each other's assumptions, assimilating the partner's skills, beliefs, values and context of action while detecting gaps between beliefs and experience, through observation, personnel transfer and job rotation (Ghosh 2004). In seeking and applying relevant knowledge, a firm will need to possess a knowledge base in the same or similar area. This similarity will allow for an understanding of the intricacies of the new knowledge as well as of its applicability to the firm's unique circumstances. Indeed, a firm will seek knowledge complementary to its own, especially when that facilitates the absorption of other knowledge (Gulati, Nohria and Zaheer 2000).

Learning with Alliance Partners

This stream of literature focuses on collective learning which relates to learning how to build an effective relationship with a partner. It is about 'learning with alliance partners' (Soekijad and Andriessen 2003). This type of learning occurs when partners jointly develop new capabilities in alliances (Hennart and Zeng 2005). Indeed, strategic alliances offer a significant potential zone for development of diversity when partners bring different paradigms, experiences and sets of skills. In an alliance a synthesis of each of the partner's skills, knowledge and paradigms may occur to create previously non-existent competencies and products (Inkpen and Dinur 1998; Ghosh 2004). Partners may generate new and more efficient joint practices (Wagner 2003). Interaction helps individuals and partnering organizations transform the tacit type of knowledge into a more explicit one. It enables not just operational learning of 'how' things are done but more importantly 'why' they are done, challenging the prevailing conditions, procedures or conceptions and leading to new shared mental models. A change in the shared mental models of partners' members is evidence of learning having taken place (Ghosh 2004). Learning generated can be used by partners to enhance their own strategy and operations. In fact, learning allows partners to revise their expectations and to gain an understanding of those of the other partners (Inkpen and Dinur 1998). Doz (1996) points out that the learning related to this stream allows the partners to readjust the initial conditions of the alliance (interface structure, task definition, partners' expectations and partners' organizational routines) and to make them increasingly effective over time. Indeed, learning allows the partners to revaluate the alliance on the basis of perceived efficiency, equity and adaptability.

In international alliances, different mental models and cultural differences complicate all types of collaboration, including learning and knowledge management (Mowery, Oxley and Silverman 1996). By contrast, culturally embedded alliance relationships, where mental models are shared, facilitate learning through frequent interactions between individuals and within groups. Channels and codes for dealing with information, shared experience and common understanding of coordination procedures represent the alliance's 'sunk' organizational capital. This later ensures joint inputs into problem solving and, as a result, collective learning (Huber 1991). Collective learning also depends on character-based trust, such as a common cultural identity. When alliance managers share common kinship and ethnic ties, character-based trust is promoted, especially when based on a small group of homogenous exchange participants. Such trust promotes collective learning between alliance partners by encouraging better knowledge about each other (Wang and Nicholas 2005).

Alliance Capability

A third part of the literature suggests that organizations can learn about alliances (Gulati, Nohria and Zaheer 2000; Soekijad and Andriessen 2003). In this stream, organizations may want to acquire knowledge that can be used for the management of alliances in general (Inkpen and Dinur 1998). In other words, firms may learn to manage cooperation, or acquire an alliance capability (Anand and Khanna 2000). The theoretical literature suggests that 'competence at managing collaborations has become a key driver of a new logic of organizing' (Powell 1998, 231). Like any other competency, the management of alliances is a skill that can be built up and which can become a significant source of competitive advantage (Kale and Singh 1999; Draulans, De Man and Volberda 2003). An alliance capability is defined as a higher-order resource which is difficult to obtain or imitate and has the potential to enhance the performance of the firm's alliance portfolio (Heimeriks and Duysters 2007). A firm's alliance capability can be seen as its ability to internalize alliance management knowledge. As a result, an alliance capability represents mechanisms or routines that are purposefully designed to accumulate, store, integrate and diffuse relevant organizational knowledge acquired through individual and organizational experience of alliances (Kale, Dyer and Singh 2002). This definition highlights the importance of the alliance experience and the quality of alliance management (De Man 2005).

Accumulation and application of knowledge form the basis for building organizational capabilities (Grant 1991). Relevant knowledge is usually built by making associations between past actions, the effectiveness of those actions and future actions. Such knowledge can arise from the firm's own experience or from similar experiences of other organizations. The knowledge underlying the alliance capability is often built over time with alliance experience (Anand and Khanna 2000; Parise and Henderson 2001). In line with previous research, this chapter defines alliance experience as the lessons learned, as well as the know-how generated through a firm's former alliances (Kale and Singh 1999; Kale, Dyer and Singh 2002). These lessons and know-how are likely to become embedded in the minds of the individuals involved. This provides a basis for an organizational routine with respect to performing a certain task or activity (Nelson and Winter 1982).

Shared experience engenders the development of 'common perspectives' (Nonaka 1994, 24), enabling a firm to absorb new knowledge more effectively (Grant 1991; Heimeriks and Duysters 2007). Most experience benefits appear to be based on learning-by-doing through repeated engagements. Repeated engagements in strategic alliances allow the firm to create codified routines, policies and procedures as well as tacit knowledge with respect to the entire range of alliance management, beginning with partner selection and alliance formation to alliance management and finally alliance termination (Rothaermel and Deeds 2006). A firm with greater alliance experience should be able to manage a larger number of alliances effectively (Dyer and Singh 1998). Involvement in more alliances helps companies to develop a tacit proficiency in managing this form of collaboration (Cohen and Levinthal 1990). Prior empirical work has produced some evidence that experience in alliances can positively affect a firm's rate of patenting, new product development and stock market value created (Anand and Khanna 2000). Other research has recently focused not only on alliance experience, but rather on how firms leverage their experience in developing an alliance capability (Kale, Dyer and Singh 2002; De Man 2005).

However, alliance experience may be a necessary but not a sufficient condition for firms to build alliance capability. It is a crude approximation of the mechanisms that lie at the foundation of building such a capability. In fact, alliance capability would also rest upon proactive efforts to accumulate and leverage alliances associated with prior experience (Kale and Singh 1999). Several activities (creating a repository or database containing the firm's entire cooperation history, informal conversations and discussions between managers, formal mechanisms such as of alliance committees, training programs, having an alliance specialist in a company, the presence of an alliance department, etc.) can help identify best practice on what was done right or wrong in similar situations in the past. These lessons facilitate, in turn, effective action and decision-making in ongoing and future alliances and stimulate the exchange of alliance best practices (De Man 2005; Kale and Singh 1999; Kale, Dyer and Singh 2002). The value of this conceptual framework is determined by empirical research. The next section describes the research method and approach, results and discussions.

RESEARCH METHOD AND APPROACH

This research is based on a case study approach. Parkhe (1993) has argued the need for qualitative research that allows us to understand the core issues underlying a theory of collaboration. In particular, Smith, Carroll and Ashford (1995) call for more longitudinal case studies that are capable of capturing the complexities and dynamics of cooperation. This research is related to a Tunisian group of companies (ALPHA) which operate in four sectors of activities: construction, joinery-cabinet work, plumbing and electricity. ALPHA has concluded two South–South JVs (Figure 4.1): the first with an Egyptian partner (STEFM), the second with a Libyan partner (STL).

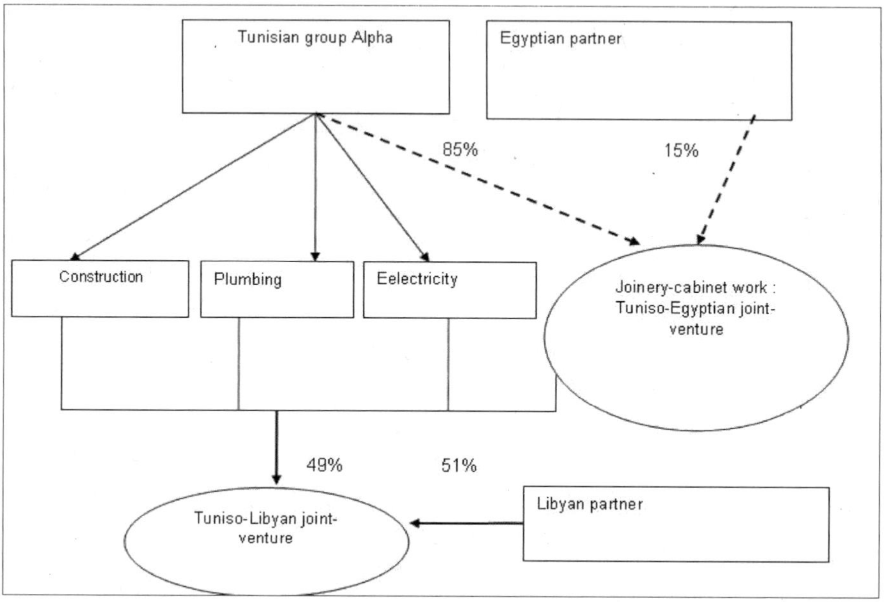

Figure 4.1 South–South cooperative agreements concluded by ALPHA.

This study provides illustrative descriptions and develops insights into the analysis of the learning types in South–South cooperation. The selection of the case was based on the following two considerations.

- An investigation related to Tunisian companies, which maintain business relationships with other companies from developing countries, showed that the majority of these relationships are limited to commercial transaction (import/export) and do not fall within the field of alliance definition.
- ALPHA has more than one South–South cooperative relation.

STEFM was approached in a longitudinal way, stretching out over a period of three years (from January 2004 to March 2007). Data collection consisted of interviews with fifteen individuals (three shareholders, three middle managers and nine operating employees). Several face-to-face and telephone interviews were conducted. The data was also collected through a set of observations. STL was studied over a period of ten months (from May 2006 to March 2007). Interviews involved individuals that were both directly and indirectly (four Tunisian managers) engaged in the JV itself and who could provide additional information on how STL has been functioning. We could not interview the Libyan managers as STL is based in Libya.

In both cases, all interviews were semi-structured. These interviews addressed topics as varied as clauses of contract, tasks distribution, motivation, the competencies internalized by each partner, existence of teamwork, periodic meetings, shift of tasks repetition and partners' expectations, duration of JV and the people who operate in two JVs. All of the notes and notable facts were recorded in a small log file or diary by the researchers, which formed an important point of reference in the analysis. Other sources of data included documents, such as buildings plans and contracts.

RESULTS AND DISCUSSIONS

This section presents insights from the two JVs of STEFM and STL and examines the learning developed by each partner.

First South–South Joint Venture: STEFM

STEFM was founded in July 1996 between ALPHA and an Egyptian partner. STEFM is active in the 'wood, cork and furnishing' industry. The production process is composed of six stages (design, cutting, hewing/sculpture, assembly, varnishing/painting and tapestry). The distribution of operating employees (Tunisian and Egyptians) according to different stages of the production process is presented in the Table 4.1.

The activity concerns the cabinet work and more precisely the sculpture and the sale of furniture. This activity lasted four years. Towards the end of 2000, it was extended to joinery. The activity is related to

Table 4.1 Repartition of Operating Employees

Stages	*Effect of operating employees*	
	Tunisians	*Egyptians*
Design	1	1
Cutting	4	4
Hewing (sculpture)	6	4
Assembly	3	2
varnishing/painting	2	1
Tapestry	4	3
Total	20	15

outsourcing for the hotels all the equipment of joinery including furniture. The reasons which led STEFM to redesign the activity include:

- the degradation of the cabinet
- the activity referring to the sculpture and appearance of new fashions
- the aim of Tunisians to take advantage from the Egyptian reputation to attract hotel promoters
- the existence of qualified Tunisian and Egyptian operating employees
- the existence of problems related to suppliers' payment

Two main periods can be identified within this JV arrangement. In the first period (1996–2000), the main activity was related to cabinet work. In the second period (2001–2007), the activity was related to joinery. Figure 4.2 shows the evolution of the financial turnover in each period.

There are several motives for the creation of the Tuniso-Egyptian JV. For the Tunisian partner, the choice to cooperate was related to the cabinet work which represented an expansion of the sculpture fashion. The lack of Tunisian competencies relating to the speciality of sculpture constituted the first reason which led the Tunisian partner to cooperate. In this activity, the Egyptians possess a high level of know-how and skills. In fact, their knowledge and reputation constituted an essential factor for the Tunisian involvement in this JV.

The Egyptians interviewed have identified different motives. Access to a new market and benefits from their competencies and their reputation

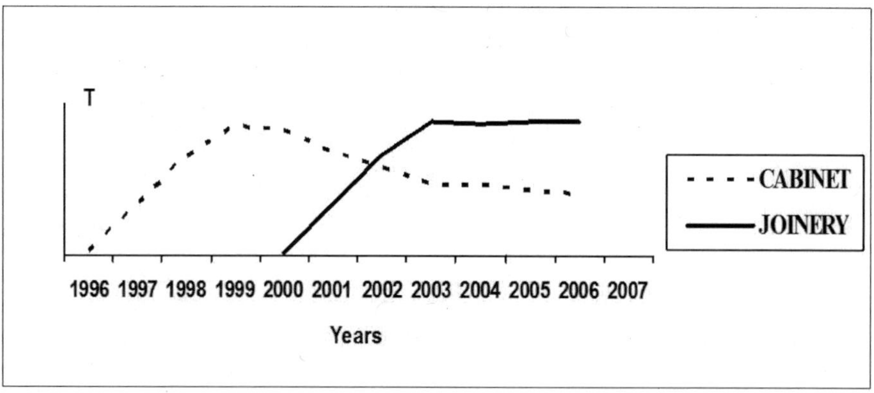

Figure 4.2 Evolution of the financial turnover of STEFM.

* T: Turnover.
Source: Financial department.

constituted the first reason to cooperate. Moreover, the relationships that the Egyptian partner maintained with various Tunisian actors (competitor, suppliers, etc.) constituted another reason that encouraged them to cooperate. By referring to the study of Darréon and Faiçal (1993) on the alliances between the large and small or medium enterprises, it can be argued that the motivations of the Tunisian and Egyptian partners are articulated around the two following motives:

- 'access to the markets' for the Egyptian partner
- reputation for the Tunisian partner, who has benefited from the Egyptian notoriety

The Tuniso-Egyptian JV has offered some opportunity for the Tunisian partner to acquire skills and improve its learning. Most of the learning realized by the Tunisian partner refers to the cabinet work. Thus, the local partner has internalized Egyptians' skills related to the speciality of sculpture. As suggested by a Tunisian manager, in four years, two Tunisian employees among six have reached a level of competence in the sculpture speciality which is equivalent to the Egyptian counterpart.

The fact that the Egyptians had mastered the majority of stages of the production process stimulated their Tunisian colleagues to internalize their skills, related mainly to cutting and varnishing. For varnishing, two Tunisian employees affirmed that the acquisition of learning has lasted five years because this knowledge is essentially tacit. A Tunisian employee argued that he has gained an understanding of Egyptian varnishing skills through interaction, continuous communication and observation. This confirms Hamel's (1991) suggestion that complex skills based on tacit knowledge can be acquired through close observation and relationship.

During the evolution of the Tuniso-Egyptian JV, the partners learned to work together effectively and jointly generated new capabilities. The tasks were distributed between the members of the two partners so as to be carried out jointly by constituting teamwork. From a managerial perspective, it is important for cooperation and knowledge sharing, in particular, to take place in groups where there is a possibility of group cohesion, active interaction, comparable knowledge levels and diversity in skills and expertise (Soekijad and Andriessen 2003). The complementarities between the specialization of the Egyptian team and the flexibility of the Tunisians have led to a good level of new and shared learning.

Some products resulting from the alliance are created jointly. For example, the design of articles is the result of the re-examination of the Egyptian series of models and collective creation of other new models. This result can be explained by the common cultural heritage which had contributed to the creation of products adapted to the customer requirements. Culture has a circular relationship with learning in that it creates and reinforces learning and is created by learning. It influences ongoing learning and adaptation within and between alliance partners (Parkhe 1993, 589).

Compared to the models that exist in the market, these prototypes are original. The difference is located at the level of sculpture and varnishing. At the end of the first period, a series of models were developed. For the second period, these series, which correspond to different periods of time and various styles (Arabesque, Sculpture, Tapestry, etc), contributed to attracting the hotel developers.

In addition, the initial partnering conditions and motivations of the JV changed, making it imperative to revaluate and readjust learning priorities. In the first period, the objective of Egyptians was to acquire knowledge about the Tunisian market. During the second period, their learning priorities shifted. Unexpected learning had been acquired. Indeed, the orientation towards joinery constituted for the Egyptians an opportunity to acquire the Tunisian partner's competencies related to this field.

Although learning in the Tunisian business environment (suppliers, customers and competitors) occurred during the two periods, Egyptian interviewees consider the knowledge related to the second period more appropriate because the JV operates in several areas in different part of Tunisia such as Hammamet, Tabarka, Mahdia, Jerba and Monastir. An example of an alliance between two multinationals, 'NAMCO' and 'Hexagon', described and tracked by Arino and De La Torré (1998), indicates that the partners made the necessary readjustments in their expectations, changed behaviour and modified the implicit contract and norms governing the alliance relationship.

Second South–South Joint Venture: STL

STL is a 49 per cent/51 per cent JV owned by ALPHA and a Libyan partner. In 2003, ALPHA participated in an invitation to tender announced by the Libyan social affairs ministry. It was about the construction of a building and a chain of five hotels including the activities of joinery, cabinet work, plumbing and electricity. As the offer was accepted in September 2004, ALPHA decided to cooperate with a Libyan company. The thinking which motivated this strategic decision include the following five main reasons, which correspond to the points highlighted by Beamish (1987), who investigated thirty-two JVs in less-developed countries.

1. The important financial capital needed for the project.
2. The access to a new market.
3. The minimization of the political risks because the Libyan legislation towards foreign investment and JV was not very clear. Through this JV, the responsibility of the Tunisian partner was related to an operating level. The Libyan partner was in charge of legal arrangements, relationships with the external environment

(competitor, suppliers, customers, etc.) and control of quality. The overall JV was jointly management by the two partners.
4. The enhancement of the reputation, skills and networks.
5. The geographical proximity.

Operating employees in STL represent an interbreeding of nationalities. This situation procures an advantage insofar as each nationality refers to the speciality of its employees. According to Hamel (1991), the difference in national origin might be crucial to learning outcomes. STL has profited from this opportunity by recruiting the qualified workers in every activity. For instance, STL recruited some Syrian and Egyptian operating employees who are respectively reputed in construction and joinery-cabinet activities. For each activity, interactions between operating employees led to some opportunities to internalize one another's technical skills and also to develop new competencies. Hamel (1991) showed that operating employees interact day by day, facilitating the sharing of knowledge between partners. The Tunisian partner also acquired knowledge about the Libyan business environment (suppliers, customers, competitors, laws, etc.). The Libyan partner internalized managerial skills. Regular meetings (one meeting per month) were organized between managers from each partner to negotiate, discuss and revise the conditions of the JV (contract clauses, tasks distribution, partners' expectations and partners' organizational routines). As stated by a Tunisian manager, 'these meetings helped us to develop interpersonal relationships and reduce the cultural and social differences'.

For Neuville (1997), these meetings can, in the long term, strengthen relationships between the partners' members and expand their relational network. By using a large-sample survey data, Kale, Singh and Perlmutter (2000) noted that regular contact and communication lead to greater sharing and learning and facilitates internalization of each others' knowledge and know-how. It also helps to learn how to work together.

Through the two JVs, the Tunisians developed a competence to manage alliances. This capability can be measured through two dimensions: alliance experience and the quality of alliance management. Powell, Koput and Smith-Doerr (1996) state that either the number of alliances a company has entered into or the number of years a company has been involved in alliances indicate the level of alliance experience. The Tuniso-Egyptian JV has lasted ten years. During this period, the Tunisian partner developed a 'capability to cooperate'. Some interviewees have suggested that working with the Egyptian partner has enabled them to improve their technical and management capability in a cross-cultural environment.

ALPHA has also developed its capability on contract management. Mayer and Argyres (2004) suggested that learning how to contract is an important component of learning how to collaborate. ALPHA's top manager reported that the Tuniso-Egyptian cooperation contract suffered from some significant drawbacks. This contract did not provide any guidance or procedures for dispute resolution. Mayer and Argyres (2004) argue that agreements on how disputes should be resolved are as important as JV relationships are complex and dynamic. In order to overcome such gaps, the negotiations for the Tuniso-Libyan contract have lasted one year. Through these negotiations, the partners have attempted to arrive at a relatively inclusive and stable set of administrative procedures for governing the JV in the future. Contracting experience might also help managers and their firms to better understand the implications of contingencies for the relationship, for the firm's performance and for its future contractual relationships.

The second element of alliance capability is the quality of alliance management. In his article which investigated whether there is a difference in the level of alliance capability between European and American companies, De Man (2005) mentioned that companies may implement a variety of management techniques to manage their alliances (internal and external alliance training, alliance database, standard partner selection process, legal experts, etc.). In the case study investigated, ALPHA dedicated four alliance managers, responsible for managing the two JVs. In their research, Draulans, De Man and Volberda (2003) argued that alliance knowledge and experience are concentrated within certain individuals. These persons named as 'alliance specialists' are involved in and responsible for multiple alliances. Another element in building up an alliance capability is hiring outside help by using 'legal experts', like Tunisian lawyers who specialize in international business.

ALPHA reinforced its capability to manage South–South alliances by cooperating with a Libyan partner. The acquisition of this skill has encouraged the company to anticipate the concluding of other cooperative agreements, in particular with partners belonging to the South. As suggested by Anand and Khanna (2000), firms with more JV experience create significant value in subsequent JVs that are formed. This is supported by Cohen and Levinthal (1990), for whom firms which have learned to learn will continue to do so at an increasing rate. In the context of this case study, both partners have reached a better understanding of the expectations and behaviours of their counterpart.

In addition to learning how to design and manage future alliances, ALPHA has developed its capability to internalize the partner's skills and improved its experience and expertise in developing and working within JVs with partners from developing countries. Although a great deal of the literature review (Hamel 1991; Doz 1996; Anand and Khanna 2000; Kale, Singh and Perlmutter 2000) focuses only on one type of learning, this research has given some evidence of interdependence between these three types of learning, as highlighted in Figure 4.3.

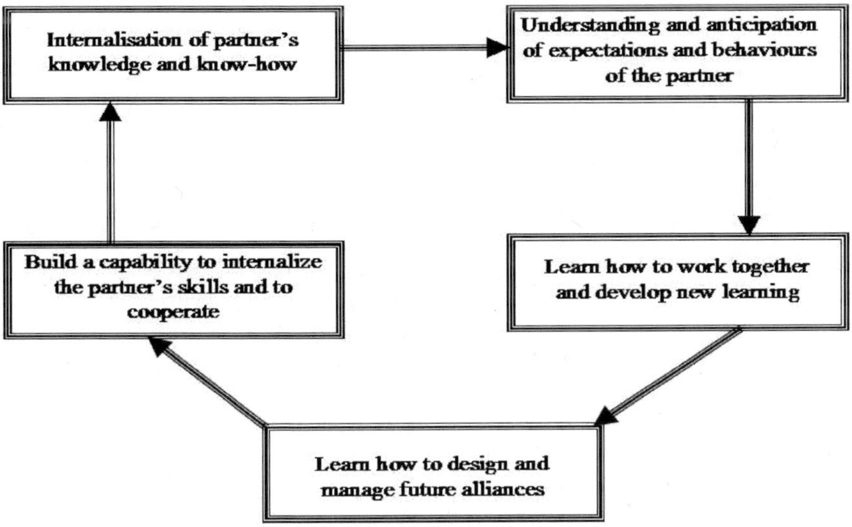

Figure 4.3 Interdependence between three types of learning.

CONCLUSION

This research has attempted to explore the emergence and interdependence of three types of learning that can be generated in South–South cooperation. These types of learning include internalization of partners' capabilities, learning how to build effective relationships with a partner and learning to manage alliances. The empirical research was based on a case study focusing on a Tunisian group of companies (ALPHA) which was involved in two JVs with partners from developing countries (Egypt and Libya).

Through the two JVs, the research has demonstrated that both partners managed to internalize each other's competencies and learned how to work together and how to develop new knowledge and know-how. Most of the learning achieved occurred at an operating level. A significant level of learning on how to manage alliances was also achieved by the Tunisian group (ALPHA).

The strategic alliance literature also illustrates some factors of learning such as trust, absorptive capacity and cooperative norms. Future research needs to explore the conditions for successful knowledge sharing and learning in South–South cooperation. Additional case studies or quantitative analysis would lead to a better understanding of the transfer of learning through South–South cooperation.

REFERENCES

Anand, B. N., and T. Khanna. 2000. 'Do firms learn to create value? The case of alliances'. *Strategic Management Journal* 21:295–315.

Argyris, C., and D. Schön. 1978. *Organizational learning*. London: Addison-Wesley.
Arino, A., and J. De La Torré. 1998. 'Learning from failure: Towards an evolutionary model of collaborative ventures'. *Organization Science* 9 (3): 306–335.
Baber, Z. 2001. 'Modernization Theory and the Cold War.' *Journal of Contemporary Asia* 31(1): 71–81.
Beamish, P. W. 1987. 'Joint ventures in LDCs: Partner selection and performance'. *Management International Review* 27 (1): 23–37.
Becker, G. S. 1964. *Human Capital: A Theoretical and Empirical Analysis, with Special Reference to Education*. New York: Columbia University Press.
Cohen, W., and D. A. Levinthal. 1990. 'Absorptive capacity: A new perspective on learning and innovation'. *Administrative Science Quarterly* 35:128–152.
Darréon, J., and S. Faiçal. 1993. 'Les enjeux des partenariats stratégiques entre grandes entreprises et PMI'. *Revue Française de Gestion* (September–October): 104–115.
De Man, A. P. 2005. 'Alliance capability: A comparison of the alliance strength of European and American companies'. *European Management Journal 23* (3): 315–323.
Delande, P. 2003. 'Partenariat Nord-Sud et la réalisation de PMI conjointe'. http://www.polotique-africaine.com/numeros/pdf (accessed 18 June 2003).
Doz, Y. 1996. 'The evolution of cooperation in strategic alliances: Initial conditions of learning processes'. *Strategic Management Journal* 17:55–83.
Draulans, J., A. P. De Man and H. W. Volberda. 2003. 'Building alliance capability: Management techniques for superior alliance performance'. *Long Range Planning* 36 (2): 151–166.
Dussauge, P., and B. Garette. 1995. *Les stratégies d'alliances*. Paris: Les Editions d'Organisation.
Dyer, J., and H. Singh. 1998. 'The relational view: Cooperate strategies and sources of interoganizational competitive advantage'. *Academy of Management Review* 23 (4): 660–679.
Etzkowitz, H. 2002. 'The triple helix of university-industry-government implications for policy and evaluation'. Working paper. http://www.sister.nu/pdf/wp_11.pdf (accessed 11 December 2008).
———. 2003. 'Learning from transition: The triple helix as an innovation system'. Paper presented at the Symposium on Knowledge Based Society: A Challenge for New EU and Accession Countries, Zagreb, Croatia, October.
Etzkowitz, H., and J. Dzisah. 2007. 'The triple helix of innovation: Towards a university-led development strategy for Africa'. *ATDF Journal* 4 (2): 3–10.
EU, A Background Paper. 2008. 'Lifelong learning for creativity and innovation'. http://www.sac.smm.lt/images/12%20Vertimas%20SAC%20Creativiy%20and%20innovation%20%20SI%20Presidency%20paper%20anglu%20k.pdf (accessed 11 December 2008).
Fiol, C., and M. Lyles. 1985. 'Organizational learning'. *Academy of Management Review* 10 (4): 803–816.
Gherzouli, K. 1997. ' Les conditions de réussite de la coopération euro-maghrébine'. *Revue Française de Gestion, Juin* (July–August): 118–131.
Ghosh, A. 2004. 'Learning in strategic alliances: A Vygotskian perspective'. *The Learning Organization* 11 (4–5): 302–311.
Grant, R. 1991. 'The resource-based theory of competitive advantage: Implications for strategy formulation'. *California Management Review* 33 (3): 114–135.
Gulati, R., N. Nohria and A. Zaheer. 2000. 'Strategic networks'. *Strategic Management Journal* 21:203–215.
Hamel, G. 1991. 'Competition for competence and inter-partner learning within international strategic alliances'. *Strategic Management Journal* 12:83–103.

Heimeriks, K., and G. Duysters. 2007. 'Alliance capability as a mediator between experience and alliance performance: An empirical investigation into the alliance capability development process'. *Journal of Management Studies* 44 (1): 25–49.
Hennart, J. F., and M. Zeng. 2005. 'Structural determinants of joint venture performance'. *European Management Review* 2 (2): 105–115.
Holmqvist, M. 2003. 'A dynamic model of intra- and interorganizational learning'. *Organization Studies* 24 (1): 95–124.
Huber, G. P. 1991. 'Organizational learning: The contributing processes and the literatures'. *Organization Science* 2 (1): 88–115.
Huelsmann, M., J. Lohmann and C. Wycisk. 2006. 'The role of inter-organizational learning and self-organizing systems in building a sustainable network culture'. *International Journal of Knowledge, Culture and Change Management in Organisations* 5 (2): 21–30.
Inkpen, A. C., and A. Dinur. 1998. 'Knowledge management processes and international joint ventures'. *Organization Science* 9 (4): 454–468.
Kale, P., J. H. Dyer and H. Singh. 2002. 'Alliance capability, stock market response, and long-term alliance success: The role of the alliance function'. *Strategic Management Journal* 23 (8): 747–767.
Kale, P., and H. Singh. 1999. *Building alliance capability: A knowledge-based approach*. Chicago: Academy of Management Best Paper Proceedings.
Kale, P., H. Singh and H. Perlmutter. 2000. 'Learning and protection of propriety assets in strategic alliances: Building relational capital'. *Strategic Management Journal* 21:217–237.
Karthik, I. 2002. 'Learning in strategic alliances: An evolutionary perspective'. *Academy of Marketing Science Review* 20:1–13.
Kogut, B., and U. Zander. 1992. 'Knowledge of the firm, combinative capabilities, and the replication of technology'. *Organization Science* 3 (3): 383–397.
Lane, J. P., and M. Lubatkin. 1998. 'Relative absorptive capacity and interorganizational learning'. *Strategic Management Journal* 19:461–477.
Larsson, R., L. Bengtsson, K. Henriksson and J. Sparks. 1998. 'The interorganizational learning dilemma: Collective knowledge development in strategic alliances'. *Organization Science* 9 (3): 285–305.
Levitt, B., and J. March. 1988. 'Organizational learning'. *Annual Review of Sociology* 14:319–359.
Matthews, J. H. 2003. 'Knowledge management and organizational learning: Strategies and practices for innovation'. Paper presented at the International Conference on Organizational Learning and Knowledge, Lancaster, June.
Mayer, K. J., and N. S. Argyres. 2004. 'Learning to contract: Evidence from the personal computer industry'. *Organization Science* 15 (4): 394–411.
Moatassime, A. 1983. 'Sud-Sud au Maghreb ?' *Revue Tiers Monde* 96 (October–December): 909–918.
Mowery, D., J. Oxley and A. Silverman. 1996. 'Strategic alliances and interfirm knowledge transfer'. *Strategic Management Journal* 17:77–91.
Nelson, R. R., and S. G. Winter. 1982. *An evolutionary theory of economic change*. Cambridge, MA: Harvard University Press.
Neuville, J. P. 1997. 'La stratégie de la confiance : le partenariat industriel observé depuis le fournisseur'. *Sociologie de Travail* 3:297–319.
Nonaka, I. 1994. 'A dynamic theory of organizational knowledge creation'. *Organization Science* 5 (1): 14–37.
Ouédraogo, A. 2003. 'Alliances stratégiques dans les pays en développement, spécificité, management et conditions de performance: une étude d'entreprises camerounaises et burkinabé'. PhD diss., HEC Montreal.
Parise, S., and J. C. Henderson. 2001. 'Knowledge resource exchange in strategic alliances'. *IBM Systems Journal* 40 (4): 908–925.

Parise, S., and L. Sasson. 2002. 'Leveraging knowledge management across strategic alliances'. *Ivey Business Journal* (March–April): 1–11.

Parkhe, A. 1993 'Strategic alliance structuring, organizational learning, and longevity in global strategic alliances'. *Journal of International Business Studies* 22 (4): 579–601.

Peteraf, M. 1993. 'The cornerstones of competitive advantage: resource-based view'. *Strategic Management Journal* 14:179–191.

Powell, W. W. 1998. 'Learning from collaboration: Knowledge and networks in the biotechnology and pharmaceutical industries'. *California Management Review* 40 (3): 220–240.

Powell, W. W., K. W. Koput and L. Smith-Doerr. 1996. 'Interorganizational collaboration and the locus of innovation: Networks of learning in biotechnology'. *Administrative Science Quarterly* 41: 116–145.

Ratten, V., and Y. Suseno. 2006. 'Knowledge development, social capital and alliance learning'. *International Journal of Educational Management* 20 (1): 60–72.

Rothaermel, F. T., and D. L. Deeds. 2006. 'Alliance type, alliance experience and alliance management capability in high-technology ventures'. *Journal of Business Venturing* 21 (4): 429–460.

Shrivastava, P. 1981. 'Strategic decision-making process: The influence of organizational learning and influence'. PhD diss., University of Pittsburgh.

Smith, K. G., S. J. Carroll and S. J. Ashford. 1995. 'Intra- and inter-organizational cooperation: Toward a research agenda'. *Academy of Management Journal* 38:7–23.

Soekijad, M., and E. Andriessen. 2003. 'Conditions for knowledge sharing in competitive alliances'. *European Management Journal* 21 (5): 578–587.

Teece D. J., G. Pisano and A. Shuen. 1997. 'Dynamic capabilities and strategic management'. *Strategic Management Journal* 18 (7): 509–542.

Voduzek, Z. 1988. 'Quelques réflexions sur la coopération économique et industrielle entre pays en voie de développement'. *Revue Tiers Monde* 115 (July–September): 961–974.

Wagner, B. A. 2003. 'Learning and knowledge transfer in partnering: an empirical case study'. *Journal of Knowledge Management* 7 (2): 97–114.

Wang, Y., and S. Nicholas. 2005. 'Knowledge transfer, knowledge replication, and learning in non-equity alliances: Operating contractual joint ventures in China'. *Management International Review* 45 (1): 99–119.

Williamson, O. E. 1975. *Markets and hierarchies*. New York: Free Press.

Yoshino, M. R., and U. S. Rangan. 1995. *Strategic alliances: An entrepreneurial approach to globalization*. Boston, MA: Harvard Business School Press.

5 Guarding the Gatekeeper

University Research Managers in a Triple Helix Environment

John Kirkland and Julie Stackhouse

INTRODUCTION

Relationships between academics, industry and governments are not new. A major characteristic of the past twenty years, however, has been the degree to which such relationships have become institutionalized and regulated. As governments have recognized the importance of the relationship, they have increasingly sought both to help its development, through legislation and the provision of incentives, and to oversee it, through evaluation and monitoring.

For universities, this changing environment has had important implications. Active engagement with government and industry is infinitely more important than ever before—both in financial terms and increasingly in terms of credibility. It is also something for which they are increasingly held to account. At an institutional or departmental level, project-based collaboration accounts for an increasing proportion of income. In systems such as the United Kingdom, extensive data is collected which contributes directly to core funding for the institution as a whole.

As the relationship has begun to matter more, so it has become more complex. This is true for all parties. The proportion of research commissioned by industry for reasons of curiosity, or an act of charity, is declining. Universities today are more likely to be undertaking research that forms part of core corporate strategy. In view of this, the results matter more. Governments, too, expect to see the results of their work disseminated more widely, even where no commercial interest is involved (Godfrey, Funke and Mbizvo 2010). Legal issues such as ownership of background and foreground intellectual property, confidentiality, disclosure and ethical clearance, not to mention the negotiation of financial terms, are no longer formalities. Each needs to be negotiated, and from a position of equal knowledge. Increasingly, each project requires individual negotiation.

Nor does the relationship end at negotiation. An increasing issue for higher education, in particular, has been the implementation of agreements. Historically, universities were highly individualistic institutions, with neither the mechanisms nor desire to regulate the research choices of their staff. Even now, the accountability of academics is primarily to their academic department, whereas negotiation of contractual terms is undertaken by a unit of the central institution. This raises the issue, for example, of who, on a day-to-day level, has accountability to the client for maintaining, say, confidentiality in a laboratory environment.

THE ROLE OF GATEKEEPER

These trends pose the question of how far the core relationships in a triple helix model can, or should, be regulated and who should be responsible for such regulation. Such issues are important in triple helix theory because gatekeepers play a role (although not exclusively) in determining who is allowed to enter the overlapping circles between universities, industry and government, and what type of activity is allowed to take place there. The issue is not only one of the formal seniority of research managers with respect to other key players, but also the expertise and personal qualities required to perform such a function, and their relationship with other key players. It is this issue that has led to the development of 'research management' as a distinct profession in the university sector. Yet detailed analysis of the background, motivation and impact of such gatekeepers has been limited.

No single definition of research management exists, but for current purposes it can be taken to mean 'an activity instituted at university level which seeks to add value to the research activity of the academic staff, without being part of the research process itself' (Kirkland 2008, 718). Traditionally, even these issues were handled within the research team itself. Even today, the ability of academic researchers to handle them is critical. For example, however much support institutions give in identifying funding bodies and presenting applications to them, the ability of academic staff remains the main single determinant as to whether funding is obtained.

Yet for the reasons of complexity and accountability cited earlier, the relationship between academics and administrators has changed. Thirty years ago, the question of how much information and control institutions *should* have regarding the work of their individual researchers was often seen in terms of academic freedom. Today it is couched in more pragmatic terms. Institutions *need* to know what their staff is doing as a matter of self-protection.

Given the financial and legal responsibilities surrounding external research, this 'need to know' has inevitably led to mechanisms that monitor, direct, restrict and encourage research output. These manifest themselves in several ways, for example, through decentralized resource allocation systems that make departments fully responsible for their success or failure to attract funds, and through university research strategies that prioritize specific areas of research. Perhaps the most visible of these developments has been the growth of the university research management office (Connell 2004).

Although precise structures vary between institutions, most research management offices combine a proactive and restrictive role. Their role is both to help increase the research output of the university and ensure that such research only takes place on terms and conditions that the university finds acceptable. In the context of a triple helix model, it is worth asking how far such offices can be regarded as having a 'gatekeeper' function.

The role of research managers as gatekeepers is restricted in three ways. The first is in the quantity of contact. Despite the huge growth of recent years, research managers are small compared with those of researchers. Research managers are usually based centrally, whereas researchers are spread throughout the institution, although an international benchmarking study (Higher Education Funding Council for England 2006) revealed examples of universities that were seeking to decentralize, particularly in Canada. In these circumstances, it is unrealistic to regard researcher managers as the gate through which all contact must pass.

The second restriction is expertise. The survey that follows reveals research management to be a graduate-level qualification, with a significant proportion of staff that have postgraduate- and even doctoral-level qualifications. Research managers do not claim to duplicate the expertise of academics, however. Some blurring of lines might exist in the development of research proposals to ensure that projects are realistic and financially viable, but, in general, research methodology remains the province of the academic.

The third (related) restriction is type of issue. Partly because of the time and expertise restrictions mentioned earlier, research managers are only effectively able to perform a gatekeeper function in specific areas. They are limited, for example, to issues that have implications for the institution, rather than those concerning the research group or individual. In many cases, their role is confined to project-based, externally funded work, as opposed to that undertaken by staff as part of core funded activities.

Within these confines, the notion of the research manager as a gatekeeper is a legitimate one. Research managers do not control, or even know about, the majority of contact between their institutions and the outside world. Much 'knowledge transfer' remains informal and unregulated. However where formalized relationships are required, research management structures now play a strong role in authorizing these, and a smaller one in managing and promoting them.

WHERE DO RESEARCH MANAGERS COME FROM?

The growth of research management systems is an international, although not yet global, response to these trends. The world's oldest professional body in the field, the United States–based Society for Research Administrators International, is over forty years old and boasts over four thousand members. In more recent years, there has been a proliferation of other national and regional bodies, and some trend towards specialization within them. The United Kingdom has two substantial bodies—the

Association of Research Managers and Administrators and Association of University Industry Research Links—and others dealing with specific areas such as science parks and university spin-off companies. In Europe, the past decade has seen the establishment of the European Association of Research Managers and Administrators, as well as a number of national associations. The Australasian Research Management Association now caters for Australia and New Zealand, whereas 2002 and 2006 saw the establishment of the Southern and West African Research and Innovation Management Associations. Such organizations appear to have penetrated the profession well, with 83 per cent of our respondents in the survey that follows reporting membership of a professional organization in the field, and 71 per cent having attended an event or training course in the past twelve months. Most parts of Asia, and to some extent South America, remain outside the trend, although individuals from these regions have taken part in international activity.

Relatively little is known, however, about research managers themselves—their backgrounds, perspectives and aspirations—and how these compare internationally. To address this need, early in 2008 the Association of Commonwealth Universities conducted the first international online survey of research managers (Stackhouse 2008). The survey attracted almost four hundred responses from practitioners in twenty-four countries. Reflecting the commonwealth nature of the sample and the strength of the profession, numbers showed some bias towards Australasia (129) and Europe—mostly UK—(127). Africa produced fifty-eight replies and the Americas—mostly Canada—fifty-three. Only ten respondents came from Asia.

The survey paints a picture of research management as a fast-moving profession. Over 75 per cent had been in their current position for under five years. Under half felt that they would remain in research management for the rest of their career, with about a third undecided. Slightly more, at just under 60 per cent, thought that they would remain in higher education for the remainder of their career, but even here almost 30 per cent were undecided. Whereas the majority had consciously applied for their position—between 75 and 85 per cent in each of Australasia, Canada and the United Kingdom—others suggested a degree of serendipity. Twenty-six per cent overall had entered the profession by research management functions being added to an existing job. In Africa, this figure was much higher, at 52 per cent, perhaps reflecting the earlier stage of development there.

In the context of triple helix theory, it is interesting to examine the extent to which those at the external interface themselves have experience of other sectors. From the evidence of the survey, it appears that this is somewhat limited. Eleven per cent had come from a position in the private sector, and 13 per cent from one in the public sector, outside higher education. The figures were slightly higher in the case of Australasia and the Americas, but even here the combined totals reached only 29 per cent and 25 per cent respectively. Perhaps surprisingly, in view of the strong recognition that research managers had strong transferable skills, even fewer anticipated that their next job would be outside the public sector. Whereas 13 per cent of Australians thought it likely that they would move into the private sector, the figure for the sample as a whole was only 9 per cent. The same proportion anticipated a move into another public sector role outside higher education.

Overall, the results suggest that rather than being a force for bringing new skills and experiences into higher education, the research management profession is gradually being consolidated within the existing administrative structure. In Africa, where research management is at an earlier stage of development and the culture of university administration generally is arguably less valued, the respondents were likely to have taken on the role as part of a senior academic position, and often on a part-time basis. Fifty-eight per cent listed their last position as being academic, compared with only 17 per cent for the sample as a whole. The same can be said for the small Asian sample, where 40 per cent had come from an academic position.

Elsewhere, the notion of a university research management profession, or research management as a discrete arm of university administration, was strong. Twenty-one per cent had come into their present job from another research management position at their institution, and 12 per cent from a research management position at another academic institution. Eighteen per cent had come from another area of university administration—although this form of mobility was significantly more common in Europe (28 per cent) than anywhere else. A similar pattern was found in future aspirations, with a combined total of 57 per cent considering that their next role would be in one of these three areas. Fifty-seven per cent also disagreed with the statement that 'research management cannot be considered as a discrete profession'. Only 28 per cent agreed, these being disproportionately from Africa (40 per cent) and the small Asian sample (78 per cent).

It also seems likely that this professional consolidation has increased over time. The overwhelming majority of those who had moved from research management positions at other institutions had done so within the last five years, whereas 23 per cent of those in their current positions for less than five years had moved from other research management jobs at the same institution. This is perhaps not surprising in the context of an expanding profession—82 per cent agreed with the statement that 'the office in which I work has expanded in the last three years'. It does, however, send an important message to those planning this expansion. It seems likely that the profession may not yet have reached its full size, and that there may still be a shortage of experienced staff. Competition for such staff may less likely be with other sectors, than between higher education institutions themselves.

WHAT DO RESEARCH MANAGERS DO?

The majority of respondents described their position as a senior- or middle-level administrative one. Except in Africa, there appeared to be strong recognition of the boundaries between the research management and the academic direction of research itself. The majority also spent all of their working time on research management activities—the overall figure of 72 per cent masking wide divisions between 91 per cent of Canadian respondents and 38 per cent of Africans.

In purely quantitative terms, it appears that most effort is devoted to negotiating and, particularly, managing and reporting on contracts, followed by the more promotional and awareness-raising aspects of the role, with less time spent on the relatively specialist field of technology transfer and commercialization.

Analysis of the skills that research managers considered necessary for their work revealed two interesting trends, both of which perhaps reinforce the notion of the research manager as a generalist. First, it was clear that respondents placed high emphasis on more personal, transferable skills alongside specific knowledge. The ability to develop good personal relationships was ranked as the most important attribute, followed closely by the ability to absorb complex information and attention to detail. Presentation, negotiation and numeracy skills, although still valued, were ranked somewhat lower. See Figure 5.1.

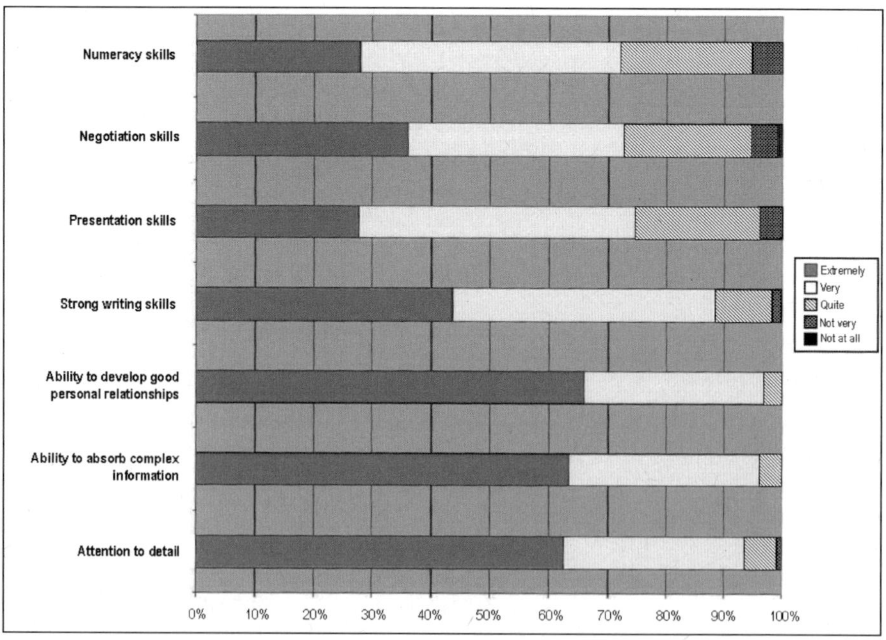

Figure 5.1 Presentation, negotiation and numeracy skills.

Second, although the specialist nature of some aspects of research management meant that few respondents felt an urgent need to increase their skill in these areas, significant proportions considered that they had some need to do so. Only 4 per cent and 6 per cent considered that they had an urgent need to improve their financial and marketing skills, but 65 per cent and 41 per cent had 'some' need. For legal knowledge, the respective figures were 13 and 55 per cent, intellectual property 14 and 62 per cent, ethics six and 40 per cent and innovation or technology transfer issues 12 and 56 per cent.

These figures may partially be explained by a natural tendency of respondents to move to the centre of any scale in their replies. They suggest a continuing demand for knowledge and training, and recognition that the research process cannot be seen as a series of unrelated stages. Thus, for example, those whose prime responsibility is for writing proposals or reporting on them to sponsors still require some knowledge of the legal, ethical and technology transfer environment in which they operate.

WHAT CHALLENGES DO RESEARCH MANAGERS FACE?

Research managers appeared divided on the future prospects for their profession. On the one hand, there was recognition that much had been achieved in recent years. Seventy per cent agreed that awareness of research management issues had risen amongst academics during the past three years, 58 per cent that research management was highly valued by the leadership of the institution and 56 per cent that it was now highly valued by academics at the institution. Respondents were also confident of their impact. Eighty-nine per cent said that research management plays an important role in improving the research reputation of the institution, and 86 per cent that it plays an important role in increasing institutional income.

Much, however, remained to be done. Despite the increased recognition mentioned earlier, only 35 per cent felt able to agree with the statement that research management was generally well resourced at their institution. Only 38 per cent agreed that career prospects for research managers embarking on their career were very good and just a bare majority of 54 per cent that research management offered good training and development opportunities. The guarded confidence that institutional leaders and academics were starting to understand research management issues did not extend to a more macro level. Only 21 per cent agreed that research management issues are widely understood by their country's government.

Despite—or perhaps because of—the expansion of recent years, their confidence in the importance of their work and increasing acceptance of their role, research managers appeared to feel under some pressure. When asked to rate the extent to which they felt challenged by specific areas, 71 per cent considered increased expectations from funding bodies to be a serious challenge. When combined with the lack of confidence in governments to understand research management issues mentioned earlier, this might be said to raise questions about the policy environment in which university research management takes place. Other issues were also recognized as serious challenges, although to a lesser extent. Sixty-four per cent considered the need to deal effectively with an increasing range of international partners as a serious challenge, 62 per cent the increasing complexity of contracts and 57 per cent the increasing complexity of ethical issues.

This external pressure can also be seen at the level of the individual. When asked to rate the barriers to their increasing expertise, to meet the needs identified earlier, those identified most strongly were a lack of time to attend (3.15),[1] and the unavailability of appropriate courses (3.05). The lack of institutional support (despite being highly rated in the Americas) and lack of motivation were notably lower.

RESEARCH MANAGERS IN THE DEVELOPING WORLD

Several chapters in this volume draw attention to the need to adapt triple helix models to particular circumstances. Chapter 8 provides an analysis of how triple helix models might be applied in Ghana, in particular, and draws attention to the importance of institutional base and culture, noting critical differences between, for example, government research institutes and universities.

In this context it is important to summarize the implications of our study for developing countries in particular. It is clear that the relationships between research managers, their institutions and the outside world in Africa and Asia differ significantly from those of other parts of the world. In part, this reflects the different stages of development of the profession, in part the different systems and cultures within which universities operate.

Perhaps surprisingly, visible awareness of formal research management structures appears to be stronger in African universities than Asian ones (Association of Commonwealth Universities 2001; ACU SARIMA 2005; Stackhouse and Day 2005). In each case, however, development is at a much lower level than in more developed regions. The majority of the small Asian sample, and over 40 per cent of the African one, agreed with the statement that 'research management cannot be considered as a discrete profession', compared with around a quarter in Europe and the Americas, and 22 per cent in Australasia. Nonetheless, 71 per cent of Africans agreed that awareness of research management issues had risen amongst academics over the past three years, a figure slightly above the global average.

In earlier analysis (Kirkland 2008), we have suggested that the growth of national research management systems can be seen in three (not necessarily consecutive) stages—those of awareness raising, building on structures at the institutional level to support these and the development environment which support these. Analysis of African universities involved in a project supported by the Carnegie Corporation of New York (Botha et al. 2007) suggest that Africa has advanced well on the first of these, and that some moves have taken place towards the third, but that the second stage—that of developing robust structures at the institutional level—remains the most problematic.

African research mangers also had a significantly different career profile from those in developed countries. They tended to have markedly higher-level academic qualifications—with 69 per cent having obtained a doctorate, compared with 40 per cent in Asia and less than 30 per cent in each of the other three regions. Fifty-eight per cent of African respondents and 40 per cent of Asians had come into their post from an academic position, compared with only 17 per cent in the total sample, although interestingly only 26 per cent of the African sample expected that their next position would be an academic one. None of the Africans had come from the private sector, and only 7 per cent expected their next post to be there.

The markedly closer relationship between research management and academic life in African universities may have implications for triple helix relationships. At first sight, it would appear to suggest that research management will be less likely to develop a different set of values to conventional academic ones. On the other hand, the relatively senior level of individuals claiming to be research managers may, in the short term, give the activity more credibility amongst academic peers. Interestingly, however, the majority of those who have made the transition do not expect to move back into a purely academic life. The extent to which this cohort adopts distinctive norms and values will be critical in the development of the profession over the next decade.

RESEARCH MANAGERS AND THE TRIPLE HELIX MODEL

The role of university research managers, and others who exist at the boundaries of university, industry and government, needs to be fully understood both by triple helix theorists and policymakers. It is, for example, important to know the balance between their promotional and regulatory roles, and thus the extent to which their role is primarily to encourage, or restrict, interaction. Another key balance is between the role of mediator, working towards consensus by encouraging greater awareness of business approaches within the institution, and that of representing the university view to the outside world. The survey reported earlier provides some important pointers in this respect.

First, the results confirm that university research management has developed as a discrete profession, confident in its abilities and developing its own structures and boundaries. As such, its members probably represent the most visible and coherent set of gatekeepers or boundary dwellers in the triple helix model. Consideration needs to be given to the identification of similar groups in industry and government, and the extent to which core administrative or policy values are shared between them. It is also important to note that expansion is not happening at an even pace between countries. Our survey revealed that in Africa, in particular, the process is a relatively new one, often been carried forward by academics for whom research management has been grafted onto existing functions. The extent to which the profession that results reflects local culture and expectations, or gravitates towards international norms, represents a further area for future analysis.

University research managers appear to regard themselves very much as part of the organizational framework of their institutions. The majority came into their positions from elsewhere in higher education, their institutions and increasingly university research management. Similarly a significant proportion (57 per cent) regard higher education, and research management in particular (41 per cent), as the most likely source of their next employment.

Given this, it might be thought likely that research managers perceive their role more as one of representing their institution to the outside world, rather than influencing behaviour within the institution. This, however, cannot be assumed. Awareness raising (presumably amongst colleagues within the institution) was cited as one of the most time-consuming functions of the research manager. The suggestion that institutional leaders and academic now understand research issues more fully also suggests that an element of influencing opinion exists. In reality, research managers may combine an element of adopting institutional values, influencing such values and, as the profession expands, developing a set of professional values of their own.

Despite a generally confident and expansionary outlook, research managers appear to recognize the weight of expectation on them. Within their institutions, most have benefited from expansion and a majority recognize that institutional leaders and policymakers increasingly understand their concerns. These views are not universal, however. A significant minority disagree and substantial numbers expect to move on. There is significant unmet demand for training, the obstacles to which include lack of time and availability of suitable sources, in many cases even more than lack of institutional support.

The external policy environment appears more problematic. Despite expansion and increasing opportunities for specialization, a majority recognize that increasing complexity has raised serious challenges. The issues of increased demands from funding bodies, and the lack of understanding of research management issues by government, suggest that these may in part be due to a lack of understanding in other parts of the research environment. The survey also showed significant differences between regions, with structures in developing countries being generally less robust, despite recent developments in Africa. There has also, to date, been a lack of such a clear structure to manage university–industry relations in the private sector.

Overall, our results suggest that the role and number of university research managers is still increasing and may continue to do so for some time yet. As such, they have the potential to be a major stakeholder group in the triple helix environment, with the potential to improve not only the effectiveness of their institutions, but the workings of the system as a whole. Whether this potential is fulfilled, however, will depend on both changes in the external environment and the extent to which institutions provide the training and career development opportunities characteristic of more established professions. Research management is very much an area in which the gatekeepers themselves will need continuing protection.

NOTES

1. Figures refer to overall average rating, for all respondents, on a 1–5 point scale, where 1 corresponds to 'not a barrier' and 5 corresponds to 'serious barrier'.

REFERENCES

Association of Commonwealth Universities. 2001. 'Research management in African universities: Report of a benchmarking seminar held in Durban, South Africa. 1–3 April 2001'. Workshop report. http://www.acu.ac.uk/resman/pdf/ResearchManagementinAfricanUniverities.pdf (accessed 16 February 2009).

Association of Commonwealth Universities, Southern African Research and Innovation Management Association. 2005. 'The demand for research management support in African universities: Report of an ACU-SARIMA seminar for African Universities, Cape Town. 19–20 February'. Workshop report. http://www.acu.ac.uk/resman/pdf/TheDemandforReserchManagmentinAfricanUnis.pdf (accessed 16 February 2009).

Botha, A. P., P. van Eldik, P. Waugaman, J. Kirkland and P. Ajai-Ajagbe. 2007. 'Report on a workshop on the Carnegie Programme to strengthen research management structures in African universities. 11, 12 May 2007'. Workshop report. http://www.acu.ac.uk/resman/pdf/CarnegiepProgrammetosupportresearchmgmt.pdf (accessed 16 February 2009).

Connell, H., ed. 2004. *University research management: Meeting the institutional challenge.* Paris: Organisation for Economic Co-operation and Development.

Godfrey, L. Funke, N. and Mbizo, C. 2010. 'Bridging the Science Policy interface: a new era for South African research and the role of knoledge brokering: South Africa Journal of Science vol. 106, no. 5–6, May/June (http://www.scielo.org.za/scielo.php?script=sci_arttext&pid=50038-235320^00003000^3&lng=en&nrm=iso).

Higher Education Funding Council for England. 2006. *International research management benchmarking programme: Report to HEFCE by the Association of Commonwealth Universities.* London: HEFCE.

Kirkland, J. 2008. 'Research management: An emerging profession in the developing world'. *Technology Analysis and Strategic Management* 20 (6): 717–726.

Stackhouse, J. 2008. 'Research management survey—recruitment and retention'. *Research Global* 20: 4–5, 7.

Stackhouse, J., and R. Day. 2005. 'Global and regional practices in university research management: Emerging trends'. *International Journal of Technology Management and Sustainable Development* 4 (3): 189–205.

6 Revisiting Agricultural Science for Development

From *Mode 1* to *Mode 2*[1]

Norman Clark

INTRODUCTION

This chapter concerns ways of rethinking the ways in which agricultural science relates to the problems of the rural poor, especially with respect to conditions in sub-Saharan Africa. For some time now relevant scientific communities have been under pressure to show national governments and donor communities that their work is indeed having some tangible impact on the resolution of poverty, levels of which are becoming less tractable across large geographical areas. The problem seems to be that the institutional patterns and norms that govern scientific behaviour (and the organizations within which science is conducted) have great difficulty relating to the general issue of poverty impact. Scientists are caught between two stools. On the one hand, they have been brought up to apply rigorous experimental procedures to precisely defined problems with results that have to meet standards of peer review. On the other hand, they are being told that such reductionism is insufficient and that unless they broaden the scope of their work they will suffer reduced support. Scientists then reply that the lacuna is not the responsibility of science as such but that of other downstream bodies better placed to deal with poverty issues, bodies such as extension agencies, NGOs, local government departments, private firms, etc. Moreover, broadening the scope of their work would dilute its value and ultimate impact.

The debate has now been going on for some time and indeed is very much to the fore in the governing Science Council of the Consultative Group for International Agricultural Research (CGIAR). Generally speaking there are two competing viewpoints. The 'traditionalists' argue that the Consultative Group (CG) system was established in light of the great breakthroughs made by the Green Revolution scientists in the 1950–1980 period. These appeared to show that concentrating scarce resources in high-quality agricultural research centres would increase the efficiency of science and lead ultimately to substantial technological results, in this case the high-yielding hybrids (HYVs) of rice, wheat and maize. The 'apostates' argue that contextual circumstances have now changed significantly such that purely technological solutions to poverty are unlikely to be reached easily. And in any case there is now so much 'donor fatigue' that it would be strategically unwise to carry on down the Green Revolution road. Science will suffer and the eventual results will be even worse from a poverty standpoint.

How can these positions be reconciled? Indeed, should they be so reconciled? This chapter takes the view that the debate is actually a microcosm of a wider debate on the nature and purpose of science. One of the easiest ways of describing this (albeit an oversimplified one) is to put the issue in terms of the *mode 1/mode 2* distinction developed by Gibbons and his colleagues (1994) over the past decade or so. The second section of the chapter makes the point by making a link between this broad distinction and the emergence of *innovation systems* as an operational metaphor for science policy analysis. The third section of the chapter then summarizes three case studies that give an empirical picture of the debate in more precise context. The fourth section draws a number of lessons that may be learned that in turn give pointers to how agricultural scientific research may be readjusted in ways that reconcile the conflicting agendas. Finally, the fifth section draws the chapter to the general conclusion that the two positions are not irreconcilable. Indeed, it argues that there is actually a continuum between the two positions, elaboration of which could not only give a new role to scientists, but actually improve both the quality of science and its applicability to the poverty agenda.

SCIENCE AND INNOVATION SYSTEMS

The distinction between *mode 1* and *mode 2* paradigms of scientific discovery was first put forward by Gibbons and colleagues in 1994[2] and subsequently revisited by some of the authors over the past decade or so. In the words of Nowotny, Scott and Gibbons (2003) its broad thesis was that:

> the old paradigm of scientific discovery (*Mode 1*) characterised by the hegemony of disciplinary science, with its strong sense of an internal hierarchy between the disciplines and driven by the autonomy of scientists and their host institutions, the universities, was being superseded—although not replaced—by a new paradigm (*Mode 2*) which was socially distributed, application-oriented, trans-disciplinary and subject to multiple accountabilities.[3]

The argument has caused considerable debate within scientific and related communities. The essential characteristics of *mode 2* knowledge may be summarized as follows:

- It is generated within the context of application and not solely through scientific experiment.

- It is essentially transdisciplinary and not solely reducible to the outputs of single disciplines.
- It is developed within and across widely different organisational forms.
- It is 'reflexive' in the sense that it is not reducible to an objective investigation of 'natural law' but is rather a dialogue between research actors and subjects.
- Quality is controlled not only by scientific peer review but also by other actors including research 'clients'.
- Issues of policy, commercialization (including intellectual property rights) and accountability are now very much to fore in corresponding science management.

Much of this debate has tended to bypass developmental agricultural science until very recently. Here the Green Revolution agenda has continued to obtain although research centres have found it increasingly hard to mobilize resources. The tradition in agricultural technology development had always been to regard national and international R&D centres as the major players. Where this failed to produce the desired results, responsibility for failure tended to be laid at the door of poor government policies or inefficient extension services.

However, nowadays it is increasingly difficult to regard publicly funded agricultural science as the only source of production/welfare improvements and thus, international food security and social well-being. Moreover, the agricultural research agenda has now expanded well beyond improving 'yields per hectare' to include issues of continued (and worsening) poverty, environmental sustainability, the multifaceted contribution of the private sector, the complementary roles of non-governmental organisations (NGOs) and community-based organizations (CBOs), the importance of farmer knowledge, the growth of relevant agribusiness and changing (national and global) macroeconomic conditions. In short, the agenda for agricultural science has arguably become much more complex and multidimensional. In particular, it is about building up knowledge on how to integrate agricultural science better with client needs and complementary capabilities, especially with relevance to poor rural communities.

In short, modern literature shows that the agenda for agricultural research has changed dramatically from the days of the Green Revolution, and with it the demands on relevant organizations. And as a recently published IAASTD Report (2009) has emphasized,[4] it is this new complex agenda that has created the need for a fresh look at science policy analysis for agriculture. Arguably agricultural R&D can no longer be left on its own to meet the new demands of the twenty-first century using the old institutional methodologies. In turn, this means new types of relationship with other stakeholders and new types of capacity on the part of scientific institutions and organizations. This does not mean any reduction in the quality of the science. Rather the reverse in fact, as a UK Parliamentary

Select Committee has pointed out in its 2004 report.[5] It implies that scientists and the organizations in which they work need to improve their capacities to undertake quality science. But to do this they also must become more aware of the wider context of their research and how this can inform the nature and purpose of what they are trying to do.

It is here that the 'innovation system' metaphor begins to play a role.[6] Initially put forward by economists such as Nelson and Freeman in the 1980s to explain the rapid economic growth of the so-called 'newly industrializing countries' (NICs) over the latter part of the twentieth century[7] its use has been extended and developed widely over the past decade.[8] Originally the concept was developed to deal with the inability of conventional economic variables (such as capital investment and R&D expenditures) to explain differential rates of economic performance and to locate the role of knowledge among a much wider range of stakeholder groups than had previously been the case.

The concept is now used as a kind of shorthand for the network of inter-organizational linkages that apparently successful countries have built up as a support system for economic production across the board. In this sense it has been explicitly recognized that economic creativity is actually about the quality of 'technology linkages' and 'knowledge flows' amongst and between economic agents. In agricultural research policy discussion is even more recent, but it is growing. One of the most widely recognized groups to develop the idea was Hall and his research team at ICRISAT in the period 1996–2005. In a series of around thirty or so case studies undertaken in South Asia and Africa they demonstrated that successful pro-poor technology development in the crop post-harvest sector depended heavily on the capacity of the whole system to change institutionally. By this they meant that both the stakeholder organizations themselves and the norms that pattern their behaviour had to be capable of (in some cases) fairly radical change. Where they did not (or could not), what were quite promising potential interventions on behalf of the poor farmer often did not succeed. Hall noted that it is often scientific organizations themselves that are most reluctant to change largely because they have been brought up in *mode 1* ways. His use of the *innovation systems* metaphor was designed, therefore, to give greater credibility to appropriate policy recommendations.

However, the concept of an innovation system is itself still ill-understood. Some[9] regard it as a potential scientific theory from which testable propositions may be deduced. Others[10] regard it as a dangerous concept that may well reduce the effectiveness of what little pro-poor agricultural research is left in poor countries. Yet others[11] regard it as an interesting heuristic device that suggests the importance of better linkages across stakeholder groups. What follows in this chapter is a brief account of three contrasting cases of technology development. One of these is historical and represents a classic *mode 1* approach. The second illustrates a mixture of both approaches. The third case is unambiguously *mode 2* both in design and actuality. The fourth section then goes on to set out what these cases seem to indicate from the standpoint of developing a more plausible theory of technology development that incorporates a systems approach.

CASE STUDIES

Livestock Diseases—The Case of Trypanosomiasis

Livestock plays a particularly important role in the livelihood of the poor farmer. It acts as a form of capital asset and stored wealth; it provides insurance against sudden loss of crops due to natural disasters like drought; it acts as a source of natural fertilizer for crop production; it provides vitamins and other nutrients not present in cereal and leguminous crops; it acts as a source of draught power and transport; and in some cases acts as a source of income for otherwise disenfranchised groups like women. At the same time it is prey to a range of diseases (mainly vector-borne) that affect its growth, health and ability to perform such functions efficiently.

Trypanosomiasis is one such disease. It is caused by unicellular protozoan parasites, termed trypanosomes, which propagate in the blood and tissue fluids of their hosts. Pathogenic species of *Trypanosoma* occur in Africa, Asia, Latin America and the Middle East, and infect, among others, man, cattle, sheep, goats and water buffalo. Some species of trypanosomes also cause sleeping sickness in humans. The susceptibility of host species differs—the disease can be either acute or chronic. Trypanosomiasis is frequently fatal in highly susceptible animals (such as *Bos indicus* Zebu cattle, and some exotic breeds), whereas in more resistant ones (including N'Dama cattle, a West African *Bos taurus* breed) the disease results in decreased productivity. Symptoms of trypanosomiasis include sporadic periods of fever, wasting, enlargement of lymph nodes, anaemia, infertility and immune dysfunction. The major trypanosome species that infect animals in Africa are *Trypanosoma congolese* and *Trypanosoma vivax*. These are mainly transmitted by tsetse-flies, although non-tsetse-transmitted forms of trypanosomiasis also occur in Africa and South America.

For decades, trypanosomiasis control had been attempted mainly through two routes—vector (tsetse-fly) control and trypanocide drugs. The former has involved a range of approaches, from tsetse habitat clearings and the use of impregnated traps, to the widespread application of insecticides and the use of the sterile male technique. Indeed in the early to mid-1980s, according to one authoritative source, 'the days of tsetse seemed numbered'.[12] Large-scale spraying at ground and aerial levels had all but eliminated tsetse from large areas of east, west and southern parts of Africa that had previously been infested. This had given enormous benefit to livestock owners, many of whom were poor farmers. In addition, a range of newer technologies such as odour-baited targets and pyrethroid-treated cattle seemed to indicate that the problem was showing every sign of getting under permanent control.[13] According to Torr et al. (2005), however, what seemed a promising (if expensive) research trajectory began to fall apart due to changing donor priorities and research policy positions. This combined with general economic decline in these countries meant that considerable ground was lost.

In the mid-1970s to late 1970s, however, one of the CG institutes, the International Laboratory for Research on Animal Diseases (ILRAD), embarked on a biological solution to the trypanosomiasis problem, namely, the discovery of a vaccine against the disease. ILRAD had been set up in the mid-1970s with a mandate to find immunological solutions to diseases affecting livestock and initially it decided to concentrate on trypanosomiasis and east coast fever, two diseases that are particularly virulent in east and southern Africa. The thinking at the time seems to have been that although there were other remedies such as those outlined earlier dealing with the vector, tsetse, it would be worthwhile concentrating on a 'solution' that promised to be much more comprehensive and cheaper in the long run.[14] Accordingly, just at the point where vector control appeared to have lost ground in the attempt to deal with the parasite itself, considerable resources were placed at the disposal of research into bovine immunology. Unfortunately, things did not turn out as anticipated; the development of a vaccine was hampered due to antigenic variation and unanticipated complexities in the research itself. The result was finally that vaccine research efforts effectively ended in the mid-1990s. The focus of trypanosomiasis research at the laboratory (now renamed the International Livestock Research Institute, ILRI) has shifted to the genetic characterisation of trypanotolerant cattle.

The issue here seems to be one of research concentration. At the time of establishment of the centre it appeared sensible to go for biological solutions to livestock diseases. We do not now know what the rationale for this was but one can hypothesize a number of factors as follows:

- a strong belief on the promise of modern biology and related technological solutions
- anticipation of scientific synergies across livestock and other diseases (e.g. malaria)—so that breakthroughs would have spin-off benefits in cognate areas
- the belief that scientific solutions had not previously had sufficient opportunities to prove themselves
- a feeling that vector control was expensive, politically sensitive and perhaps environmentally destructive
- strong signals from the donor community
- economic cost factors

For whatever reason, however, this centre decided not to act as a general animal health centre but to focus on a 'high science' trajectory. With the failure of the vaccine project, farmers are still reliant on expensive trypanocide drugs. National veterinary services have been much reduced mainly due to macroeconomic constraints whereas capacities for vector control are by now much diminished. In recent years the research centre in question has begun to reposition itself in the former mode and has been developing projects of a broader perspective in relation to livestock problems for the poor.

Provision of Fodder for Livestock: The Case of Groundnut[15]

One of these areas is the provision of fodder for livestock. A problem that particularly affects poor farmers is the economic acquisition of sufficient amounts of fodder of the necessary quality to enable livestock to serve its livelihood function. Accordingly, in the early part of 2002 DFID funded ILRI to undertake collaborative research in India and Nigeria to improve fodder provision. The geographical focuses were Andhra Pradesh in India and Kaduna State in Nigeria although there seems to have been no clear rationale behind this choice of regions. The emphasis was to be on testing and evaluating 'already proven' fodder technologies through scaling them out to the ultimate client sector—the poor farmer in both regions. The overall project consisted of a number of sub-projects in both countries.

One of these investigated groundnut haulms, which are an important fodder for cattle in mixed farming systems in the semi-arid zones of Andhra Pradesh and other states in India. In this case a new variety had been developed in on-farm trials resulting in higher yields of both grain and haulm. The project started by scaling the new 'technology' out to farmers through participatory rural appraisal (PRA), rapid rural appraisal (RRA), focus group discussions (FGDs) and field-days. This was followed by a sample survey conducted with sixty farming households participating from three villages in the district. The purpose was to understand and characterise the livestock-livelihoods-fodder scenario in a farming systems perspective. Another objective was to explore how far the prevailing crops and cropping systems supported the fodder requirement of cattle across different seasons of the year. The three villages chosen were representative of several typical features: red loamy soils, predominantly rain-fed agriculture and groundnut-based cropping. They also had a majority of poor households.

The survey indicated that given the low rainfall and virtual monocropping of groundnut in the area the improved variety would enhance the livelihoods of crop–livestock farmers. Focus groups and workshops were then held to operationalise scaling out procedures. It was at that point that project staff began to realize that there were many other factors preventing the technology being adopted. For the poor farmer there was no reliable source of seed supply, especially for farmers in remote rural areas. Reliance on the private sector was hampered by poor information on prices and resultant exploitation by traders. Unstable rainfall patterns, inability to source fertilizer and poor extension provision also played their part in disenfranchising precisely the client sector that had been targeted. In addition, there were many more stakeholder groups in the picture whose activities influenced possibilities for scale out. Examples are producer associations, government departments at state and district level and NGOs of various types. The response of project staff was then to undertake further studies into why this should have been the case but it gradually became clear that the total system was much more complex than had initially been realized.

A second sub-project was established in Nigeria as a means of using existing knowledge of forages and fodder options to improve the livelihoods of smallholder farmers. In this case a key component was building partnerships with 'non-technical groupings'. Many farmers across Africa endure animal feed shortages during the year, especially during the dry season. In order to alleviate this shortage the project was hoping to initiate a project with a variety of stakeholders to identify promising fodder options for dissemination across Nigeria. As in India, forages were seen as a relevant technology in West Africa although there has always been debate on their use, partly stemming from some reluctance on the part of farmers to adopt it. In this sub-project the methodology adopted was to identify key component best practice technologies through meetings of major international and local research institutes, to build partnerships with local stakeholder groups (including NGOs) and then to scale out the most promising technologies in a series of extension activities including all stakeholders.

Again things did not go as expected. At the beginning of the project the emphasis was on working with partners to research the adoption of specific fodder technologies and impacts on livelihoods following a single model i.e. demonstration plots and farmer-to-farmer dissemination of germ-plasm and practices. However, de Haan,[16] while reviewing this project, paints a picture of dissonance among partners due to patchy attendance at meetings, dominance of technical partners and poor information flows. Uptake was also slow. The focus then changed (over a three-year period) to one of working more directly with key stakeholder groups, particularly seed suppliers, and to trying out new information mechanisms to improve uptake. But progress was slow. In the seeds area a key problem was that the private companies were unwilling to invest in new varieties which did not have a guaranteed market and for which they knew nothing of the demand. It was also clear that a potential way of linking farmer demand to the private sector was through input suppliers. In this case the project signed an agreement with the private seed company to underwrite seed production in 2005 and purchase any unsold seed from the 2006 growing season. In fact, 4.7 tons were produced; seeds of all but one variety sold out and in 2006 the company expanded its seed production activity from 4.7 to 19 tons.

Andhra Pradesh Netherlands Biotechnology Programme (APNLBP)

The Andhra Pradesh Netherlands Biotechnology Programme (APNLBP) was established in the mid-1990s with Dutch bilateral aid. It was part of a larger international program in four countries designed to develop biotechnology capacity for the rural poor but unlike most such programs it built something approaching an innovation systems flavour right from the beginning.[17] The first phase of the program started on 1 November 1995 and concluded by 31 March 2002 with a total budget of Rs.155 million. On the basis of a satisfactory performance evaluation, the program was extended for another five years up to 31 March 2007 with a total budget of Rs.275 million. Thus the total duration of the program has been more than eleven years with a total outlay of Rs.430 million. During the period between 1995 and 2005 the program established about seventy-five research projects with a total commitment of about Rs.300 million in association with a number of research organizations, NGOs and state government departments in Andhra Pradesh.

It has been the strong belief of the Dutch public policy since the early 1990s that the potential of agricultural biotechnology can help redress problems of food insecurity in developing countries provided these countries are empowered to design their own technologies to suit their local conditions. The Indian program focused on Andhra Pradesh, one of the states where Dutch Development Cooperation is strong. After several years of preparation the substantive program began in November 1995. From the beginning its unique feature was that it should follow an 'interactive bottom-up' (IBU) approach, an approach based on the principles of participatory technology development (PTD). All projects were to be formulated on the basis of local needs assessment and priority setting, to which end users, researchers, policy makers, government and NGOs should all be involved. In addition, a central principle was to be constant interaction between farming communities and scientists in the process of technology development and adaptation. These interactions would be used to combine indigenous knowledge of people with both tacit stakeholder and modern scientific knowledge.

Developed by a group of researchers at the Free University of Amsterdam in the late 1980s,[18] the IBU approach placed emphasis on identifying needs and priorities for interventions based on end-user demands. These should be formulated by all stakeholders interacting with each other in multidisciplinary fora until common ground is achieved regarding what the priorities are and how they should be tackled. Equally, power relations should be such that no one group could dominate any other. Basic scientists, for example, would not be in a position to dictate procedures, but on the contrary would be subject to constant scrutiny from other participatory groups. Moreover, the process would be interactive in the sense that constant monitoring and adaptation would take place throughout the life of projects. Using this process a multidisciplinary team consisting of natural scientists, social scientists, extension workers, administrators and NGO representatives participated in a local 'need assessment survey'. This led to intensive discussions and deliberations in prioritizing specific areas for intervention in dry land agriculture. The output of this survey resulted in a base document for designing the entire program and defining the priority areas in a priority-setting workshop wherein different stakeholders participated and deliberated. Its broad objectives were as follows:

1. To promote application of biotechnologies relevant to small-scale agricultural producers and processors in Andhra Pradesh in such a way as to contribute to sustainable agricultural production, taking into account in particular the position of target groups such as women and poor farmers.
2. To develop appropriate biotechnologies through research activities that focus on identified priority problems.
3. To conduct supportive activities required to ensure development and adoption of biotechnologies including training, transfer of technology activities, workshops and information dissemination.
4. To strengthen capacities of local organizations in Andhra Pradesh to develop and transfer biotechnologies and conduct analysis in the field of technology assessment.
5. To promote the adoption of biosafety measures and to contribute to discussions on issues of intellectual property where appropriate.

The program has since focused on four priority areas that affect primarily the livelihoods of poor farming communities. These are (a) agroforestry and horticulture, (b) food crops, (c) oil seeds and (d) animal production and health. It focuses mainly on a few selected villages in two districts close to Hyderabad. All technologies developed through this program are being tried initially in these villages and their impact is being assessed. The plan is that proven technologies will then be propagated in other parts of the state. Scaling out has been subcontracted to two research stations, one in each district. Each station has been funded to enable the purchase of laboratory equipment necessary to pursue simple interventions associated with second-generation biotechnology, such as tissue culture. The program is managed by a small unit based at a local university and reports to a cross-stakeholder 'policy committee' that meets frequently for project decision and monitoring purposes.

Several features should be noted about the APNLBP as follows:

1. The decision to operate through NGOs (eighteen in all) as an interface with farmers was taken as a deliberate measure to ensure that scientific institutions did not dominate project selection and to enhance acceptability of interventions. A good example here was that of biocontrol agents where the benefits tend to accrue only after a period of time (compared to the impact of chemical agents).
2. Only a relatively small proportion of projects (around 23 per cent as of 2005) involved gene transfer technology because the objective was

to choose projects that had a balance between immediate impact and long-term benefits (in some cases two separate projects were sanctioned that dealt with the same problem).

3. As a deliberate policy a series of projects were put in place to enhance the livelihoods of relatively disadvantaged groups such as women and unemployed youth. Examples here are projects on vermiculture, mushroom cultivation and poultry husbandry.
4. Consistent interactions between farmers and scientists have taken place throughout the program with two-way biannual visits.
5. Considerable spread effects have taken palace through 'word of mouth'. There is a clear homology here with the aims of farmer field schools.
6. The quality of science undertaken at the laboratory level was just as high as normal.
7. Decision-making procedures at the 'top level' are taken by cross-disciplinary technical groups that also have representation from relevant government departments and NGO stakeholders.
8. The donor agency has kept an unusually low profile throughout the lifetime of the program (nearly ten years in all). There were no 'log frame' deliverables specified in pre-project management and even most of the periodic external reviews were managed by Indian national personnel and organizations. It seems likely that withdrawal of the donor (due soon) will not affect the capacity of the program both to continue and to act as a 'model' for similar interventions both in India and overseas.

DISCUSSION

The case studies outlined in this chapter were selected deliberately to show that there is a continuum of approaches towards technology development. The first case study shows how the approach to dealing with a problem was seen in terms of a technological fix (vaccine discovery) that would be led by scientific research. The second case study indicates that the new technology could not readily be scaled out and therefore the need for 'applications engineering' had become necessary. But even here the underlying presumption was that, provided the proper context could be engineered, the final results would be beneficial. However, even where the project made a deliberate effort to work with non-technical partners, progress was slow. The third case illustrates an approach that has broken with tradition in a fundamental sense, although at this stage we are unsure how successful the model will be. Essentially, the APNLBP put an early premium on specific developmental outcomes and enrolled laboratories as only one element in their strategy.

How then should science administrators proceed? One thing that has finally become clear is that the knowledge system that underlies economic production is much more complex than had previously been accepted. And at this point it is useful to return to the analysis outlined in the second section of this chapter because in all three cases it is clear that the relationships are highly complex. The first point to make is that all three cases reflect the general tension that always exists between *mode 1* and *mode 2* views about the proper conduct of science. The traditional linear and reductionist view is that science should focus on precisely defined experiments that are capable of giving robust conclusions that may be validated by peer review prior to publication of results. These are driven very much by ruling scientific disciplines and on the whole cross-disciplinary interaction is viewed as compromising rigour. Similarly, although such results would then be exploited by client groupings scientists as a community are discouraged from relating closely with client sectors because of similar possibilities. In the first case, therefore, the building up of detailed knowledge of the biology of trypanosomes combined with free discourse among livestock scientists would act as a precursor to discovery of a vaccine that would play a major role in eliminating this deadly disease. But how that vaccine should be deployed when found was not felt really to be a job for the scientific community, except perhaps in an advisory sense.

Conversely the *mode 2* position is essentially an interactive one. Scientists see their job more as reflecting social goals and prefer to carry out research and development to achieve these even if this means engaging with complexity and an inability to adhere to disciplinary norms. Motivations may also be financial. Driven perhaps by scarce resource constraints, scientists (and the organizations within which they function) are conscious that donors have wider agendas that concern social welfare goals and are willing to take approaches that go beyond the unwritten rules of *mode 1* thinking. In practice there is a continuum between the two normative positions and the second case study on fodder clearly reflects this. To begin with the scientists saw their role as largely one of technology development because the key research leading to the better variety had already been done. It was only when they encountered the reality of the scale out problems that they felt the need to change their approach.[19]

To take the point a little further in an *innovation systems* sense, whereas the 'inputs' to the project are similar to those of the old style, the 'outputs' are much more complex. They include not only 'knowledge' outputs (robust research results) but also developmental outputs such as welfare improvements among target groups (in this case, poor farmers). They include also intermediate outputs such as stakeholder identification, strength of interactions among stakeholder groups, institutional bottlenecks, etc., many of which cannot be reduced easily to quantitative indicators. We are therefore dealing with a 'complex simultaneous equation' scenario where singling out cause and effect is very hard, to say the least.

One of the earliest people to explore the properties of dynamical systems independent of the context of application was Rosen.[20] Rosen is a systems ecologist whose original background lay in the analysis of complex biological systems, but who has also tried to show that similar metaphors are applicable to the dynamical behaviour of socio-economic systems as well. He takes the view that conventional science works best with physical systems. Conversely, in the case of living systems (ecologies) that are creative, and as a result experience evolution of internal structures, we need an experimental approach that allows us to capture the inevitable indeterminism of their behaviour. Such an approach needs above all to avoid constraining our analysis of the system under investigation by presuppositions about its 'true nature' and, in particular, that of imagining that it can be modelled deterministically.

In formal terms, Rosen (1987) puts the issue quite neatly in terms of ecological creativity. Imagine an ecosystem in which there are species each having either a predator–prey or a symbiotic relationship with its fellows, and we wish to construct some kind of model of that system. The first conclusion we shall immediately come to is that to model such an ecosystem in terms of deterministic relationships amongst its constituent species would not be a very useful exercise. Rosen shows that this is because determinate systems in general are merely limit cases of a wider class of complex system whose relationships are informational but not exact. For example, in our predator–prey model the relationship between predator and prey can be pictured as the quantity uij where:

$$\text{uij}(\text{x1}\ldots\ldots\text{xn}) = \delta/\delta\text{xj}(\text{dxi}/\text{dt}) \quad (1)$$

xi, xj are examples of various species x1, x2. . . . . . . xn co-existing in a given ecological space and t = time. The uij represent values giving the rate of change of the production of xi as a result of a change in the production of xj over time. If the two species are symbiotic, uij will be positive, but if xj is a predator on xi then uij will be negative. Rosen shows that the equation set (1) is indeterminate—it cannot be solved for specific values of the uij. However, under specific limiting conditions (1) can approximate to a conventional dynamical system which takes the form:

$$\text{dxi}/\text{dt} = \text{fi}(\text{x1, x2, x3,}\ldots\ldots\text{xn}) \quad (2)$$

It is then a determinate system. However, no living system could survive if it behaved according to the relationships outlined in (2), simply because unpredictable and creative behaviour would be impossible by definition. All species would be wiped out. The point is that for a system to be alive and to evolve creatively, its behaviour must be relatively indeterminate. It follows that to try to model it deterministically is tantamount to turning it from a living system into a dead system, and conversely that we need to develop models which somehow capture the essential creativity of living systems.

Now, if this is true with natural ecologies it must be even more the case with socio-economic systems because these are both more complex and subject to much more rapid evolution. In addition whereas the 'information' relevant to predator–prey relationships is relatively simple (species x very quickly learns whether species y is friend or foe and can take appropriate action), that for creative socio-economic systems is infinitely more complex. That being the case, it is likely that the capacity to assimilate, process and use relevant information in a rapidly changing context will be crucial in any development context. My argument is that such a capacity cannot depend solely, or even largely, upon hierarchically related organizations of a *mode 1* kind. On the contrary, because those most likely to know most about any given part of the system are either those who live and work in it *or* those who have some (recently acquired) expert knowledge about it, it is *their* knowledge that will generally play the most important role. It should immediately be apparent that this is precisely what obtains in an innovation system. The evolution of new technological options depends on a series of relationships involving a wide variety of stakeholder groups. In the case of the fodder project, the scaling out of the new variety did not work precisely because there were unanticipated constraints. These constraints presumably could have been anticipated but even if they had been they would still have required some *ex ante* research to provide developmental guidance.

But how is this to be done? In another article Rosen attacks the problem by focusing on the lack of clarity (fuzziness) surrounding many of the concepts associated with development planning. He argues that not only have analysts no clear idea about what constitutes 'development' (they are often actually incompatible and contradictory), 'even among those who happen to share the same views as to the ends of development, there are similarly incompatible views as to the means by which the ends can be attained'.[21] Similar views have been expressed by Thompson (1993). But, Rosen goes on, if 'so many distinct and contradictory views . . . can be held by (so many) able people . . . [then] . . . a first step in dealing with such concepts is to try to identify and remove the source of the fuzziness'.[22] And an important source in his view is that different analysts 'live intellectually' in a variety of different 'analytical worlds' where the underlying concepts used often have different meanings to each analyst. Sometimes, unfortunately, they are not fully understood by the analysts themselves even in terms of their own 'world'. However, if we begin to separate out these worlds and characterise them unambiguously, then we shall have made an important first step in clearing up much of the confusion.

Rosen's method is to abstract from the Newtonian mechanics which underlie most models of system behaviour (including system dynamics), and in particular to introduce the notion of 'anticipation'. Now a moment's thought will show that there are therefore an infinite number of possible planning models to choose from because every analyst will have his/her own 'world' in mind in relation to the 'real system' that happens to be under investigation. In fact, Rosen does not take his own analysis much beyond this point, but what he does say is suggestive in the broad sense that by couching the problem in terms of *anticipatory* systems he opens up much more clearly the nature of the potential confusions that arise typically in development policy and planning. Each analyst, or perhaps it is better to concentrate on each *analytical ideology*, has his/her (its) own social construction of reality which guides the policy questions asked and the answers that are ideologically acceptable. Rosen's argument is that at least the recognition of this fact should help to clear up (some of) the intellectual baggage that obscures the policy process.

We all know, of course, examples of such ideological confusion. Economists are well known for seeing any particular issue in terms of resource allocation and the price mechanism. What usually differentiates them from a policy standpoint is the position they take up on 'market failure' in relation to that issue. Those on the 'right' tend to want to believe that a 'market solution' is the better option, whereas those on the 'left' will tend to favour intervention on the part of some central authority to correct 'market failure'. Their 'worlds' are predetermined in this narrow, ideological sense. Similar differentiation occurs across disciplines. For example, on issues of environmental degradation, physical geographers will not see the problem in terms of the efficiency of the price system but will tend rather to concentrate on purely physical processes such as climatic changes and soil and water stress. Political scientists will eschew both in favour of an analysis concentrating upon power structures. Finally, it should be noted that such social construction goes well beyond intra- and interdisciplinary battles; it relates more widely to strongly held views on the part of many powerful and highly motivated pressure groups. As Thompson (1993) has pointed out in graphic detail, the passion with which rival adherents typically cling to their respective 'worlds' can potentially waste many billions of dollars. It is in this profound sense that policy analysis is not a costless exercise.

From my point of view, however, there are two important inferences to be drawn from this discussion:

1. If development policy is to be successful there has to a way of allowing permanent model revision. Not to do so (and therefore to rely on one model) is to open the door once again to confusion amongst the analysts (and even more so, those further down the hierarchy) whose 'worlds' will continue to vary and to evolve even though the model itself does not.
2. Information is a crucial component of the whole exercise because necessary (and continuous) model revision will depend on the accuracy, speed and general efficiency of information flows. And it is for this reason that 'information search', sometimes called 'policy research', has become a significant factor in modern development discourse.

It is perhaps the third case study that provides some potential answers. The planners of the APNLBP took an early decision to think in *mode 2* terms. That is, by identifying all potential stakeholders in the innovation system and by asking them to be directly involved in priority setting right from the beginning they were in a much stronger position to anticipate constraints. In addition, by maintaining continuous review and monitoring of project development they were able to 'troubleshoot' where unexpected problems arose. They were also in a position to hedge against such problems and to build up trust among groups. In the second case study, of course, the question is rather different. Having started out as a *mode 1* project, how is it possible to retrieve ground and ensure that the wider system begins to benefit? Here perhaps what is needed is to rely more strongly on heuristic research design principles that depend upon site experience and learning. Thus, it seems clear that in this case access to seed supplies at the right time and at costs that reflect poor farmer purchasing power appears to have been a crucial constraint. It follows that the project organizers could then design a comparative study (or studies) into ways to overcome this. They would clearly need to interact with key stakeholder groups in setting such a study (or studies) up and very probably also arrange relevant workshop activities to change focus as needed. Something along these lines was done at various points of the APNLBP.

A similar view reached in a rather different way is that of Stephen Biggs (2006), who uses three project case studies to show the importance of learning from positive experiences even where these were not initially anticipated by project designers. His conclusions are first 'that we can do more to learn from and support positive situations where poverty reduction and social inclusion are already taking place [and] second, [that] effective innovations in policy and institutional arenas are, by their very nature, generally location and time specific'.[23] It follows that policy interventions to deal with poverty alleviation should always be open to modification as circumstances change. This will often happen as local stakeholders themselves change behaviour to act in innovative ways. And usually this is because their own local knowledge dictates what should happen. The trick for the sensitive policy analyst is to be aware of this and to tailor interventions to allow as much supportive flexibility as possible.

CONCLUSIONS

This chapter has been concerned with how to reposition agricultural science in ways that make it more accessible to the needs of poor farmers' livelihoods. The case studies cited appear to show how by widening the canvas of scientific communities in project selection, research and dissemination not only can the pro-poor relevance of R&D be enhanced, but at the same time there should be no diminution of the scientific quality of the research itself. However, when it comes to scaling results out to wider stakeholder client groups it is important to recognize the complexity of the whole innovation system and to take appropriate action right from the start of programs. Here it is vital to recognize that we are normally dealing with systems that are indeterminate and evolutionary. And this is so because of the inevitable impact of new information over time. Accordingly, ways need to be found of improving research design such that interventions not only have a good basis in natural science, but that they also have a reasonable expectation of useful final impact on poor communities.

The APNLBP shows us one way of doing this that emphasizes constant dialogue among stakeholder groups, but it is certain that there will be many other possibilities. In all cases, however, analysts and decision-makers must recognize that they are not living in a Newtonian world where system parameters are fixed and therefore determinate. On the contrary, innovation systems are constantly changing in structure and applicability. All we can say hopefully are that system parameters are evolving more slowly than system variables and that we can keep abreast of this by means of appropriate dialogue activity and related research design. It is my firm belief that by taking on board the insights of analysts like Rosen there should be no problem linking a *mode 2* approach to science that at the same time ensures that the quality of laboratory research is, if anything, enhanced. If this is so, it should certainly be possible to enable such activity to make, as it should, a much greater impact on the lives of the poor in developing countries.

NOTES

1. An earlier version of this chapter was delivered at the INNOGEN Annual Conference on Genomics for Development: The Life Sciences and Poverty Reduction, 5–6 September 2006, Regent's College, London.
2. See Gibbons et al. (1994).
3. See Nowotny, Scott and Gibbons (2003, 194).
4. See Scoones (2008).
5. See DFID (2004).
6. I prefer to see the concept as a metaphor rather than a theory. For more detailed analysis of its applicability to third world agriculture, see Clark (2002).
7. See Freeman (1987) and Nelson (1993), for example.
8. See, for example, Oyeyinka (2005).
9. See, for example, Arnold and Bell (2001, 288).
10. See, for example, Dalrymple (2005).
11. See, for example, Bezanson and Oldham (2005).
12. See Torr et al. (2005, 1).
13. And in the same paper the authors argue that cheap mechanisms for tsetse control have now been refined to a stage that were they carefully applied, trypanosomiasis could well be effectively if not completely eliminated.
14. Although it is doubtful whether any formal cost-benefit analysis was undertaken to justify such a conclusion.

15. This section has drawn heavily from Prasad et al. (2006).
16. See de Haan et al (2005).
17. For more details, see Clark, Hall and Yoganand (2002) and Clark, Reddy and Hall (2006).
18. See Bunders and Broerse (1991).
19. A related point, however, is that not only is there often a cognitive disjunction on the part of scientists, but similar ambiguities often obtain within donor agencies. Thus many donors still maintain quite strict rules regarding reporting on outcomes. They will often require scientists to specify what the eventual 'research outcomes' will be and often also ask for this to be measured in quantitative terms. Hence the *ex ante* use of 'log frames'. The result is that proposers often feel under pressure to revert to *mode 1* styles of work and to dress up their work as 'scientific' with the aid of the appropriate jargon.
20. See also Kline (1985), who was writing about the non-linear properties of innovation systems at roughly the same time.
21. See Rosen (1974, 245).
22. Rosen (1974, 246).
23. See Biggs (2006, 14).

REFERENCES

Arnold, E., and M. Bell. 2001. 'Some new ideas about research for development'. *Danish Ministry of Foreign Affairs: Partnership at the leading edge: A Danish vision for knowledge, research and development* (April): 288.

Barnett, A. 2006. *Journeying from research to innovation: Lessons from the DFID's CPHP 'Partnerships for Innovation'*. Brighton: Policy Practice Ltd.

Bezanson, K., and G. Oldham. 2005. 'Donors should take an "innovation" rather than a "research" approach to designing scientific and technological aid programmes'. *SciDev.Net*, 10 January.

Biggs, S. D. 1990. 'A multiple source of innovation model of agricultural research and technology promotion'. *World Development* 18 (11): 1481–1499.

———. 2006. 'Learning from the positive to reduce rural poverty: Institutional innovations in agricultural and natural resources research and development'. Paper prepared at the Impact Assessment Workshop organized by the CGIAR System-wide Program on Participatory Research and Gender Analysis for Technology Development and Institutional Innovation and the International Maize and Wheat Improvement Centre, Mexico, 19–21 October.

Bunders, J. F. G., Broerse, J. E. W., (eds), 1991. "Appropriate Biotechnology in Small-scale Agriculture: How to reorient research and development," C. A. B. International, Wallingford, UK.

Clark, N. G. 2002. 'Innovation systems, technology assessment and the new knowledge market: Implication for third world development'. *Economics of Innovation and New Technology* 11 (1–2): 353–368.

Clark, N. G., A. Hall and B. Yoganand. 2002. 'New science, capacity development and institutional change: The case of the Andhra Pradesh-Netherlands biotechnology programme'. *International Journal of Technology Management and Sustainable Development* 2 (3): 53–67.

Dalrymple, D. G. 2005. '"Innovation system": Old wine in a new bottle?' *SciDev.Net*, 13 April.

De Haan, N. C., P. Bezkorowajnyj, D. Romney and O. Olufajo. 2006. 'Feeding livestock through partnerships'. *KM4D Journal* 2 (3): 123–135.

DFID. 2004. 'The use of science in UK international development policy, House of Commons Science and Technology Committee, thirteenth report of Session 2003–2004'. The Stationery Office, 26 October.

Freeman, C. 1987. *Technology and economic performance: Lessons from Japan*. London: Pinter.

Gibbons, M., C. Limoges, H. Nowotny, S. Schwartzman, P. Scott and M. Trow, eds. 1994. *The new production of knowledge*. London: Sage.

Hall, A. J., ed. 2004. *Innovations in innovation*. Patancheru, India: ICRISAT.

Hall, A. J., G. Bockett, S. Taylor, M. Sivamohan and N. G. Clark. 2001. 'Why research partnerships really matter: Innovation theory, institutional arrangements and implications for developing new technology for the poor'. *World Development* 29 (5): 783–797.

Hall, A. J., N. G. Clark and J. H. Crouch. 2004. 'The evolving culture of science in the consultative group of international agricultural research: Concepts for building a new architecture of innovation in agri-biotechnology'. In *Innovations in innovation*,ed. A. J. Hall, 135–162. Patancheru, India: ICRISAT.

Hall, A.J., N. G. Clark, R. V. Sulaiman, M. V. K. Sivamohan and B. Yoganand. 2000. 'New agendas for agricultural research in developing countries: Policy analysis and institutional implications'. *Knowledge, Policy and Technology* 13:70–91.

Kline, S. J. 1985. 'Innovation is not a linear process'. *Research Management* (July–August): 36–45.

Nelson, R. R., ed. 1993. *National innovation systems: A comparative analysis*. New York: Oxford University Press.

Nowotny, H., P. Scott and M. Gibbons. 2003. 'Mode 2 re-visited: The new production of knowledge'. *Minerva* 41 (2): 179–194.

Oyeyinka, O. 2005. 'Systems of innovation and underdevelopment: An institutional perspective'. UNU/INTECH Discussion Paper Series, Maastricht, the Netherlands.

Prasad, V. L., P. G. Bezkorowajnyj, S. N. Nigam and D. Romney. 2006. *Participatory varietal selection to multiple actor orientation—a case study of groundnut in Anantapur*. Andhra Pradesh: ILRI.

Rosen, R. 1974. 'Planning, management, policies and strategies'. *International Journal of General Systems* 1:245–252.

———. 1987. 'On complex systems'. *European Journal of Operational Research* 30:129–134.

Scoones, I. 2008. 'Global engagements with global assessments: The case of the International Assessment of Agricultural Knowledge, Science and Technology for Development (IAASTD)'. IDS Working Paper No. 313, November.

Thompson, M. 1993. 'Good science for public policy'. *Journal of International Development* 5 (6): 669–679.

Torr S. J., J. W. Hargrove and G. A. Vale (2005). "Towards a Rational Policy for Dealing with Tsetse." Trends in Parasitology 21(11): 537–541.

UN Millennium Project. 2005. *Innovation: Applying knowledge in development: Task force 10 on science, technology and innovation*. London: EARTHSCAN.

Part II

Operationalising the Triple Helix System and Bridging the Gaps

7 The Triple Helix Paradigm in Korea and Taiwan

A Test for New Forms of Capital

Matthew A. Shapiro

INTRODUCTION

Successful collaborative research and development (R&D) output is a function of how partnerships are initiated, so it is particularly important that we examine the nature of the relationships between government research institutes (GRIs), universities and firms. Specifically, it is considered here whether researchers tend to partner repeatedly with cross-sector research entities. Cross-sector or public–private R&D collaboration is increasingly applied and necessary in Korea and Taiwan, given the complexity of new technologies. Such advancements typically require interdisciplinary skills, large financial investments, testing and experimental and production facilities (Yusuf 2003), which is reflected in these two countries' policies related to R&D and science and technology (S&T). In this way, this chapter (as well as Chapter 4) is essentially a response to a triple helix–based claim which strongly emphasizes the benefits of relations between private research entities, universities and GRIs. For Korea and Taiwan, this presents a unique opportunity to determine whether traditional methods of networking dominate, or whether R&D trumps culture.

For other countries approaching a similar level of technological capacity, this chapter provides evidence of East Asia's industrial success beyond the underlying macro-level differences.[1] The conclusions offered here confirm triple helix–based claims that new forms of relationship-based capital arise through public–private R&D collaboration. Korea and Taiwan are largely homogenous in these findings, despite the strength of traditional network structures in these two countries. These traditional structures were expected to limit the formation of new relationships.[2] There is, thus, support for the application of policies calling for public–private R&D collaboration in other technology-oriented countries, such as India and China.[3] This line of research also bridges the divide between the triple helix paradigm, the sociologically based research on the East Asian organizational structure (Hamilton 1990; Park 2000; Hamilton and Biggart 1988; Biggart 1990) and East Asian political economy (Fields 1995).

Theoretically, this chapter offers a deeper look at the linkages between the three parts of the triple helix structure than is generally presented in the literature. These three helices between the government, universities and private firms offer interesting conclusions but offer little in terms of differences between subgroups. *Subgroup* refers here to the different types of research entities within the public and private research sectors. The private sector, for example, is comprised of a range of different size firms, each providing unique qualities to the triple helix dynamic. Further, when standard triple helix linkages are drawn between universities, firms and the government, GRIs are not specifically identified and positioned in the schematic. Knowledge creation occurs in the context of a fluid and evolving community and formal organizations are poor methods for learning (Powell, Koput and Smith-Doerr 1996). It is, thus, all the more relevant to identify and examine the sources of innovation where firms, universities and GRIs intersect. To test for this, GRIs must be included in the triple helix paradigm, as should delineations between different size firms.

Methodologically, triple helix studies are typically conducted at the case study level, and comparative research of the magnitude attempted in this chapter is not the standard. Indeed, there are a number of stylized facts which are ripe for quantitative analysis. Using the unique KORTAI R&D data set, in combination with interview results between the author and research directors in Korea and Taiwan, this chapter offers a response to a number of program-, sector- and country-level questions which are often raised on a case-by-case basis, as the subsequent literature review shows. With appropriate techniques and a sufficient sample size, generalizations could be tested with confidence. Here, the contexts of such tests are instances of public–private R&D collaboration resulting from a specific policy structure which targets innovations with long-term potential and commercializability. Presently, the East Asian developmental state is functioning to facilitate the interaction and collaboration between the public and private research sectors. The mechanisms utilized for this effect include the continued use of government funding and R&D subsidization, as well as reinforced targeting of joint R&D, the sharing of research facilities, more effective tax incentivization and a bolstered intellectual property rights (IPR) regime (Chung 2004). These are the minimal requirements for other countries attempting to implement a similar structure.

The structure of this chapter is designed to present a full analysis of the Korean and Taiwanese cases, test the aforesaid aspects of the triple helix paradigm and make propositions about the viability of transference of best practice techniques. To that effect, the second section briefly reviews the triple helix–related literature and presents background details on the Korean and Taiwanese cases. The third section introduces the relevant variables from the KORTAI R&D data set and presents summary statistics to validate a number of key assumptions. The fourth section outlines the empirical specification and also presents the results and robustness checks from this model. Additional data on the sources of pre-existing and new relationships is also presented here. The fifth section concludes this chapter with particular reference to the influence of traditional network structures in these two countries.

LITERATURE REVIEW AND CASE DESCRIPTION

Despite the breadth of literature on the triple helix phenomena, a thematic breakdown by country/region, method or theory reveals a case study focus with little attention to within-region comparative analysis. As well, studies of the triple helix paradigm in developing countries are not in great supply. In this chapter and throughout the whole book, attention is drawn to the micro-level institutions at work with respect to the Korean and Taiwanese cases.

A survey of the related literature also reveals an overall absence of simultaneous assessments of both the public and private research sectors. There is also an overwhelming focus on the European phenomenon, recently presented in an analysis of increased entrepreneurship at a single Belgian university (Van Looy et al. 2006) and in an earlier study of diffusion through an analysis of technology transfer offices at German universities (Kruecken 2003). Colombo and Delmastro (2002) generate micro-level data in a comparison of firms located in Italian science parks, and Schartinger et al. (2002) also gathered micro-level data for their study of the complexity of knowledge interactions between Austrian universities and firms. The data of these last two studies, it should be noted, is similar in nature to the content of the KORTAI R&D data set, which is the source of the data used here. Marques, Caraca and Diz (2006) also provide valuable insights with their case study of a Portuguese university's effectiveness in promoting innovation and entrepreneurship throughout the local area.

Beyond Europe, the focus turns to Latin America and East Asia with few exceptions. These studies also fail to simultaneously consider multiple components of the triple helix paradigm, but they do provide deep insight into the particulars of individual countries, such as the studies of Brazil and Mexico (Chapters 2, 3 and 11), Indonesia (Chapter 9), Malaysia (Chapters 10 and 13) and India (Chapter 12).

In addition, Casas, Gortari and Santos (2000) show how the presence of key scientific and engineering fields in Mexico's Bajio region exhibit relationships along the lines of the triple helix model. Mirroring the aforementioned Portuguese university case study, Bernasconi (2005) emphasizes the absence of the triple helix paradigm in a case study of Chile's Pontificia Universidad Catolica, detailing the transition from a teaching to a research orientation, in spite of shortages in government funding. The value of these developing country–based studies has led to explicit calls in support of the triple helix paradigm, such as Saad and Zawdie's (2005) detailed progression of Algeria's post-independence industrialization. This is echoed by

Etzkowitz and Leydesdorff (2000), who survey related events in Europe, Latin America and Asia, particularly the government's involvement in altering the relationship between knowledge producers and users.

Turning now to the most relevant case-specific literature, Korea and Taiwan are typically presented comparatively with other countries having triple helix qualities. Park, Hong and Leydesdorff (2005) compare Korea and the Netherlands based on 'knowledge infrastructure'. They conclude that Korea's scientific and technological output is greater than that of the Netherlands, measured by webometric, scientometric and technometric indicators, but offer little in terms of policy prescription. Etzkowitz and Brisolla (1999) also study Korea comparatively, using Korea with Brazil as proxies for their respective regions. The authors ultimately make a connection between technology-bolstering policies and the international political economy, concluding that intervention in technology policies is no guarantee of success.

This chapter, thus, extends the aforementioned micro-institution-based literature with a largely unexplored aspect of the triple helix paradigm, all the while focusing on the Korean and Taiwanese cases. The research question under consideration is a response to an untested hypothesis by Etzkowitz (2003), which states that new social arrangements and channels of interaction are needed if industry and government are joined by universities in knowledge-based economies. Earlier studies allude to this (Faulkner and Senker 1994), concluding that cooperation between universities and private firms is based on personal contact.[4] Others have found that interactions between university and firm researchers occur through a dense network of interpersonal relationships (Dierdonck, Debackere and Engelen 1990),[5] whereas still others have determined that the source of innovation-based relationships is captured through the personal contacts of research institution employees (Fritsch and Schwirten 1999).

The Korean and Taiwanese cases provide an ideal opportunity to test for relationship-based capital and advance the triple helix concept to developing countries, as Chapter 12 and the remaining chapters in this volume attempt to do. The triple helix in Korea and Taiwan is rooted in the 1960s and 1970s, during which imitation was the source of rapid industrialization. This imitation, or 'reverse engineering', of existing foreign technologies required minimal investment in R&D for the production of simple products. Chapter 2 provides a more detailed description and discussion of the rapid industrialization of Korea. Reverse engineering rarely occurs in a vacuum, and multilevel interactions among firms, universities and public R&D institutes were required (Kim and Nelson 2000).

RESEARCH QUESTIONS AND DATA

A test of Etzkowitz's (2003) new relationship-based capital claim is bound to a number of related research questions which are intended to advance the overall understanding of the triple helix structure. The additional issues included here address subgroups of the public and private research sectors. Private sector subgrouping is divided between small and medium enterprises (SMEs) and large firms. Ernst (2000) notes that the chaebol-dominant industry structure in Korea was accompanied by 'octopus-like diversification' into many unrelated industries. Such over-diversification minimized specialization, which some claim actually hindered the accumulation of knowledge (Ernst 2000). As such, SMEs can now be viewed as the vehicle through which ideas and technologies germinate. In Taiwan, on the other hand, SMEs have always been the dominant firm model. To test for this difference, cross-sector subgroup collaborative tendencies will be included as control variables.

Growth in the research efforts of SMEs has some attributes of a structural change. Kim (2001) identifies the Asian crisis of 1997 as a factor in the growth of SME innovation in Korea. The chaebols reduced R&D investment following the crisis, prompting an SME-based upsurge. The number of new firms in Korea also increased from one hundred (pre-crisis) to more than seven thousand in June 2000, as a result of post-crisis layoffs by the chaebols. This growth in SMEs and new start-ups is also affected by the targeting of public funds, either by using the new technology as bank loan collateral, subsidizing R&D personnel or providing technical information and services (Chung 2004).

Collaborative tendencies between public and private groups are captured in the KORTAI R&D data set. Exactly, collaborative tendencies measure the percentage of R&D collaboration done by the respondent with the opposite sector, or with a subgroup of the opposite sector.[6] Looking first at aggregate patterns in Tables 7.1 and 7.2, the pattern of collaboration in both Korea and Taiwan is similar, as public sector respondents collaborate more with the private sector (45.2 and 30.9 per cent, respectively) in both countries than private sector respondents collaboration with the public sector (34.4 and 17.3 per cent, respectively). The only major distinction between Korea and Taiwan in this respect is in terms of scale, as both public and private respondents in Korea engage in cross-sector R&D collaboration approximately 50 per cent more than Taiwanese respondents.

Table 7.1 Collaborative Tendencies: Aggregate Level, Korea

Source	*Collaborator*	*Percentage*
public sector	private sector	45.2%
private sector	public sector	34.4%

Note: Percentages do not add up to 100 because of possible overlap in collaborative efforts.
Source: KORTAI R&D database.

Table 7.2 Collaborative Tendencies: Aggregate Level, Taiwan

Source	*Collaborator*	*Percentage*
public sector	private sector	30.9%
private sector	public sector	17.3%

Note: Percentages do not add up to 100 because of possible overlap in collaborative efforts
Source: KORTAI R&D database.

Collaborative tendencies between each subgroup and the subgroup of the opposite sector, presented in Tables 7.3 and 7.4, show relatively similar patterns across countries. GRIs in both countries collaborate more with SMEs than with large firms, whereas private firms in both countries collaborate evenly with GRI and university partners. There are only three major distinctions between Korea and Taiwan. First, the magnitude of private sector collaboration is higher in Korea, which is consistent with Tables 7.1 and 7.2. Second, GRIs collaborate slightly more with SMEs in Korea than in Taiwan, although the overall pattern is similar in both countries. Third, with regard to the university subgroup of the public research sector, there is virtually no difference in collaborative patterns with SMEs and large firms in Korea (at around 44 per cent), whereas universities collaborate more with SMEs than with large firms in Taiwan. This last point is not an entirely robust finding, however, given that the sample size for the Taiwan university subgroup is not very large. On the whole, however, it can be concluded that SMEs are the preferred collaborative R&D partner for GRIs in both countries, and that private respondents are relatively indifferent as to whether they collaborate with GRIs or with universities.

Table 7.3 Collaborative Tendencies: Subgroup Level, Korea

Source	*Collaborator*	*Percentage*
GRI	SME	54.5%
	large firm	36.8%
university	SME	44.3%
	large firm	44.0%
private sector	GRI	32.7%
	university	36.1%

Note: Percentages do not add up to 100 because of possible overlap in collaborative efforts.
Source: KORTAI R&D database.

Table 7.4 Collaborative Tendencies: Subgroup Level, Taiwan

Source	*Collaborator*	*Percentage*
GRI	SME	47.8%
	large firm	33.6%
university	SME	47.5%
	large firm	31.7%
private sector	GRI	17.6%
	university	17.0%

Note: Percentages do not add up to 100 because of possible overlap in collaborative efforts.
Source: KORTAI R&D database.

The measures capturing new and pre-existing relationship-based capital, also available in the KORTAI R&D data set and detailed in the following, are designed to show whether new forms of capital created through triple helix–structured collaborations have a greater effect upon research output than pre-existing forms of capital. Again, our test is based on Etzkowitz's (2003) claim that the triple helix generates new forms of capital, which is the logical result of having a new, dynamic research-based relationship with an entity from the opposing sector.

From the KORTAI R&D data set, new forms of capital generated through public–private R&D collaboration are measured by the percentage of collaboration done with partners from previous projects (*new_relations*).[7] This proxy for new, relationship-based capital is delineated by a discrete, time invariant value from zero to ten, measuring percentage values from zero to one hundred in increments of ten. The KORTAI R&D data set also accounts for eight different reasons for such repartnering, all based on a seven-point Likert scale response (7 being greatest): a lack of other qualified partners (*noother*); a stipulated funding condition (*fundstip*); a shared commitment (*sharecom*); a lack of tension (*lacktension*); ease of communication (*easecom*); complementarity in knowledge (*compknow*); the presence of trust (*trust*); and expected commercialization (*expcom*). Interviews between the author and project managers in Korea confirm that these reasons all have potential relevancy.

The effects of new forms of capital are held up here in comparison to the effects of pre-existing forms of relationship-based capital.[8] The variable for pre-existing capital is measured as the percentage of public–private R&D collaboration originating from pre-existing connections—personal ties (*old_relations*)—between the respondent and collaborators from the opposing sector. The variable is formulated by a number from zero to ten, representing percentage values from zero to one hundred in increments of ten.[9] As before, the KORTAI R&D data set also includes five dummy variables capturing the sources of the various forms of personal ties: university-based ties (*sameuni*), former university laboratory-based ties (*sameunilab*), former private firm–based ties (*samefirm*), ties through working on multiple previous projects (*sameproj*) and ties from meeting at a conference (*sameconf*).[10]

A conceptual framework outlining the possible interactions between these potential new forms of capital and research output is presented on the right-hand side of Figure 7.1, whereas the interactions between pre-existing forms of capital and research output are shown on the left-hand side of Figure 7.1.[11] The dependent variable is R&D project output, measured by (a) the number of patents strictly from cross-sector R&D collaboration and (b) the number of total patents excluding cross-sector R&D collaboration. Consistency among the predicted effects upon these two dependent variables will confirm that there is no difference in the generation of both types of patents. On the other hand, use of both measures enables one to perform a counterfactual test of the new-versus-pre-existing hypothesis: if new capital is truly important for collaboration in the triple helix structure,

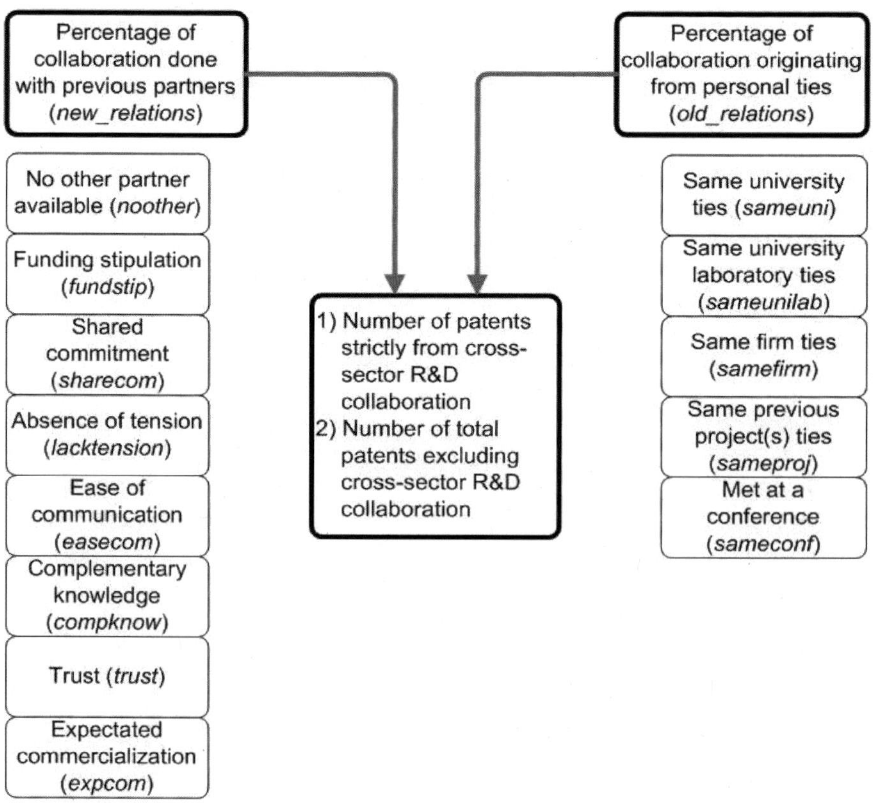

Figure 7.1 Tracing the effects of personal ties and repartnering.

the effects of repartnerships (*new_relations*) should be greater for collaborative patents than for non-collaborative patents.

As evidence of the growing importance of public–private R&D collaboration as a source of R&D output, Tables 7.5 to 7.8 and Figures 7.2 to 7.5 show the time trends for the average number of patents by subsector via collaboration and excluding collaboration, respectively. This effectively illustrates that output is growing across all subgroups, with a couple of anomalies. Whereas these tables and figures show important changes over time, the subsequent empirical analysis is entirely static in nature (i.e. patent measurements will reflect data only for 2005), as all of our explanatory variables are time invariant.[12]

On the whole, the three subsectors in these two countries have followed similar trends, although there are a couple of anomalies. For the Taiwanese case, the average number of collaborative patents for universities is much higher after 2002, although this can be attributed to a couple of outlying cases, not to mention the small sample size for the Taiwan university subgroup. Another pattern of divergence among the three groups over time

Table 7.5 Average Number of Patents through Cross-Sector R&D Collaboration: By Subsector: Korea

	1997	*1998*	*1999*	*2000*	*2001*	*2002*	*2003*	*2004*	*2005*
GRI	.022	.022	.022	.022	.130	.043	.244	.370	.348
university	.000	.075	.150	.225	.075	.150	.150	.250	.500
firm	.075	.150	.100	.125	.175	.100	.225	.325	.500

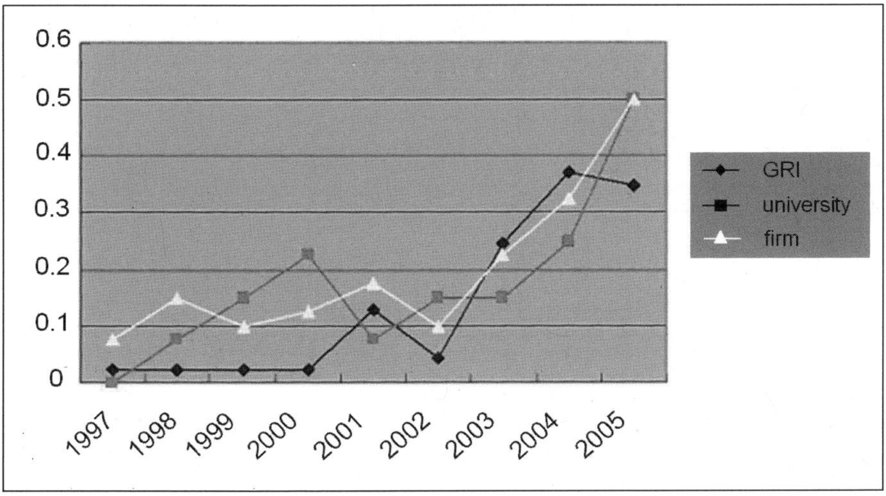

Figure 7.2 Average number of patents through cross-sector R&D collaboration: By subsector: Korea.

is the increase of average non-collaborative patents by Korean GRIs from 2001. Whereas this does not necessarily portray an image of a focus on public–private R&D collaboration in Korea, it does confirm that patenting overall has grown in emphasis. GRIs in Korea are simply not limiting their efforts to cross-sector collaboration.

Table 7.6 Average Number of Patents through Cross-Sector R&D Collaboration: By Subsector: Taiwan

	1997	*1998*	*1999*	*2000*	*2001*	*2002*	*2003*	*2004*	*2005*
GRI	.000	.000	.045	.227	.023	.023	.114	.205	.477
university	.000	.000	.000	.000	.000	.250	.500	1.170	1.330
firm	.000	.000	.009	.009	.017	.148	.052	.113	.530

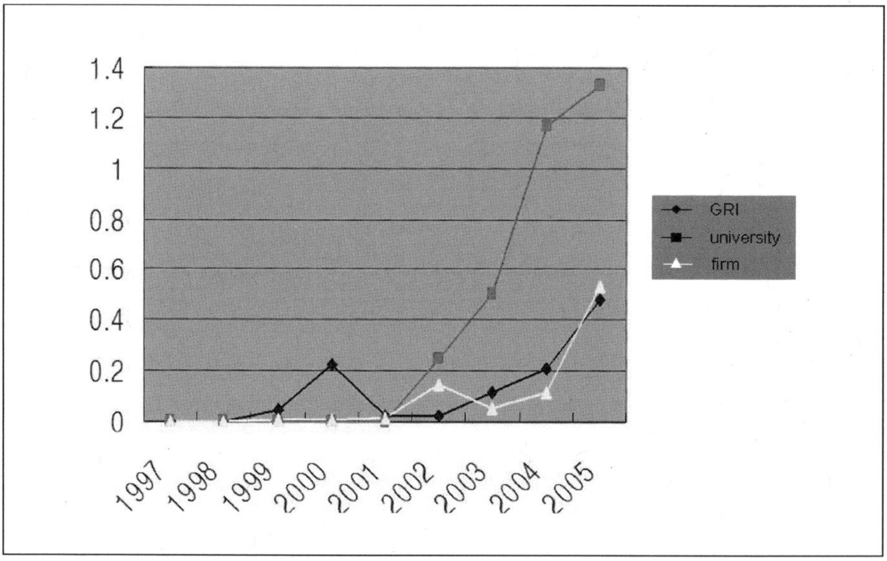

Figure 7.3 Average number of patents through cross-sector R&D collaboration: By subsector: Taiwan.

Table 7.7 Average Number of Total Patents Excluding Cross-Sector R&D Collaboration: By Subsector, Korea

	1997	*1998*	*1999*	*2000*	*2001*	*2002*	*2003*	*2004*	*2005*
GRI	0.978	1.809	1.915	2.191	2.383	3.021	3.891	6.532	8.340
university	0.175	0.450	0.725	0.675	1.000	1.125	1.350	1.350	1.750
firm	0.025	0.075	0.050	0.200	0.231	0.615	0.745	0.692	1.462

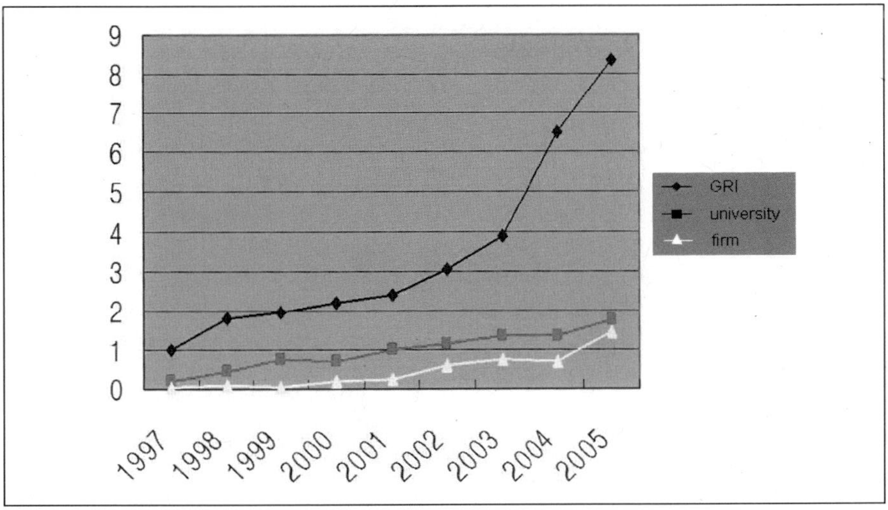

Figure 7.4 Average number of total patents excluding cross-sector R&D collaboration: By subsector, Korea.

Table 7.8 Average Number of Total Patents Excluding Cross-Sector R&D Collaboration: By Subsector, Taiwan

	1997	*1998*	*1999*	*2000*	*2001*	*2002*	*2003*	*2004*	*2005*
GRI	1.070	1.116	1.116	1.860	0.302	1.023	0.977	1.581	1.405
university	0.000	0.000	0.000	0.000	0.000	0.500	0.333	1.083	1.917
firm	0.026	0.061	0.052	0.783	0.435	0.491	0.596	0.947	1.588

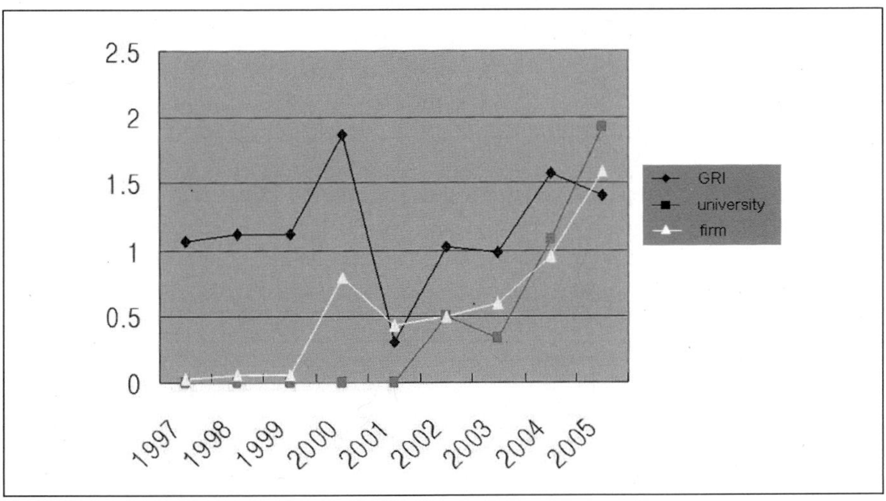

Figure 7.5 Average number of total patents excluding cross-sector R&D collaboration: By subsector, Taiwan.

EMPIRICAL SPECIFICATION AND RESULTS

The first task, the comparing of new forms of capital arising from the triple helix with pre-existing capital, is done through a comparison of the *new_relations* and *old_relations* coefficients' separate effects upon research output. The first general model, thus, is:

$$Y_{ij} = X_{ik} + \varepsilon_i \,, \tag{1}$$

Y is the number of patents generated by respondent i in 2005, and j denotes the nature of the patent (either accounting only for collaborative projects or excluding such projects altogether). X represents the relationship-based capital discussed at length in the preceding sections. For respondent i, k is either new capital (*new_relations*) or pre-existing capital (*old_relations*), and ε is the error term for respondent i.

Whereas the simple model in Equation (1) considers either new or pre-existing relationship-based capital (given subscript k), Equation (2) tests for their simultaneous effects[13]:

$$Y_{ij} = X_{1i} + X_{2i} + \varepsilon_i \tag{2}$$

The results of the specification offered in Equation (2) are intended to provide a deeper understanding of the possible effects of relationship-based capital, in contrast to the restrictions of Equation (1).

Among the possible techniques to study the effects of new versus pre-existing forms of capital in a triple helix construct, the ordinary least squares (OLS) statistical analysis is preferred, given the nature of our dependent and explanatory variables. For those cases in which j is comprised strictly of collaborative patents, OLS results are presented in Table 7.9. Columns 1 and 2 reflect Equation (1), whereas column 3 represents Equation (2). For those cases in which j is comprised strictly of non-collaborative patents, which was determined by taking total patents less collaborative patents, OLS results are presented in Table 7.10. Robustness checks beyond standard methods include the inclusion of a number of control variables, as well as a *new_relations*old_relations* interaction term. All of the results from these checks are available in Appendices 7.2 through 7.7 and do not present any significant challenges to the results offered in Tables 7.9 and 7.10.[14]

The results for the basic hypothesis are presented in Table 7.9. First and foremost, there are no statistically significant differences between Korea and Taiwan or between the public and private sectors. Pre-existing relationship-based capital (*old_relations*) is positive and significant when accounted for individually (column 2), as is new relationship-based capital (*new_relations*; column 1). Both variables continue to remain positive and statistically significant when considered simultaneously, which is expected to most accurately reflect the reality for public and private researchers in these two countries.

Table 7.9 OLS Results for New and Pre-Existing Capital's Effects upon Collaborative Output

	(1) patcol	(2) patcol	(3) patcol
new_relations	0.121**		0.105**
	(0.0365)		(0.0371)
old_relations		0.1000**	0.0782*
		(0.0365)	(0.0369)
taiwan_dummy	0.126	0.0117	0.0316
	(0.224)	(0.229)	(0.227)
private_dummy	-0.00635	0.0299	0.0424
	(0.222)	(0.224)	(0.222)
Constant	-0.00936	0.112	-0.221
	(0.225)	(0.217)	(0.245)
Observations	299	299	299
R-squared	0.037	0.026	0.052
F	3.789	2.619	3.997

Standard errors in parentheses
+ $p < .10$, * $p < .05$, ** $p < .01$, *** $p < .001$

The results of Table 7.10 present regression output for the case in which the dependent variable is non-collaborative patenting. As it was mentioned earlier, this applies a counterfactual test for the primary hypothesis, in that an expected negative effect of new relationship-based capital on non-collaborative patents reaffirms its importance in terms of collaborative patent output. As Table 7.10 shows, these expectations were not met. In fact, the effect of pre-existing relationships (*old_relations*) upon these two forms of patenting seems to support the conclusion that personal ties are most important for public–private R&D collaboration. Pre-existing personal ties have a positive (and significant) effect on collaborative patenting, but they have a negative (and insignificant) effect on non-collaborative patenting. New relationship-based capital in the form of previous participation, on the other hand, has positive effects for both collaborative and non-collaborative output. This dual importance is attributed to the increased output potential arising from repartnerships between the public and private research sectors. Indeed, in tandem with the results of Table 7.9, the results of Table 7.10 confirm that new relationships are a valuable source of increased patenting, regardless of whether the patenting is done with research entities of the opposite sector. Still, the fact that pre-existing relationships function primarily to generate public–private R&D collaborative output is a conceptual challenge, requiring further investigation.

Table 7.10 OLS Results for New and Pre-Existing Capital's Effects upon Non-Collaborative Output

	(1) *nocollpat*	(2) *nocollpat*	(3) *nocollpat*
new_relations	0.724+ (0.415)		0.785+ (0.424)
old_relations		-0.127 (0.414)	-0.291 (0.422)
taiwan_dummy	-0.346 (2.542)	-0.146 (2.604)	0.00387 (2.595)
private_dummy	-3.145 (2.520)	-3.420 (2.546)	-3.326 (2.536)
Constant	3.840 (2.557)	7.127** (2.466)	4.628+ (2.803)
Observations	299	299	299
R-squared	0.017	0.007	0.019
F	1.728	0.736	1.413

Standard errors in parentheses
+ p < .10, * p < .05, ** p < .01, *** p < .001

Tables 7.11 and 7.12 are variations on these earlier results, confirming that new relationship-based capital (*new_relations*) has a positive effect for each sector.[15] The coefficient is not always statistically significant, as it is in the aggregate case presented in Tables 7.9 and 7.10, but the pattern is consistent. At this subgroup level, Table 7.12 attempts to account for the effects of collaborative tendencies upon collaborative and non-collaborative patenting output. This is, thus, a direct attempt to empirically apply some of the triple helix's theoretical shortcomings. For the public subgroups (columns 1 and 3 of Table 7.10), collaboration with SMEs (*coll_sme*) and with large firms (*coll_large*) positively predicts patents, but is not highly statistically significant. (These variables are all assigned values from zero to ten, measuring the percentage values from zero to one hundred in increments of ten.) For the private subgroups (columns 2 and 4 of Table 7.12), the only statistically significant result is the percentage of collaboration with universities (*coll_uni*), which has a positive and significant effect on collaborative patents.

A number of robustness checks help establish whether there are persistent cross-national and cross-sectoral differences. To accommodate the hypothesis that new relationship-based capital arises from triple helix constructs, such as public–private R&D collaboration, the percentage of collaboration done with partners from previous projects (*new_relations*) is the dependent variable and, given that larger values of previous participation correspond with higher percentages of participation, an ordered logit model is adopted.[16] By country and sector, age of respondent is shown to have a negative impact upon the degree

Table 7.11 OLS Results for New and Pre-Existing Capital's Effects upon Collaborative and Non-Collaborative Output, by Sector Subgroups

	Public sample (1) *patcol*	*Private sample* (2) *patcol*	*Public sample* (3) *nocollpat*	*Private sample* (4) *nocollpat*
new_relations	0.0340	0.156**	1.448+	0.326
	(0.0453)	(0.0563)	(0.847)	(0.336)
old_relations	0.0235	0.116*	-0.555	-0.156
	(0.0448)	(0.0563)	(0.839)	(0.336)
taiwan_dummy	0.199	-0.0991	4.110	-4.164+
	(0.253)	(0.375)	(4.739)	(2.243)
Constant	0.211	-0.414	1.468	5.571*
	(0.260)	(0.399)	(4.871)	(2.382)
Observations	143	156	143	156
R-squared	0.014	0.091	0.025	0.029
F	0.656	5.073	1.177	1.539

Standard errors in parentheses
+ p < .10, * p < .05, ** p < .01, *** p < .001

Table 7.12 OLS Results for New and Pre-Existing Capital's Effects upon Collaborative and Non-Collaborative Output, by Sector Subgroups

	Public sample (1) *patcol*	*Private sample* (2) *patcol*	*Public sample* (3) *nocollpat*	*Private sample* (4) *nocollpat*
new_relations	0.0122	0.154**	0.992	0.415
	(0.0480)	(0.0568)	(0.895)	(0.342)
old_relations	0.0324	0.103+	-0.389	-0.111
	(0.0453)	(0.0562)	(0.844)	(0.338)
coll_sme	0.0403		0.0823	
	(0.0401)		(0.748)	
coll_large	0.0550		1.339+	
	(0.0433)		(0.808)	
coll_uni		0.175*		-0.514
		(0.0841)		(0.506)
coll_gri		-0.0687	-0.389	
		(0.0737)	(0.443)	
taiwan_dummy	0.230	0.142	4.773	-5.787*
	(0.254)	(0.413)	(4.743)	(2.486)
Constant	-0.154	-0.774	-3.072	8.237**
	(0.373)	(0.491)	(6.959)	(2.954)
Observations	143	156	143	156
R-squared	0.028	0.118	0.045	0.044
F	0.800	4.002	1.293	1.394

Standard errors in parentheses
+ p < .10, * p < .05, ** p < .01, *** p < .001

of repartnering in all cases except for the Korean public sector; i.e. the older the researcher, the less likelihood of repartnership. It is found that repartnering is positively affected by job experience for Korea, particularly the public subgroup. For Taiwan, however, years of job experience negatively predicts the tendency for repartnership with former public–private R&D collaborators. Another distinction between Korea and Taiwan is the predicted effect of different research emphases upon the propensity to repartner. In Korea, basic research emphasis is more likely to generate repartnering in Korea, but not in Taiwan. In both countries, there is evidence that applied research emphasis is more likely to generate repartnering in the private sector than in the public sector. This is consistent with assumptions about differing research emphasis between research sectors. Complementarity between sectors, however, has a uniform positive, significant predicted effect for all subgroups.

The richness of the KORTAI R&D data set allows us to examine details of relationship-based capital even further. With reference to Figure 7.1, the sources of new and pre-existing capital enable us to understand how and why each subgroup in Korea and Taiwan is repartnering or utilizing personal ties with collaborators. Rather than base this examination on the summary statistics for the various sources of repartnering and personal ties, weights are assigned. That is, *new_relations* is weighted by each reason for repartnering to indicate the precise amount of impact: (*new_relations*$_i$ x Z_{ni}) where Z_n is the *n*th reason for repartnering for respondent *i*. In this way, if a reason is strong but repartnering in weak, the reason is given less weight. The rankings of these weighted reasons are presented in Tables 7.13 and 7.14 for Korea and Taiwan, respectively.

Table 7.13 Rankings of Weighted Reasons for Repartnering: Korea

Rank	*GRI*	*university*	*firm*
1	trust 27.91	trust 27.42	trust 28.41
2	expcom 26.35	easecom 24.36	expcom 26.04
3	noother 25.32	compknow 24.00	compknow 25.17
4	compknow 25.26	expcom 22.19	easecom 24.45
5	easecom 24.47	noother 21.14	noother 21.69
6	fundstip 21.88	fundstip 20.69	sharecom 21.66
7	sharecom 22.15	sharecom 19.83	fundstip 21.17
8	lacktension 19.56	lacktension 18.42	lacktension 19.37

Table 7.14 Rankings of Weighted Reasons for Repartnering: Taiwan

Rank	*GRI*	*university*	*firm*
1	trust 29.43	trust 21.50	compknow 26.49
2	expcom 26.83	easecom 21.50	trust 26.32
3	compknow 26.06	expcom 21.25	expcom 25.36
4	sharecom 23.89	compknow 19.75	easecom 23.24
5	easecom 22.78	sharecom 14.75	sharecom 19.15
6	fundstip 18.77	fundstip 14.00	fundstip 18.68
7	noother 15.64	lacktension 11.71	noother 16.39
8	lacktension 12.00	noother 9.71	lacktension 14.45

Consistent results across all three sub-samples include the dominance of *trust* and the relative unimportance of *fundstip* and *lacktension*. The low ranking of *fundstip* is an indicator that other repartnering forces are at work. A subtle distinction between these two countries is that *sharecom* lacks relative emphasis in Korea whereas, in Taiwan, *noother* is ranked among the bottom three reasons for repartnering. What is unmistakable is that trust is of great importance for both of these countries and across all research groups, particularly given that trust outranks expected commercialization of results (*expcom*) for the private sectors in Korea and Taiwan.

A similar weighted ranking system is also constructed for the nature of personal ties, although the dichotomous nature of the variables prevents a similarly scaled measurement. The Likert scale response for *old_relations* is weighted with each dummy variable (i.e.[*old_relations* x *sameuni*],[*old_relations* x *sameunilab*], etc.), generating the mean values for each relevant group, which are then ranked. What is most notable among these results for the Korean case, presented in Table 7.15, is the consistent, high ranking of *sameproj* for all Korean groups. For the Taiwanese case, presented in Table 7.16, personal ties arising from attending the same conference is ranked highest, for GRIs and universities. Taiwanese firms (Table 7.16) are the only exception, as personal ties arising

Table 7.15 Rankings of Weighted Source of Personal Ties: Korea

Rank	*GRI*	*university*	*firm*
1	*sameproj* 1.96	*sameproj* 2.25	*sameproj* 2.08
2	*sameconf* 1.26	*sameconf* 1.70	*sameconf* 1.20
3	*sameunilab* 1.09	*sameuni* 1.18	*sameuni* 0.88
4	*sameuni* 0.38	*samefirm* 0.85	*sameunilab* 0.48
5	*samefirm* 0.33	*sameunilab* 0.85	*samefirm* 0.18

Table 7.16 Rankings of Weighted Source of Personal Ties: Taiwan

Rank	*GRI*	*university*	*firm*
1	*sameconf* 3.36	*sameconf* 4.12	*sameuni* 1.91
2	*sameproj* 2.07	*sameproj* 2.25	*sameunilab* 1.41
3	*sameuni* 1.36	*sameunilab* 1.83	*sameconf* 1.36
4	*sameunilab* 1.27	*sameuni* 1.75	*sameproj* 1.33
5	*samefirm* 0.91	*samefirm* 0.00	*samefirm* 1.07

from university (*sameuni*) and university-lab (*sameunilab*) are ranked highest. This is the likely result of the strong university-based firm incubation programs in Taiwan.

Based on the weighted rankings of the source of personal ties in Tables 7.15 and 7.16, one can conclude that researchers in Korea and Taiwan are largely establishing collaborations not through university or firm connections, but through conferences and work on previous projects. This ranking pattern of personal ties actually corresponds with repartnering, lending additional support for the 'new capital' hypothesis of Etzkowitz (2003). The overall impression from the high ranking of personal ties based in former projects (*sameproj*) particularly shows that repartnerships are widespread.[17]

CONCLUSION

Theoretical and methodological contributions have been offered here in an attempt to bolster an understanding of the triple helix paradigm in Korea and Taiwan. This discussion has confirmed earlier claims which detail the benefits of the triple helix structure, particularly the increase of new forms of capital (Etzkowitz 2003). There are a number of qualifiers included here, however, which make it clear that caution must be exercised when breaking down the components of the paradigm on a number of levels.

The evidence provided in this discussion supports the 'new capital' hypothesis of Etzkowitz (2003) with regard to the triple helix structure. Two different dependent variables are applied to compare the effects of this new, relationship-based capital upon the triple helix and related methods of R&D. The results clearly show that there is a positive impact of new, relationship-based capital on patenting, and that there are homogenous effects between countries and research sectors.[18] What is most notable among the OLS results from the model presented in Figure 7.1 and Equations (1) and (2) is that personal ties—pre-existing relationship-based capital—are important for public–private R&D collaborative output but not for non-collaborative output. This does not necessarily invalidate the importance of new capital, but it does show that personal ties are most important in terms of their effects on collaborative output. As it has been shown that personal ties most strongly reflect work on previous projects, this result does not challenge the 'new capital' hypothesis. Indeed, it provides further validation.

From a policy perspective, it would appear prudent to facilitate those opportunities for public and private research entities to work together, such as the funding programs included in the KORTAI R&D data set. Based on the ranked and weighted reasons for repartnering, R&D funding policies were found to be less important than other reasons for repartnering. This does not rule out the possibility that policy-directed cross-sector R&D collaboration is still a necessary catalyst. In other words, other reasons for repartnering become salient for the researcher over time. In this way, policies have a direct effect for the initial partnership and an indirect but still important effect for repartnering. There are no grounds for cutting public funding, by any means.

Returning now to the distinct patterns of networking which are evident in East Asian countries such as Korea and Taiwan, there is a distinction, albeit hazy, between the effects of new, relationship-based capital and the effects of 'traditional' forms of networking. Traditional organization structures are typically comprised of familial and alumni ties (Biggart 1990), where work-related rewards depend on how long and well one serves the organization rather than individual achievements, and employees are hired through personal recommendation, family background, education and examination scores (Rozman 1992). It was precisely the measurement of these characteristics which was attempted through the data on the sources of personal ties. Given that the weighted sources of personal ties reveal that university, university laboratory or firm ties are relatively less important than relationships which are established through conference attendance and/or former projects[19]traditional networking structures in Korea and Taiwan are superseded by the new relations highlighted in the triple helix paradigm. In addition, the importance of trust as a reason for repartnering represents a major transition from traditional social norms, given that R&D partnerships are done beyond traditional conventions. It has been determined, for example, that Korea has a relatively low level of trust in comparison to other developed nations, such as Japan and Germany (Kim 2000). Some claim that this causes low R&D productivity because of poorly coupled links between research entities (Kim 2000). In this context, there is considerable added value from applying the triple helix paradigm to R&D efforts.

Returning to the robustness checks, the effect of age on the propensity to repartner provides some insight into a change R&D model in Korea and Taiwan. In line with von Hippel (1988), who emphasizes that the duration of a relationship is positively linked to the degree of knowledge shared between partners, the age of the researcher diminishes the likelihood of repartnerships. If age were a factor, one could assume that older researchers would have a more established network of colleagues with whom s/he could engage in public–private R&D collaboration. Ultimately, this is further confirmation of the claim that S&T trumps culture in Korea and Taiwan.

There is still room, however, to interpret the preceding results as conforming to traditional norms, in the context of the developmental state. Most discussion of the East Asian–Western distinction is couched in the idea of patrimonialism, where rule by a patriarch and administration by a personal staff have been the key mechanisms to maintaining social order, subordinating the self and exercising individual self-control (Biggart 1990). In this context, the heavy involvement of the state is the logical result of an indigenous, patrimonial framework. Despite research which correlates the diminishing function of these traditions with increases in growth (Park 2000), the patriarchal function of the government is still evident as it provides R&D funding in order to advance the country's S&T levels.

APPENDICES

Appendix 7.1 Variable Notation

Variable name	*Variable description*
old_relations	Percentage of PPRD collaboration originating from personal ties (data points are from zero to ten, representing percentage values from zero to one hundred)
sameuni	Personal ties based on university ties dummy variable
sameunilab	Personal ties based on former university laboratory ties dummy variable
samefirm	Personal ties based on former private firm ties dummy variable
sameproj	Personal ties based on multiple previous projects dummy variable
sameconf	Personal ties based on meeting at a conference dummy variable
new_relations	Percentage of PPRD collaboration done with partners from previous projects (data points are from zero to ten, representing percentage values from zero to one hundred)
noother	Repartnering affected by a lack of other qualified partners(seven-point Likert scale response, seven being greatest)
fundstip	Repartnering affected by funding stipulation(seven-point Likert scale response, seven being greatest)
sharecom	Repartnering affected by a shared commitment(seven-point Likert scale response, seven being greatest)
lacktension	Repartnering affected by a lack of tension(seven-point Likert scale response, seven being greatest)
easecom	Repartnering affected by ease of communication(seven-point Likert scale response, seven being greatest)
compknow	Repartnering affected by complementarity in knowledge(seven-point Likert scale response, seven being greatest)
trust	Repartnering affected by presence of trust(seven-point Likert scale response, seven being greatest)
expcom	Repartnering affected by expected commercialization(seven-point Likert scale response, seven being greatest)
coll_large	Percentage of PPRD collaboration done with large firm partners (data points are from zero to ten, representing percentage values from zero to one hundred)
coll_sme	Percentage of PPRD collaboration done with SME partners (data points are from zero to ten, representing percentage values from zero to one hundred)
coll_uni	Percentage of PPRD collaboration done with university partners (data points are from zero to ten, representing percentage values from zero to one hundred)
coll_gri	Percentage of PPRD collaboration done with GRI partners (data points are from zero to ten, representing percentage values from zero to one hundred)

Appendix 7.2 Robustness Checks

	(1) patcol	(2) patcol	(3) patcol	(4) patcol	(5) patcol	(6) patcol
new_relations	0.121** (0.0365)		0.105** (0.0371)	-0.0123 (0.0505)		-0.0487 (0.0595)
old_relations		0.1000** (0.0365)	0.0782* (0.0369)		-0.0416 (0.0486)	-0.0663 (0.0572)
new_relations* old_relations				0.0279*** (0.00746)	0.0308*** (0.00721)	0.0384** (0.0118)
taiwan_dummy	0.126 (0.224)	0.0117 (0.229)	0.0316 (0.227)	0.00262 (0.222)	0.0386 (0.223)	0.0359 (0.223)
private_dummy	-0.00635 (0.222)	0.0299 (0.224)	0.0424 (0.222)	0.0393 (0.218)	0.0228 (0.218)	0.0152 (0.218)
Constant	-0.00936 (0.225)	0.112 (0.217)	-0.221 (0.245)	0.0854 (0.222)	0.139 (0.211)	0.301 (0.289)
Observations	299	299	299	299	299	299
R-squared	0.037	0.026	0.052	0.081	0.083	0.085
F	3.789	2.619	3.997	6.452	6.635	5.436

Standard errors in parentheses
+ $p < .10$, * $p < .05$, ** $p < .01$, *** $p < .001$

Appendix 7.3 Robustness Checks (Industry Controls Included)

	(1) patcol	(2) patcol	(3) patcol	(4) patcol	(5) patcol	(6) patcol
new_relations	0.133*** (0.0373)		0.114** (0.0379)	-0.0111 (0.0514)		-0.0542 (0.0612)
old_relations		0.109** (0.0372)	0.0848* (0.0376)		-0.0486 (0.0499)	-0.0772 (0.0595)
new_relations* old_relations				0.0300*** (0.00759)	0.0339*** (0.00743)	0.0425*** (0.0122)
taiwan_ dummy	0.140 (0.232)	-0.0118 (0.239)	0.0278 (0.236)	0.0120 (0.229)	0.0612 (0.232)	0.0610 (0.232)
private_ dummy	0.115 (0.236)	0.154 (0.239)	0.182 (0.236)	0.177 (0.231)	0.155 (0.232)	0.141 (0.232)
Constant	-0.0877 (0.309)	-0.0236 (0.315)	-0.364 (0.330)	-0.0513 (0.301)	0.0382 (0.304)	0.215 (0.364)
Observations	299	299	299	299	299	299
R-squared	0.059	0.046	0.076	0.108	0.111	0.114
F	1.382	1.061	1.666	2.466	2.538	2.419

Standard errors in parentheses
+ $p < .10$, * $p < .05$, ** $p < .01$, *** $p < .001$

Appendix 7.4 Robustness Checks (Industry, Respondent Controls Included)

	(1) patcol	(2) patcol	(3) patcol	(4) patcol	(5) patcol	(6) patcol
new_relations	0.142***		0.125**	-0.00919		-0.0533
	(0.0403)		(0.0408)	(0.0553)		(0.0662)
old_relations		0.116**	0.0922*		-0.0493	-0.0778
		(0.0408)	(0.0409)		(0.0537)	(0.0644)
new_relations* old_relations				0.0321***	0.0361***	0.0446***
				(0.00824)	(0.00798)	(0.0132)
taiwan_dummy	0.176	0.0221	0.0667	0.0578	0.104	0.104
	(0.267)	(0.274)	(0.270)	(0.262)	(0.264)	(0.264)
private_dummy	0.306	0.341	0.378	0.404	0.387	0.382
	(0.311)	(0.315)	(0.310)	(0.304)	(0.304)	(0.304)
Constant	-0.787	-0.729	-1.203	-0.688	-0.544	-0.298
	(0.879)	(0.892)	(0.892)	(0.856)	(0.861)	(0.914)
Observations	273	273	273	273	273	273
R-squared	0.070	0.054	0.088	0.122	0.125	0.127
F	1.123	0.862	1.360	1.962	2.013	1.939

Standard errors in parentheses
+ p < .10, * p < .05, ** p < .01, *** p < .001

Appendix 7.5 Robustness Checks

	(1) nocollpat	(2) nocollpat	(3) nocollpat	(4) nocollpat	(5) nocollpat	(6) nocollpat
new_relations	0.724+		0.785+	0.558		-0.0175
	(0.415)		(0.424)	(0.587)		(0.690)
old_relations		-0.127	-0.291		-1.039+	-1.048
		(0.414)	(0.422)		(0.563)	(0.664)
new_relations* old_relations				0.0347	0.198*	0.201
				(0.0867)	(0.0836)	(0.136)
taiwan_dummy	-0.346	-0.146	0.00387	-0.500	0.0272	0.0263
	(2.542)	(2.604)	(2.595)	(2.575)	(2.585)	(2.590)
private_dummy	-3.145	-3.420	-3.326	-3.088	-3.466	-3.468
	(2.520)	(2.546)	(2.536)	(2.528)	(2.527)	(2.533)
Constant	3.840	7.127**	4.628+	3.958	7.305**	7.363*
	(2.557)	(2.466)	(2.803)	(2.577)	(2.448)	(3.356)
Observations	299	299	299	299	299	299
R-squared	0.017	0.007	0.019	0.018	0.026	0.026
F	1.728	0.736	1.413	1.333	1.969	1.570

Standard errors in parentheses
+ p < .10, * p < .05, ** p < .01, *** p < .001

Appendix 7.6 Robustness Checks (Industry Controls Included)

	(1) nocollpat	(2) nocollpat	(3) nocollpat	(4) nocollpat	(5) nocollpat	(6) nocollpat
new_relations	0.761+ (0.426)		0.836+ (0.437)	0.655 (0.604)		0.0237 (0.717)
old_relations		-0.175 (0.425)	-0.351 (0.433)		-1.143+ (0.584)	-1.131 (0.697)
new_relations* old_relations				0.0220 (0.0891)	0.208* (0.0870)	0.204 (0.143)
taiwan_dummy	-0.349 (2.656)	-0.173 (2.727)	0.116 (2.718)	-0.443 (2.687)	0.275 (2.711)	0.276 (2.716)
private_dummy	-3.220 (2.699)	-3.705 (2.733)	-3.500 (2.723)	-3.175 (2.710)	-3.703 (2.711)	-3.697 (2.721)
Constant	7.158* (3.529)	10.78** (3.590)	8.299* (3.801)	7.185* (3.536)	11.16** (3.564)	11.08** (4.267)
Observations	299	299	299	299	299	299
R-squared	0.027	0.017	0.030	0.028	0.036	0.036
F	0.616	0.380	0.618	0.574	0.768	0.714

Standard errors in parentheses
+ p < .10, * p < .05, ** p < .01, *** p < .001

Appendix 7.7 Robustness Checks (Industry, Respondent Controls Included)

	(1) nocollpat	(2) nocollpat	(3) nocollpat	(4) nocollpat	(5) nocollpat	(6) nocollpat
new_relations	0.845+ (0.460)		0.892+ (0.469)	0.613 (0.648)		-0.0177 (0.775)
old_relations		-0.0711 (0.464)	-0.244 (0.471)		-1.104+ (0.628)	-1.113 (0.754)
new_relations* old_relations				0.0491 (0.0966)	0.225* (0.0933)	0.228 (0.155)
taiwan_dummy	0.459 (3.045)	0.428 (3.111)	0.748 (3.100)	0.279 (3.070)	0.938 (3.090)	0.938 (3.096)
private_dummy	-3.562 (3.544)	-4.019 (3.583)	-3.753 (3.568)	-3.411 (3.561)	-3.733 (3.552)	-3.735 (3.560)
Constant	4.632 (10.02)	9.129 (10.15)	5.734 (10.25)	4.783 (10.03)	10.28 (10.06)	10.36 (10.70)
Observations	273	273	273	273	273	273
R-squared	0.051	0.039	0.052	0.052	0.060	0.060
F	0.808	0.603	0.776	0.775	0.903	0.852

Standard errors in parentheses
+ p < .10, * p < .05, ** p < .01, *** p <. 001

NOTES

1. See Westphal (1990) for a discussion on this point.
2. This chapter addresses head-on many of the issues which are presented in Evans (1997), which bridges the topics of social capital and state-led development.
3. This is true especially for China, given its shared Confucian influence with Korea and Taiwan.
4. For Faulkner and Senker (1994), this contact is a function of scientific publications.
5. These findings are particular to the Belgian case.
6. Responses are based on a 0 to 10 scale, representing increments of 10 from 0 to 100 per cent.
7. See Appendix 7.1 for complete variable notation.
8. The two measures of relationship-based capital—*new_relations* and *old_relations*—do not always equal 100 per cent because of overlap in these two categories. Controls for such overlap are provided in the subsequent analysis.
9. Although the measures of new and pre-existing capital may not be wholly exclusive of each other, an examination of the reasons for repartnering and the sources of personal ties as a robustness check confirms that this is not a problematic issue.
10. Again, these five categories were selected as a result of the content of interviews held between the author and research directors in Korea.
11. It should be noted that the two sides of Figure 7.1 express new and pre-existing forms of capital as operating separately upon research output, but they can also have simultaneous as well as interactive effects. These concerns are treated in a series of robustness checks in the subsequent analysis.
12. The rationale for not having time varying explanatory variables is that questionnaire respondents were not expected to accurately recall the amount of previous participation or personal ties over time.
13. A robustness check for the combined interactive effect of new capital and pre-existing capital was also conducted, although the results do not have a significant impact. See Appendices 7.2 through 7.7 for details of these robustness checks.
14. From the KORTAI R&D data set, controls are included for industrial sector, age of respondent, respondent's years worked at present job and respondent's years worked in present industry.
15. The only difference between the specifications in Tables 7.11 and 7.12 is the dropping of the private dummy variable.
16. The statistical output of these results is not included here, and the results described are in log-odds form.
17. Despite this overlap between *new_relations* and *old_relations*, there are still clear differences between these two measures, given the opposite signs of *old_relations* for collaborative and non-collaborative patents.
18. Robustness checks show that there are a number of subtle differences between the Korean and Taiwanese cases, but these have a negligible effect on patenting output.
19. The only exception here is the high ranking of university ties for the Taiwanese private sector, albeit the weighted score is not high, in relative terms.

REFERENCES

Bernasconi, A. 2005. 'University entrepreneurship in a developing country: The case of the P. Universidad Catolica de Chile, 1985–2000'. *Higher Education* 50:247–274.

Biggart, N. W. 1990. 'Institutionalized patrimonialism in Korean business'. *Comparative Social Research* 12:113–133.

Casas, R., R. De Gortari and M. J. Santos. 2000. 'The building of knowledge spaces in Mexico: A regional approach to networking'. *Research Policy* 29:225–241.

Chung, S. C. 2004. 'OECD science, technology and industry outlook 2004: Country response to policy questionnaire'. http://www.oecd.org/dataoecd/30/60/34242958.pdf (accessed 17 December 2009).

Colombo, M. G., and M. Delmastro. 2002. 'How effective are technology incubators? Evidence from Italy'. *Research Policy* 31:1103–1122.

Dierdonck, R. V., K. Debackere and B. Engelen. 1990. 'University–industry relationships: How does the Belgian academic community feel about it?' *Research Policy* 19:551–566.

Ernst, D. 2000. 'Catching-up and post-crisis industrial upgrading: Searching for new sources of growth in Korea's electronics industry'. In *East-West Center Working Papers*, 1–27. Washington, DC: East-West Center.

Etzkowitz, H. 2003. 'Innovation in innovation: The triple helix of university-industry-government relations'. *Social Science Information* 42:293–337.

Etzkowitz, H., and S. N. Brisolla. 1999. 'Failure and success: The fate of industrial policy in Latin America and South East Asia'. *Research Policy* 28:337–350.

Etzkowitz, H., and L. Leydesdorff. 2000. 'The dynamics of innovation: From national systems and "mode 2" to a triple helix of university-industry-government relations'. *Research Policy* 29:109–123.

Evans, P. B., ed. 1997. 'State–society synergy: Government and social capital in development'. University of California International and Area Studies Digital Collection, Research Series #94. http://escholarship.org/ur/item/1f0812j7#page-3 (accessed 17 December 2009).

Faulkner, W., and J. Senker. 1994. *Knowledge frontiers: Public sector research and industrial innovation in biotechnology, engineering ceramics, and parallel computing.* Oxford: Clarendon Press.

Fields, K. J. 1995. *Enterprise and the state in Korea and Taiwan.* Ithaca, NY: Cornell University Press.

Fritsch, M., and C. Schwirten. 1999. 'Enterprise–university co-operation and the role of public research institutions in regional innovation systems'. *Industry and Innovation* 6:69–83.

Hamilton, G. G. 1990. 'Patriarchy, patrimonialism and filial piety: A comparison of China and Western Europe'. *British Journal of Sociology* 41:77–104.

Hamilton, G. G., and N. W. Biggart. 1988. 'Market, culture, and authority: A comparative analysis of management and organization in the Far East'. *American Journal of Sociology* 94:S52–S94.

Kim, L. 2000. 'Korea's national innovation system in transition'. In *Technology, learning and innovation: Experiences of newly industrializing economies*, ed. L. Kim and R. R. Nelson, 335–360. Cambridge: Cambridge University Press.

———. 2001. 'The dynamics of technological learning in industrialisation'. *International Social Science Journal* 53:297–308.

Kim, L., and R. R. Nelson, eds. 2000. *Technology, learning and innovation: Experiences of newly industrializing economies.* Cambridge: Cambridge University Press.

Kruecken, G. 2003. 'Mission impossible? Institutional barriers to the diffusion of the "third academic mission" at German universities'. *International Journal of Technology Management* 25:18–33.

Marques, J., J. Caraca and H. Diz. 2006. 'How can university-industry-government interactions change the innovation scenario in Portugal? The case of the University of Coimbra'. *Technovation* 26:534–542.

Park, H. W., H. D. Hong and L. Leydesdorff. 2005. 'A comparison of the knowledge-based innovation systems in the economies of South Korea and the Netherlands using triple helix indicators'. *Scientometrics* 65:3–27.

Park, T.-K. 2000. 'Confucian values and contemporary economic development in Korea'. In *Culture and economy: The shaping of capitalism in Eastern Asia*, ed. T. Brook and H. V. Luong. Ann Arbor: University of Michigan Press. pp. 125–136.

Powell, W. W., K. W. Koput and L. Smith-Doerr. 1996. 'Interorganizational collaboration and the locus of innovation: networks of learning in biotechnology'. *Administrative Science Quarterly* 41:116–145.

Rozman, G. 1992. 'The Confucian faces of capitalism'. In *Pacific Centruy: The emergence of Modern Pacific Asia,* ed. M. Borthwick, 308–316. Boulder, CO: Westview Press.

Saad, M., and G. Zawdie. 2005. 'From technology transfer to the emergence of a triple helix culture: The experience of Algeria in innovation and technological capability development'. *Technology Analysis and Strategic Management* 17:89–103.

Schartinger, D., C. Rammer, M. M. Fischer and J. Frohlich. 2002. 'Knowledge interactions between universities and industries in Austria: Sectoral patterns and determinants'. *Research Policy* 31:303–328.

Van Looy, B., K. Debackere, J. Callaert, R. Tijssen and T. Van Leeuwen. 2006. 'Scientific capabilities and technological performance of national innovation systems: An exploration of emerging industrial relevant research domains'. *Scientometrics* 66:295–310.

Von Hippel, E. 1988. *The sources of innovation.* New York: Oxford University Press.

Westphal, L. E. 1990. 'Industrial policy in an export propelled economy: Lessons from South Korea's experience'. *Journal of Economic Perspectives* 4:41–59.

Yusuf, S. 2003. *Innovative East Asia: The future of growth.* New York: World Bank and Oxford University Press.

8 Mobilizing for Development

Putting the Triple Helix into Action in Ghana

James Dzisah

INTRODUCTION

There is much more similarity than differences in the development trajectory of most sub-Saharan African states. The post-colonial development landscape is littered with uncoordinated development programs, inadequate funding, corruption, undemocratic governance and lack of monitoring and evaluation of programs to ascertain their impact. The multifaceted nature of the problems, notwithstanding, the prescriptions from both internal and external actors in most instances is the same: mobilize institutions and resources in disparate directions without structures to promote and ensure coherent interactions. It is not unusual to see government-controlled research institutes with competent scientists far removed from the research activities of universities, industry and other stakeholders. Universities are hardly part of the development discourse, and in the rare cases that they are called upon, they are consigned to the performance of peripheral roles.

In a knowledge economy, development has moved from a single organizer like the Schumpeterian entrepreneur (Schumpeter 1934, 1954) to sources outside of the economy, in particular, the creative recombination of relatively interdependent yet separate knowledge-based institutional spheres. Separating the spheres, to an extent, facilitates the emergence of new initiatives, a degree of functional integration, bringing together diverse actors, encourages idea creation and the accumulation of relevant resources for implementation (Dzisah and Etzkowitz 2008). The manner in which the institutional spheres are combined in a resourceful and efficient way to aid innovation and development is referred to as the triple helix (Etzkowitz 2008). The significance of the triple helix transformation is that the unitary notion of development, in which central government reserved initiative for itself, industry carried out technology transfer and local application and the university limited itself to training human resources, is superseded. In the triple helix development, government devolves decision-making to collaborations with regional and local authorities and other actors. Industry engages in endogenous innovation as well as transfer. Universities play an innovative role in society, active in translational research, entrepreneurial training and community development, as well as traditional tasks.

These nascent transformations have fundamentally changed the development landscape, making universities critical actors and, in some regimes, the catalyst to the knowledge-based economy (Dzisah 2007). Consequently, finding a significant niche in the global economy requires that Ghana pay more attention to knowledge production, mobilization and use. This calls for heavy investment in infrastructure, education, research and a coherent innovation and development strategy. These are enormous challenges, but the institutional structures are already in place. The missing element, as discussed in more details in Chapters 12 and 14 of this volume, is the ability to mobilize triple helix actors to generate new ideas and resources to accelerate development.

DEVELOPMENT THINKING AND PRACTICE

The dominant models in the context of Africa's development tend to overly focus on factors to be blamed for the continent's lack of significant development. In general, these models envisioned a process of socio-economic growth wedged in a circular flow with each particular phase predicated on preceding ones (Toye 1993; Rostow 1960). Therefore, apart from the complicated nature of the development terrain, in both theory and practice, development was synonymous with modernization, which underlies assumptions and strategies of major international development organizations (Baber 2001). This notion has taken root in Africa's development discourse, in spite of some notable shifts, the move away from the exclusive focus on infrastructure, capital inputs and technology to institutions, processes, management, education and knowledge (Pieterse 2001; World Bank 1997, 1998).

In both content and meaning, this reorientation has led to the incorporation of such critical dimensions as gender, environmental sustainability, capacity building and social capital (Kothari and Minogue 2002; Pieterse 2001).These transformations, especially in science, technology and knowledge, enabled the World Bank to claim that half of the considerable difference in economic growth between South Korea and Ghana over the past four decades was attributable to the investment in the right kinds of knowledge (World Bank 1998).

In spite of this, the assumption in the educational policy formulations of the World Bank has been that mass primary and secondary education should precede the extensive development of tertiary educational capabilities. This is not to belittle the Bank's contribution to higher education. Universities produce new scientific and technical knowledge through research and advanced training and serve as conduits for the transfer, adaptation and dissemination of knowledge globally generated (Becker 1964). The World

Bank is also aware that the contribution of universities to economic growth may increase with levels of technology, especially as countries achieve universal primary and secondary education (World Bank 1995).

The problem, however, is that universities feature less prominently in the priorities and strategies of the World Bank. Based on the World Bank and IMF prescriptions, basic education in Ghana is increasingly 'pitched against supposedly ineffective and unnecessary higher education' (Manuh, Sulley and Budu 2007, 28). This policy underlies the stages of growth model where economic development is a discontinuous and dialectical process until a take-off stage of self-sustained advancement is reached (Rostow 1960). In contemporary knowledge-based development, any model that sought to entirely separate the site of knowledge production from application (Gibbons et al. 1994) is not only outdated but irrelevant to Ghana's development needs. Mobilizing for development requires a coherent triple helix strategy that will generate a new kind of synergy and interaction among university-industry-government and civil society.

THE TRIPLE HELIX AND DEVELOPMENT

A triple helix of university-industry-government as relatively equal interdependent and interacting institutional spheres is increasingly becoming the requisite basis for innovation and development in a knowledge-based society. The emergence of a double helix of government–industry relations—a bi-institutional model of society—represents the great transformation of the nineteenth century (Polanyi 1957). On the one hand, the market became the organizing principle of social relations; on the other, government moderated exchange relations to ensure a living wage. The emergence of a triple helix of innovation, with the university as a key player, is the great transformation of the late twentieth and early twenty-first century. In a triple helix regime, the university assumes a leading role as a critical element and source for the recombination of resources and the formulation of new ideas for innovation (Etzkowitz and Dzisah 2007).

The triple helix operates on the premise that the university plays an enhanced role in development in concert with government and industry, the two traditional leading institutional spheres. This makes the model globally applicable as higher education institutions are virtually everywhere and their flexible nature allows them to fill a variety of roles, well beyond traditional goals. Traditional missions of teaching and research embed a knowledge transfer capability in any society. In the training of human capital for all sectors of society, the university, through its alumni, provides the basis for enhanced interaction. However, Ghanaian universities have largely focused on teaching, as a result of their role in colonial or post-colonial technology transfer regimes, where attention was directed at importing technology rather than encouraging endogenous innovation even when research capacities were developed. The form and content of education and curricula most often mirror the prevailing concept of development, which in most instances is underwritten by donor agencies. The goal of the educational system was to turn out clerks to monitor and record in basic accounting terms the purchase of traditional agricultural export commodities, missionaries to engage in proselytizing activities and officers for the colonial civil service (Dzisah and Etzkowitz 2008).

There have been some notable changes initiated over the years. New institutions were founded, others were oriented towards the needs of the prevailing context and yet others sought to expand upon their original foundations. It is encouraging that most higher education institutions in Ghana, particularly the University of Cape Coast, have taken steps to become very proactive in developing new courses and specialized domains that can contribute to the developmental needs of Ghana. The problem that still remains is that these institutions do not work in tandem with other triple helix actors, especially industry. There is, therefore, the need to utilize a triple helix development approach as a focal point for Ghana's socio-economic development.

A triple helix development model contrasts with others that place greater emphasis on 'state-led, market-led or community-led development' (Kothari and Minogue 2002, 13). Although these other approaches are laudable, they omit a critical agent of knowledge-based development, a university that is capable of undertaking socio-economic development initiatives in cooperation with teaching and research. The triple helix development model focuses on creating intermediary mechanisms that play a broader role than in developed environments. They not only fill the gaps between industry and university and between discovery and application, but also in some instances they substitute for weak or missing actors.

A triple helix development model encompasses the following trends: (a) the transition from an industrial society to a knowledge-based society in which knowledge-producing institutions, like universities, potentially play a greater role in innovation and development; (b) the supercession of large-scale physical technologies that mandate bureaucratic forms of organization to increasingly flexible smaller-scale high technologies that can be utilized by smaller-scale organizations; (c) the emergence of polyvalent knowledge in such areas as biotechnology, computer science and nanotechnology, which is at one and the same time theoretical and practical, capitalizable and publishable; and (d) the rise of new university formats that incorporate a classic ivory tower focus on discipline development with a culture of entrepreneurship, innovation and technology transfer (Dzisah and Etzkowitz 2008).

In the triple helix regime, greater emphasis is placed on interactions, linkages and collaborations. This new way of development thinking revolves around knowledge actors, strengthens diversity and represents a radical departure from the conventional development models. In traditional development thinking, the three institutional spheres—university, industry and government—were separated, with universities, in particular, excluded from playing any developmental role apart from performing their traditional function of teaching and manpower training. The new triple helix development thinking refocuses the development field, making it possible for technology transfer to play a residual role in support of the development of indigenous technological capability (Saad and Zawdie 2005).

GHANA'S VICIOUS CYCLE: ECONOMIC GROWTH AND DECLINE

The structure of Ghana's economy is very similar to other sub-Saharan African countries. It is an economy dominated by the export of raw natural resources and agricultural products. Ghana, after obtaining political independence from Britain, sought to play economic development catch-up. In the early post-colonial period, strenuous effort was made to acquire the best possible advice. Based on the advice of the United Nations economic advisor, Sir Arthur Lewis, Ghana restricted its activities to investment in infrastructure. This resulted in an increase in the volume of investment from 15 per cent to about 21 per cent of gross domestic product (GDP) and a moderate growth of GDP at the rate of 1.5 per cent per annum (Okonjo 1986). In line with most modernization prescriptions, investment in direct economic production was left to foreign capital, which was invited to participate through such policy incentives as tax holidays.

Ghana failed to attract the necessary foreign investors and faced increasing financial difficulties. It became evident that modernization prescriptions had to be either retooled or abandoned. More detrimental to the future of the country was the depletion of its foreign reserves without any meaningful increase in productive capacity (Shillington 1992). The government, as a consequence, shifted focus and moved toward centralized corporate socialism and embarked on a massive industrialization drive combined with agricultural expansion. This policy and ideological shift resulted in great infrastructural developments, such as a national airline, hydro-electric dam, fishing and shipping harbour and a shipping line. These accomplishments came at a cost; the country committed itself to unsustainable debts (Shillington 1992). Successive governments tried to manage the situation, but in the 1980s the International Monetary Fund (IMF), the World Bank and other Western aid donors persuaded the country to adopt, in return for a new wave of loans, a Structural Adjustment Program (SAP). An estimate of the external loan component required to finance the SAP for seven years was about $1.8 billion (Hutchful 1989).

The SAP started in 1983 under the military regime that endeared itself initially to the public through socialist slogans. The regime was, however, forced by serious liquidity problems to volte-face and accepted the conditions of the IMF and the World Bank. The core of the SAP was a World Bank–assisted project directed at export rehabilitation in the cocoa, timber and mining sectors. The objective was to expand Ghana's export base so that more raw agricultural and mineral products could be exported. The World Bank insisted on the implementation of SAP in Ghana, whilst at the same time noting that debt stress and economic deterioration in the poorer zones of the world stemmed from a failure to copy the newly industrializing countries' strategy of export diversification. In the words of the World Bank, 'declining export volumes, rather than declining export prices, account for Africa's poor export revenues' (McMichael 2000, 126).

The SAP deepened Ghana's dependence on the export of raw minerals and agricultural commodities and failed to bring its 'external balances into sustainable equilibrium without changes in the composition and function of the balance of payments' (Hutchful 1989, 114). This primary exports expansion proved to be a shaky basis for adjustment. It undermined food production and manufacturing sectors of the Ghanaian economy. The imbalance in the Ghanaian economy continues to be a problem years after the completion of the SAP. For instance, by the end of 2001, the rate of economic growth was 4.5 per cent with unemployment reaching 20.3 per cent (Institute for Social, Statistical and Economic Research [ISSER] 2001; Manuh, Sulley and Budu 2007).

In spite of strong pressures from public sector workers and local entrepreneurs for living wages, the IMF claimed victory, pointing to the monetary, fiscal and budgetary deficits that were contained (Hutchful 1989). With new borrowing, much of it in the form of expensive short-term IMF credits, the country's external debt service burden was about 90 per cent of total government revenue in 2000. IMF credit increased from $46 million in 1970 to $740 million in 1992. By the close of 2000, Ghana's total external debt stood at about $6.2 billion (ISSER 2007, 107–168). These indicators put Ghana, according to the World Bank's criteria, into the category of highly indebted low-income countries (HIPC) whose debt burden is expected to prove unsustainable. Not surprisingly, in March 2001 the government declared the country as highly indebted and poor and sought relief under the World Bank's HIPC initiative. The HIPC initiative, like its SAP predecessor, involves debt reduction for highly indebted poor countries and was linked to the acceptance of 'sound policies' dictated by the international community (Hoogvelt 2001).

Through prudent microeconomic policies the government managed to reach the HIPC decision point and received a significant boost through the cancellation of most of its external debts. In spite of the massive debt relief,

Ghana's economy, although making great strides, is still dominated by agricultural and raw minerals. For instance, the agricultural sector employed about 57 per cent of the adult working population. It contributed about 39.6 per cent to national output in 2000, about 39.8 per cent in 2003, 40.3 per cent in 2004 and about 39.3 per cent in 2006. In terms of growth rates, the agricultural sector recorded about 2.1 per cent in 2000, 6.1 per cent in 2003, 7.5 per cent in 2004 and about 5.7 per cent in 2006. The industrial sector grew at the rate of 3.7 per cent in 2000, 5.2 per cent in 2003, 5.6 per cent in 2004 and 6.2 per cent in 2006. The services sector, comprising the universities, grew at a constant rate of about 4.7 per cent in 2002, 2003 and 2004 and about 6.5 per cent in 2006 (ISSER 2007, 107–168).

GHANA'S SCIENCE AND TECHNOLOGY SYSTEM AND DEVELOPMENT PLANNING

Ghana has had a long history of national development planning. It was alleged that Ghana completed the first development plan in the world: the Sir Gordon Guggisberg Plan in 1919. This was more of a public investment program than a comprehensive development plan, but it nonetheless provided the framework for the first attempt at developing the then Gold Coast up to the 1920s. This very first plan was developed by the colonial administration without the participation of the people. It was implemented largely by the administrative service (Frempong 2006).

Ghana has since been guided by several socio-economic development plans. Two years after political independence, a Research Act was promulgated, paving the way for the establishment in 1959 of the National Research Council (NRC) to coordinate science and technology (S&T) activities. The ambit of the governance of S&T was expanded through the creation in 1962 of the Ghana Academy of Sciences, which gradually took over the mandate and functions of the NRC. The process of change and transformation continued with the establishment in 1968 of the Council for Scientific and Industrial Research (CSIR) as the overseer of S&T policy and governance. Although other actors, such as the Ghana Academy of Arts and Science, Ghana Atomic Energy Commission, the Ghana Standards Boards and the National Council on Tertiary Education, operate within the S&T system, the director general of the CSIR acts as the *de facto* official government S&T advisor (Frempong 2006).

In spite of this, until 1981 S&T policy objectives and strategies were captured in general national development plans (see Table 8.1). In these plans, apart from human resource development, universities were largely peripheral. The establishment of the Ministry of Industries, Science and Technology in September 1979 brought an important boost to the institutionalization of science and the coordination of scientific activities (Frempong 2006).

Table 8.1 Snapshots of Various Development Plans in Ghana

Name of Plan	*Period*
The First Ten-Year Development Plan (condensed into a five-year plan)	1951–1956
The Consolidation Development Plan	1957–1959
The Second Development Plan	1959–1964
The Seven-Year Development Plan	1963/1964–1969/1970
The Two-Year Development Plan	1968/1969–1969/1970
The One-Year Development Plan	July 1970–June 1971
The Five-Year Development Plan	1975/1976–1979/1980
Ghana Technological Plan	1980–1982
The Economic Recovery Program/SAP	1983–1986
The National Development Policy Framework: Long-Term Development Objectives: Ghana-Vision 2020	1995–2000
National S&T Policy Document	2000
Ghana Poverty Reduction Strategy 1&II	2003–2005/2006–2009

Source: National Strategies for Sustainable Development (2001).

Ghana's development malaise continued as the 1981 Ghana Technological Plan was shelved following a change in government in 1982. In recent times, the National Development Policy Framework: Long-Term Development Objectives: Ghana-Vision 2020 was also abandoned after a democratic change of government in 2000. Apart from the S&T policy launched in 2000, not many high-profile and strategic S&T programs have been articulated in Ghana. The national S&T policy has the objective of developing infrastructure to support industry and other sectors of the economy to meet the needs of Ghanaians (Frempong 2006). Unfortunately, the strategic plan to implement the policy had yet to be adopted before the government was democratically changed. This is mostly due to the lack of a triple helix strategy to design, implement and coordinate knowledge-based development.

In addition, the steps taken to significantly reorganize ministerial portfolios, including the total change of name of the substantive ministry of S&T to Ministry of Environment, Science and Technology (MEST) in 2001, were short-lived. Even before any meaningful analysis was undertaken, the MEST was dissolved completely in 2006 with the Ministry of Education and Sports assuming responsibility for science (Frempong 2006). The ministerial reorganization and restructuring separated technology from science and made no attempt to mobilize the triple helix for development. This was, however, reversed after the change in government.

Consequently, the new government has circulated a fresh STI draft plan (2009) just as this chapter is being finalized. Whereas it is premature to critically analyse the impact that this will have in mobilizing the triple helix actors into a cohesive development unit, the fact that the document is being solely written by politicians gives one cause for concern. Experiences in other countries, developed and developing, have shown us that unless the triple helix actors come together and identify areas of cooperation, distinctness and creativity there will not be any sustained development. As it stands, Ghana's STI draft plan, like those before, will likely follow a similar predictable fate and become a white elephant after much fanfare and waste of resources in heralding it through the corridors of power. The draft plan in its current form is more about a ministerial power grab and control of the potential S&T budget. There is no concerted attempt in the plan to harness the triple helix actors into cooperative and productive knowledge-based development with core roles for the universities. In fact, it is clear that reorganizing S&T as a distinct entity could have enhanced the role of universities in the socio-economic development of Ghana. The Ghanaian universities must become the core development actors because their orientation and lifespan extend beyond political regimes.

RESEARCH AND DEVELOPMENT (R&D) IN GHANA

Ghana's research and development (R&D) actors and structures are not different from most other countries. The R&D system is dominated by higher education institutions and public and private research institutes. The higher education sector, comprising five public universities, ten polytechnics, researchers and technicians of these public higher education institutions, the staffs of the CSIR and GAEC, forms the core of the scientific workforce in Ghana. At the end of 2006, there were 84,078 students enrolled in the public universities, 9,497 in the private universities and 22,305 in the polytechnics (ISSER 2007).

However, the increase in enrolment is not matched by the training and retention of research and teaching staff at the public universities. At the end of 2004, the publicly funded universities had 260 professors across all disciplines, 492 senior lecturers and 1,183 lecturers (Manuh, Sulley and Budu 2007). In the public research institutes, such as CSIR and GAEC, the full-time equivalent researchers were estimated to be about two thousand (Frempong 2006). It is clear that radical changes are needed if Ghanaís triple helix pieces are to be mobilized for effective development. This is necessary because science and research has always been a major weakness of Ghanaís tertiary education system. Whereas there have been some indications of significant research undertakings and outcomes in the past, the situation has worsened since the 1980s. A major reason for the low research activity in universities has been the limited importance of graduate programs (Aryeetey 2005; Enos 1995; Dzisah 2006).

In recent times, much of the research that occurs in the public universities is contract research for organizations and pockets of well-funded research programs by foreign higher education institutions and philanthropic foundations. For instance, University of Ghana is part of the consortium led by the London School of Hygiene and Tropical Medicine and four other African universities undertaking a $40 million Bill and Melinda Gates Foundation grant for the treatment and control of malaria (Manuh, Sulley and Budu 2007).

UNIVERSITIES AND DEVELOPMENT

Ghanaian universities lack resources and lecturers are overburdened with teaching; however, some innovative changes are taking place. Among the traditional Ghanaian universities, the Kwame Nkrumah University of Science and Technology (KNUST), has a long history of working with local mechanics at the nearby 'Suame Magazine'. It is actively involved with the Cocoa Processing Company, which provides the University's Chemistry Department with both experience and income while helping a major national industry with research and development (Sawyer 1994).

However, it is the establishment of the University of Development Studies (UDS) that heralded in the search for difference and the realization that development is best attained through institutional interaction and the ability of universities to be entrepreneurial. The UDS was the first university in Ghana that adopted socio-economic development in addition, to teaching and research from the onset by tasking its Faculty of Integrated Development Studies with producing graduates who are well grounded in the realities of the surrounding poor communities. The Faculty of Medicine also incorporates the medical needs of the poor into the course content and is responsive to medical needs of its community. More recently, a community nursing program was started, in collaboration with the Okanagan University College of Canada (Manuh, Sulley and Budu 2007).

In spite of this, the contested nature of development came to the fore when these 'relevant contexts' and the corresponding course content were challenged by both students and faculty. The UDS, however, reaffirmed its core mandate of giving renewed vigour to the idea of a 'development university'. In the pursuit of this ideal, the UDS utilized a third-trimester approach, in which students take regular course work in two trimesters and then devote a third trimester to field practice, placements and community service. Faculty are expected to participate in these activities that both enhance teaching and address real-life development challenges of poor rural Ghanaians (Manuh, Sulley and Budu 2007).

By simultaneously drawing in expertise from other institutional spheres and infusing these spheres with new ideas and projects, universities can take their traditional education and training mission of human capital development a step further by bringing on board the task of regional economic development. The UDS approach is a classical example of how the triple helix model can serve as an effective and efficient innovation strategy in Ghana. By incorporating community service in the curriculum, both the students and the faculty at UDS engaged simultaneously in research with few development agencies in the regions to design, manage and evaluate development projects and programs. UDS has become directly engaged in the national planning of Northern Ghana's poverty reduction strategies (Manuh, Sulley and Budu 2007).

FINANCING S&T IN GHANA

The major factor affecting R&D in Ghana is funding and allocation of resources. A greater proportion of the government funds (about 97 per cent) supports administrative and personal emoluments with international donor support estimated at about 25 per cent of overall research funding (Gogo 2004). In 2000, an Act of Parliament (Act 581) established the Ghana Education Trust Fund (GETFund) through a 2.5 per cent increase in the value-added tax (VAT) to support education. In 2003 the GETFund was allocated ¢224 billion for its programs and for the first time was given an additional amount of ¢28 billion for faculty R&D, as well as ¢32 billion to support the national student loan scheme. However, it has come to light that the government is in arrears in its payment to support the fund in spite of collecting the tax from consumers.

In addition to the establishment of the GETFund, a credit facility from the International Development Association was taken as a seed for a Teaching and Learning Innovation Fund for tertiary institutions aimed at improving teaching and learning facilities and programs as well as postgraduate training. Prospects for supporting technological innovation could be derived from the Venture Capital Fund, whose bill was passed in 2004 to address inadequate capital problems of the private sector especially in the micro, small and medium enterprises (Frempong 2006). However, the overall support for R&D has always been inadequate. For example, in 2004, 81 per cent of the government's budgetary allocation to the CSIR was for personal emoluments with only 9 per cent earmarked for research activities (Frempong 2006).

As with other R&D activities, these budgetary allocations are donor dependent. Donor support is an indispensable component of R&D funding in Ghana. It accounted for about 48 per cent of total R&D budget in 2004 (Frempong 2006; Gogo 2004). Donor support for Ghana's R&D activities is not sustainable as most of the projects cease once the donor funding ends. Ghana's funding for R&D in most of the 1990s and 2000s was under 1 per cent (Frempong 2006). Thus, in the instances that donor funds failed to come in, most of the operational funds were either delayed or not released at all. For instance, in 2003, only 74 per cent of budgetary allocation for that year was released. However, this was an improvement of about 15 per cent to 23 per cent over the previous years.

There is no credible data on the level of contribution by the private sector in Ghana towards R&D activities; however, all plausible indications are that such contributions, if any at all, might be very low, because some of the private companies still have the wrong impression that R&D funding as a public activity must be supported by the government. More importantly, most of the private companies are small scale and also too financially weak to adequately support R&D activities. They are, however, likely to adopt and exploit scientific results to bolster their economic activities if they come with little or no cost to them.

Significantly, a triple helix framework could alleviate some of these problems as university, industry and government interactions offer a platform to shift resources around. This is even critical if one considers the fact that the few large companies, such us Guinness Ghana Limited, Unilever, Anglo-Gold and other multinationals, which could possibly support R&D in the country mostly resort to laboratories of their parent companies for scientific support (Etzkowitz and Dzisah 2007).

FINDING THE RIGHT SOCIO-ECONOMIC DEVELOPMENT MIX

The mobilization and utilization of the triple helix development model in Ghana could result in the following: a decrease in production costs, increased efficiency and output, improved quality of products and packaging, increased employment and income, improved production capacity of local industries and enhanced competitiveness of local firms in the global market. However, Ghana (like most developing countries) finds it difficult to move beyond the government helix. This results in a lack of democratic governance and a culture of corruption as people are unwilling to give up power, especially if moving to other helices is not seen as a viable alternative.

In developed countries, this movement across helices is sometimes viewed as creating conflicts of interest due to the close association of roles in different spheres. The bright side is that through institutional cross-fertilization, each helix is infused with new ideas and perspectives (Dzisah and Etzkowitz 2008). As knowledge is transformed into capital, persons from any triple helix sphere become potential entrepreneurs and firm founders (Etzkowitz 2008). The outcome of this transmogrification is that universities are no longer limited to their teaching and research functions but are extending these activities into the realm of shaping organizations in entrepreneurial education and incubation programs. Thus, rather than only serving as a source of new ideas for existing firms, universities are combining their research and teaching capabilities in new formats to become a source of new firm formation, especially in advanced areas of science and technology. Consequently, universities may establish incubator projects, to teach groups how to operate an organization and launch businesses and cooperative projects (Dzisah and Etzkowitz 2008).

The entrepreneurial university adds the capitalization of knowledge to the university's traditional role in the production and dissemination of knowledge. The entrepreneurial academic model comprises more than what goes on in the university, it includes extra-academic as well as transformed academic elements. A reverse linear model is based on identifying technological needs in industry or government, utilizing the resources of the university to provide solutions. In this model, a service organization is created to identify and meet those needs. This model is based on organizing the human capital resident within the university. It provides a means of capitalizing this human capital in order to provide funds for developing academic capacities (Etzkowitz 2006).

The triple helix model provides a professional capacity to seek external work and organizes groups of academics to provide the solutions. It translates consulting activity from an informal individual activity into an organized group activity. There are signs of the mobilization of the triple helix for development in Ghana. However, there is the need to strengthen S&T institutions and universities to take on the challenge of knowledge-based development. The days of finding comfort in only pointing out the colonialist roots of all our socio-economic problems without reflecting upon our own complexities, in the evolving contemporary social structures, should be a thing of the past (Baber 2001). In an increasingly knowledge-based global economy the success stories of countries with less or similar natural resource endowments must galvanize Ghana into action.

CONCLUSION: PUTTING THE TRIPLE HELIX INTO ACTION

The functional differentiation of institutions in the early modern era is being displaced by integration and hybridization of functions in post-modernity. This process begins from a different starting point and ushers in a secular triple helix. An open civil society paves the way for triple helix actors to organize and overcome blockages to the transformation of knowledge into innovation. This transformation is making universities even more central to the innovation process, and is superseding many functions of the industrial enterprise. This transition reinforces the global role of universities; it brings with it the development of less specialized universities with a wider interdisciplinary scope, and the birth of a new kind of scientist who integrates knowledge and innovation.

The emerging role of the university in creating an industrial sphere, supported by government actions, augurs a new relationship among the institutional spheres. As the institutional spheres of university and industry are enhanced, a more equal relationship with government is possible. It is clear that a continuous flow of science to the economy does not need to be achieved slowly through traditional stages of development-based models but can occur more rapidly by expanding and reorienting universities. As equal interacting spheres develop, the pace of knowledge-based development is quickened.

The lack of interaction among the development actors explains the failure of development in Ghana despite the massive government investment and huge donor funding. It is clear that Ghana needs to mobilize its triple helix actors, especially its universities, for development. This demands that policies take on post-partisan modes and universities be made to take their critical core developmental mantle.

REFERENCES

Aryeetey, E. 2005. 'The problems of African policy research institutions'. Paper presented at the International Conference on African Economic Research institutions and Policy Development: Opportunities and Challenges, Dakar, January.

Baber, Z. 2001. Modernization Theory and the Cold War. 'Journal of Contemporary Asia.' 31(1): 71–81.

Becker, G. S. 1964. Human Capital: A Theoretical and Empirical Analiysis with Speical Reference to Education. New York: Columbia University Press.

Dzisah, J. 2006. 'Information and communications technologies and development in Ghana'. *Science, Technology and Society* 11 (2): 379–396.

———. 2007. 'Institutional transformations in the regime of knowledge production: The university as a catalyst for the science-based knowledge economy'. *Asian Journal of Social Science* 35 (1): 126–140.

Dzisah, J., and H. Etzkowitz. 2008. 'Triple helix circulation: The heart of innovation and development'. *International Journal of Technology Management and Sustainable Development* 7 (2): 101–115.

Enos, J. L. 1995. *In pursuit of science and technology in sub-Saharan Africa: The impact of structural adjustment programmes*. London: Routledge.

Etzkowitz, H. 2006. 'The new visible hand: An assisted linear model of science and innovation policy'. *Science and Public Policy* 33 (5): 310–320.

———. 2008. *The triple helix: University-industry-government innovation in action*. London and New York: Routledge.

Etzkowitz, H., and J. Dzisah. 2007. 'The triple helix of innovation: Towards a university-led development strategy for Africa'. *ATDF Journal* 4 (2): 3–10.

Frempong, G. 2006. 'Improving S&T development and policy making in Ghana and other Africa countries: Lessons from United Kingdom'. *ATDF Journal* 3 (4): 12–26.

Gibbons, M., C. Limoges, H. Nowotny, S. Schwartzman, P. Scott and M. Trow. 1994. *The new production of knowledge*. London: Sage Publications.

Gogo, O. J. 2004. 'Towards effective management and funding of research, science and technology in Ghana'. Paper presented at the first national forum on Harnessing Research, Science and Technology for Sustainable Development in Ghana, Accra, Ghana, September.

Hoogvelt, A. 2001. *Globalization and the postcolonial world: The new political economy of development*. Baltimore, MD: The Johns Hopkins University Press.

Hutchful, E. 1989. 'From revolution to monetarism: The economics and politics of the adjustment programme in Ghana'. In *Structural adjustment in Africa*, ed. B. K. Campbell and J. Loxley, 92–131. New York: St. Martin's Press.

Institute for Social, Statistical and Economic Research. 2001. *The state of the Ghanaian economy in 2000*. Accra: Wilco Publicity Services Ltd.

———. 2007. *The state of the Ghanaian economy in 2006*. Legon: ISSER.

Kothari, U., and M. Minogue, eds. 2002. *Development theory and practice: Critical perspectives*. Houndmills: Palgrave.

Manuh, T., G. Sulley and J. Budu. 2007. *Change and transformation in Ghana's publicly funded universities: A study of experiences, lessons and opportunities*. Oxford and Accra: James Currey and Woeli Publishing Services.

McMichael, P. 2000 *Development and social change: A global perspective*. Newbury Park, CA: Pine Forge Press.

National Strategies for Sustainable Development. 2001. 'Status review of national strategies for sustainable development in Ghana'. http://www.nssd.net/country/ghana/gh0403.htm (accessed 12 May 2008).

Okonjo, C. 1986. 'The concept of development'. In *Rural development in Ghana*, ed. Charles K. Brown, 1–25. Accra: Ghana Universities Press.

Pieterse, N. 2001. *Development theory: Reconstructions and reconstructions*. London: Sage.

Polanyi, K. 1957. *The great transformation*. Boston, MA: Beacon Press.

Rostow, W. 1960. *The stages of growth: A non-communist manifesto*. Cambridge, MA: Harvard University Press.

Saad, M., and G. Zawdie. 2005. 'From technology transfer to the emergence of a triple helix culture: The experience of Algeria in innovation and technological capability development'. *Technology Analysis and Strategic Management* 17 (1): 89–103.

Sawyer, A. 1994. 'Ghana: Relations between government and universities'. In *Government and higher education relationships across three continents: The winds of change*, ed. G. Neave and F. A. van Vught, 22–53. Oxford: Pergamon for the IAU Press.

Schumpeter, J. 1934. *The theory of economic development*. Cambridge, MA: Harvard University Press.

———. 1954. *History of economic analysis*. New York: Oxford University Press.

Shillington, K. 1992. *Ghana and the Rawlings factor*. London: Macmillan Press.

Toye, J. 1993. *Dilemmas of development*. Oxford: Blackwell Publishers.

World Bank. 1995. *Priorities and strategies for education: A World Bank review*. Washington, DC: World Bank.

———. 1997. *World development report*. New York: Oxford University Press.

———. 1998. *World development report*. New York: Oxford University Press.

9 Bridging the Gaps in the Triple Helix

A Case Study Based on the Challenge of the Indonesian Experience

Dessy Irawati

INTRODUCTION

The phenomenon of the triple helix system has been recognized widely in developed countries (Etzkowitz and Mello 1994; Etzkowitz et al. 2007; Turpin, Sullivan and Deville 1993; Shinn 1997; Leydersdoff and Van den Basselaar 1998). The triple helix model has emerged from the needs of universities to work closely with the industry (i.e. double helix) in order to improve the knowledge spillovers (Marshall 1920) and maintain the sustainable development of the industry–university integration. This synergy needs to be supported by the government sphere as it plays the role of policymaker providing the necessary tools to encourage the local region. The tools could be based on innovative policy or incentives for the university and industry to develop their research and development (R&D) activities.

Nevertheless, the new triple helix paradigm has been recognized as a 'new' concept for some developing countries. In the Indonesian context, triple helix has been known as a non-linear mechanism to promote innovation and science policy along the joint partnerships in the 1990s. However, the progress has been considered to be relatively slow due to lack of communication and understanding among the actors involved. There have been attempts to implement this concept in order to bridge the gaps between small and medium enterprises (SMEs), large enterprises (LEs), local authorities (LAs), university and government. More emphasis has been placed upon links with and support for SMEs, so that they can play a key role in generating employment, economic growth and a more equitable development in Indonesia.

Indonesian SMEs are valued for several reasons, including their significant contribution to the economic activity in terms of the number of enterprises and employees as well as outputs. In 2006, there were about forty-nine million SMEs in Indonesia. They accounted for more than 99.98 per cent of all enterprises, employed 85.4 million people or 96.2 per cent of the total workforce and contributed to 53.3 per cent of total gross domestic product (GDP). When the Asian economic crisis hit Indonesia in 1997, SMEs weathered the crisis better than LEs.

This chapter discusses the essential stages required to establish a robust synergy between the three different actors: the university, the industry (both SMEs and LEs) and the government within the Indonesian context. A greater emphasis is placed upon the role of university in providing help for SMEs together with the government and other institutional developing agencies. This chapter also explores the promotion of SMEs by a clustering approach due to the fact that Indonesian SMEs are scattered across the country.

THE SURFACING OF THE TRIPLE HELIX THESIS

The triple helix system helps to create positive synergy among the three different actors through knowledge spillovers. The model engages the university as the centre of excellence with its academic-based research and development activities, the industry as the provider of the customer demand based on its commercial activities and the government as a policymaker. The integration of these different actors lies at the heart of the triple helix system that ideally increases knowledge spillovers in the region; thus increasing the competitive advantage of economic development, either regional or national.

The change in emphasis from a sole concentration on production and dissemination of knowledge to technology transfer and the formation of firms places the university in a new alignment with the productive sector. The modern university, which combines teaching and research, emerged in the early nineteenth century. The emergence of the industrial research laboratories and the 'scientification' of industrial production created a labour market for experimentally trained academics.

Etzkowitz and Kemelegor (1997) argue that university and industry have, up to now, been relatively separate and distinct institutional spheres. The role of the government in relation to these two spheres is moving in apparently contradictory directions. Government offers incentives, encouraging academic institutions to go beyond performing the traditional functions of cultural memory, education and research, and make a more direct contribution to wealth creation.

Accordingly, the current necessity to combine external with internal sources of innovation has revised the role of the industrial laboratory within firms by reducing both their scale and their scope (Rosenbloom and Spencer 1996). Also, the unit of analysis has increasingly become technological systems, viewed as networks of agents, interacting in a specific technology area as individual firms externalize many of their innovation activities (Hughes 1987).

Within these changed conditions, universities are increasingly viewed as actors in national and regional innovation systems; distinct boundaries are being eroded and replaced by a web of ties. Academic institutions are increasingly internalizing and decentralizing intellectual property

(IP) management and technology transfer activities, inserting the university as an entity in between the faculty and its industrial interlocutors. As the university acquires an industrial penumbra, industry takes on some of the values of the university, sharing as well as protecting knowledge. New institutional infrastructures, combining inputs from diverse sources, are merging among individual firms for the purpose of generating and diffusing technological knowledge flows into a cluster of firms (Carlsson and Stankiewicz 1991; Egelhoff and Haklisch 1991). Therefore, research groups within firms are increasingly becoming elements of research joint ventures and long-term strategic alliances, bringing them into contact with government laboratories and university research groups to achieve a common strategic goal.

Such a 'hands-on' strategy, however, requires a greater science and technology policy capacity on the part of the state, industry and academia, because the judgements of the level and type of intervention in particular areas become more critical (Etzkowitz and Kemelegor 1997). Therefore, the central issue is the synergy among the three different actors in societies, reflecting different traditions of political economy and different levels and types of economic development, including the macro- and microeconomics of each particular country.

CLUSTER AS CONCEPTUAL STRATEGY AND ITS BENEFIT FOR INDONESIAN SMEs: COMPLEMENTARY APPROACH FOR IMPLEMENTING TRIPLE HELIX

A cluster is a geographically proximate group of interconnected companies and associations of institutions in a particular field, linked by commonalities and complementarities (Porter 1998). Porter (1998) argues that the geographic scope of cluster ranges from a region, a state or even a single city nearby (or neighbouring countries), which also relates to the distance over which informational, transactional incentives and other efficiencies occur.

Thus, a cluster encompasses an array of linked industries and other entities important to competition which include suppliers of specialized inputs, such as components, machinery and services, as well as providers of specialized infrastructure. Additionally, many clusters include governmental and other institutions (e.g. universities, think tanks, vocational training providers, standards-setting agencies, trade associations) to facilitate specialized training, education, information, research and technical support.

Clusters occur in many types of industries, in smaller fields, even in some local industries, such as restaurants, car dealers and antique shops. They exist in large and small economies, in rural and urban areas, and at several geographic levels (e.g. nations, states, metropolitan, regions and cities). Moreover, clusters arise in both advanced and developing economies, although clusters in advanced economies tend to be far more developed (Porter 1998). Moreover, Porter (2000) explains that firms, either small or large, within clusters are able to more clearly and rapidly perceive new buyer needs. By contrast, an isolated firm faces higher costs and steeper impediments to assembling insight as well as a greater need to create knowledge in-house (Audretsch and Feldman 1996; Harrison, Kelley and Grant 1996; Jaffe, Trajtnberg and Henderson 1993).

Nevertheless, SMEs, particularly in developing countries such as Indonesia, are having difficulties capturing those opportunities that require products with better quality and prices, good after sales, larger production quantities and homogeneous product standards (Tambunan 2005; Hill 2001). Many enterprises experience difficulties in achieving economies of scale and are confronted with significant obstacles to internalizing functions.

Accordingly, for SMEs in Indonesia, clustering is believed to offer opportunities to engage in a wide array of domestic linkages between users and producers and between the knowledge-producing sector (universities and R&D institutes) and the goods and services producing sectors of an economy. Clustering allows SMEs to grow in manageable steps by sharing the costs and risks through collaboration.

Knowing and understanding clusters can generate value to regions only if that knowledge leads to actions that grow economies and raise standards of living. Unfortunately, there is no single recipe for less-favoured regions to follow that will meet the needs of all clusters, which embody many types of systematic relationship and kinds of industries. The choices regions can make depend on many factors which can include geographical factors, stages of development, resource constraints, special societal needs, cluster priorities, market imperfections and local preferences.

Cooperation within a cluster provides participant companies with advantages that can be derived from factors which include the presence of suppliers of raw materials, components, machinery and parts; the presence of workers with sector-specific skills; and the presence of workshops which make or service the machinery and production tools (Tambunan 2005). Clustering of enterprises makes it easier for government, LEs, universities and other development-supporting agencies to provide cheaper and better quality services. The cost of these services could be very high for individual enterprises in dispersed locations (Tambunan 2005). In the case of Indonesia, where the agglomerations of SMEs are located in rural areas, it is more beneficial for SMEs to form into clusters.

More important than the benefits of clusters in current productivity is the role of clusters in innovation and productivity growth. According to Porter (2000), cluster participation offers many potential advantages in their capacity to innovate and upgrade their competitiveness compared to an isolated location. Porter adds that some of the cluster characteristics that enhance the current productivity are important to innovation.

Porter (2000) explains that firms, whether small or large, within a cluster are able to more clearly and rapidly perceive new buyer needs. Thus, firms in a cluster can benefit from the concentration of firms with buyer knowledge and relationships, the juxtaposition of firms in related industries, the concentration of specialized information-generating entities and buyer sophistication. Cluster firms can often discern buyer trends faster than isolated competitors. Silicon Valley–based and Texas-based computer companies, for example, plug into customer needs and trends quickly and effectively and with an ease hard to match elsewhere (Porter 2000).

The era of globalization and increased competition impacts the ability of SMEs to improve their productivity and efficiency. It also requires from SMEs a greater ability to adapt and to be more flexible with regard to markets, products, technology, management and organization. In addition, the era of globalization creates greater market opportunities. However, the small size and the lack of resources often act as significant obstacles for SMEs in their efforts to achieve economies of scale, access financial services and to integrate functions such as training, technology innovation, market intelligence and logistics. SMEs are often unable to capture opportunities that require products with better quality, homogeneous standards, larger production quantities and a regular supply.

As many experiences in developed countries have shown, clusters can provide a powerful means for overcoming the obstacles facing SMEs and succeeding in an ever more competitive global market. Clusters offer participants advantages in perceiving new technological, operating or delivery possibilities. Participants can be exposed to richer insight into evolving technology, component and machinery availability and service and marketing concepts. By contrast, the isolated firm faces higher costs and steeper impediments to assembling insight as well as a greater need to create knowledge in-house (Audretsch and Feldman 1996; Harrison, Kelley and Grant 1996; Jaffe, Trajtnberg and Henderson 1993). Clusters can offer a variety of benefits to enterprises as well as to the wider economy.

THE THREE ELEMENTS IN INDONESIAN HIGHER EDUCATION: PERSPECTIVE FROM UNIVERSITY IN INDONESIA

Universities have a unique role to play in this emerging environment, particularly in the economic growth of the communities in which they exist (Gnuschke 2001; Hill 2001). Gnuschke (2001) believes that no great city exists without a great university where a centre of economic activity and intellectual activity go hand in hand to support each other. In contrast to the past, when university scholars were discouraged from becoming entrepreneurs, a new generation of university faculty is now closing the gap between the world of academic scholarship and economic activity.

Therefore, the evolving university as centre of intellectual property can be seen as an engine for economic growth and prosperity (Gnuschke 2001).

Additionally, according to Gnuschke (2001), some universities have obviously responded by recognizing the value of campus-based intellectual activity, engaging in business partnerships with the commercial activities of their faculty and focusing their research activities on issues that have both academic and commercial applications.

There are, however, some universities which are locked in traditional structures that have not encouraged and have even discouraged the development of entrepreneurial activities. They seem to fear that the traditional academic and research mission of the university either will be tainted or overrun by the for-profit component of entrepreneurial adventures (Gnuschke 2001). Nevertheless, the truth probably lies somewhere in between these extremes as they can be complementary. Entrepreneurial activities can help to provide funding for both applied and theoretical research in order to balance and effectively manage the competing interests of for-profit or traditional academic activities. For that reason, all entities, both universities and their partners, will benefit from the joint activities in which they are involved.

In view of that, Indonesian higher education must act in response to this change not simply by improving quality through proper implementation of existing standards, but by placing a greater emphasis on reviewing the relevance of the standards that have so far been used. This implies the need to question the belief that the quality of higher education can be assured by meeting the criteria and following the procedures established by the Directorate General of Higher Education (DIKTI).

The Indonesian university has three essential roles in terms of delivering education across the region. The roles are called TRI DHARMA PERGURUAN TINGGI or The Three Elements of Indonesian Higher Education, which entail education, research and community service. The program called the TRI DHARMA PERGURUAN TINGGI is aimed at improving the quality of the university's role in advancing the education in Indonesia alongside the industry sector and the government.

The university and TRI DHARMA will work together with the industry as a business partner to establish productive business networks (either on the small or medium scale). The government encourages the university and the industry to keep up this work by providing relevant policies and incentives (i.e. government loans). These policies and incentives are explained in the following.

Education—Research/Study

The purpose of improvement in the teaching–learning process is to increase the university's effectiveness in accommodating the development of both the students' and the faculty members' skills in learning and implementing management concepts and tools. Research helps to produce teaching materials, which could complement the materials used for academic programs (Soehendro 1998). Therefore, the system is expected to develop the expertise and commitment to produce teaching–learning materials, including cases, exercises/games/management laboratory, as well as innovation in methodology (Peniwati 2000).

Education—Community Service

The activities in community service should generate a sense of priority for specific knowledge relevant for the students to master, as well as improvisation and innovative applications of management concepts (Peniwati 2000). Therefore, education is also accountable for developing skills in simplifying and applying knowledge based on context, as well as communication skills required for delivering training, consulting and facilitating the process of organizational change (Soehendro 1998).

Community Service—Research/Study

Peniwati (2000) believes that carrying out community service on business terms would better ensure relevance to the needs of the community as well as high-quality work of the faculty members. As a result of this cooperation and sharing, these services can be provided at a lower cost. Therefore, activities in community service would identify topics or problems that need to be studied, as well as data and information, which should be used to conduct such studies. The research outcome must be disseminated to the community.

Nevertheless, the implementation of TRI DHARMA has merely been restricted to SMEs in non-high-technology and the non-high-capital sectors. An example is in the automotive sector, whereby SMEs which produce auto components mostly work together with the industry without the direct intervention of the university. The reason is the lack of communication and less networking between SMEs and the university. In Indonesia, SMEs which lack access to a university prefer to collaborate directly with their customers, such as large enterprises, in terms of marketing and product development. Indonesian universities are considered to be very interested in SMEs and non-traditional sectors and industries. This is why the automotive sector has established learning groups to upgrade the technical skills for all their suppliers.

Knowing that there is an additional demand for SMEs in entering non-traditional industries, starting in the late 1990s, the university has been trying to be more sensible and approachable by both SMEs and LEs. An example is the program that has also been undertaken by Astra Automotive with help from Toyota Indonesia. This program was aimed at providing technical training assistance to SMEs, particularly those that serve Toyota's group, both vendors and subcontractors. To roll out its program, coordination has been established with local governments, state-owned enterprises, private companies and universities aimed at developing nine Business Development Centres that assist 1,560 SMEs in several provinces. These provinces include Jabotabek area, West Java, Tegal, Jogjakarta, East Java, Bali, Nusa Tenggara Barat, Borneo and Sumatra.

Funding was also made available to support this program. Since 1991, a total of Rp.101 billion has been used to help SMEs expand their businesses from automotive components, motorcycle spare parts, heavy machinery, rubber and plastic manufacturing to handicrafts and house construction, as well as paint and medical equipment manufacturers.

Subsequently, Astra Automotive and joint academia from the university in the Java region have set up another educational program with support from Toyota. This program is known as the Astra Manufacturing Polytechnic (Polman Astra). Polman Astra is aimed at developing effective learning concepts and ensuring that its graduates are employable. Polman Astra has developed close links with industry, particularly with the Astra Group and with the group's customers and suppliers. For example, Polman Astra collaborated with United Tractors to launch its new Heavy Equipment Program. It also collaborated with Astra Nissan Diesel Indonesia and sent some of its staff and alumni to Nissan Diesel in Japan. The school works closely with various national and regional government and non-government organizations, including the Ministry of Education, the Japan Vocational Ability Development Association (JAVADA) and the Association for Overseas Technical Scholarship (AOTS), which has appointed Polman Astra as its Indonesian Centre of Excellence in Manufacturing.

Polman Astra has developed five core programs with a curriculum oriented toward the needs of the industry. These programs include manufacturing machining technology, manufacturing process and production technology, information system for manufacturing, automotive technology and combined engineering technology (including electronic, information, mechanical and control system engineering). These different programs have been delivered with the support of laboratories and the latest machine and engineering software.

CLUSTER STRATEGY FOR SMES IN INDONESIA: PERSPECTIVE FROM GOVERNMENT TO STRENGTHEN TRIPLE HELIX

The development of SMEs competitiveness constitutes a clear and consistently stated Indonesian government priority. This priority features prominently in most key government documents, such as the five-year plans (Repelita), the Broad Outlines of Government Policy (GBHN) and many official statements (Hill 2001). In addition, the government has set up a program known as Indonesian Knowledge and Technology Vision 2025 or VISI IPTEK 2025, in which the roles of LEs and SMEs are expected to make significant contributions to the economy, science and technology. The main objective of this program is to strengthen the national development and prosperity of the Indonesian people. Since then, the Indonesian government has been trying to engage industry/business and universities to help out in achieving VISI IPTEK 2025.

SMEs are seen as a vehicle for promoting *pribumi* (indigenous Indonesian) business and therefore as a means of asset redistribution along the ethical characteristics of the country. It is most likely, due to ethnic relations being more responsive in Indonesia than in any other East Asian country, that the gap between official pronouncements and implementation with regard to SMEs is the widest (Berry, Rodriguez and Sandee 2001). This is different from the standard economist's approaches to policy intervention, which emphasize market-oriented solutions as the key to rapid economic development (aside from specific justifications for intervention). As a result, it cannot be assumed that the policies drawn up for larger industrial units will necessarily apply to SMEs.

Additionally, international experience suggests that an effective SME sector is conducive to rapid industrial growth and a flexible industrial structure (Berry, Rodriguez and Sandee 2001). A particular strand of this argument emphasizes the importance of a well-developed SME sector in underpinning the key electronics, machine, goods and automotive industries through the establishment of subcontracting networks.

This explains the current interest in SMEs in Indonesia. This type of firm appears to have weathered the economic crisis of 1997–1998 better than larger industrial units. This statement appears to be true both for intra-country comparisons (i.e. large and small firms within a given country) and across economies (e.g. the Korea–Taiwan comparison; see Tambunan 2005).

SME clusters can be found in all provinces, and most of them are located in rural areas. The clusters were established naturally as traditional activities of local communities whose production of specific products have long been proceeding and the workers have special skills in making such products (Tambunan 2005). Clusters of batik, the traditional Indonesian textile, within the district of Java Island (i.e. Yogyakarta, Pekalongan, Surakarta and Tasikmalaya) are examples that have long been in existence. In 2003, most of the SMEs were concentrated in the Java region with 69.05 per cent, compared to the Sumatera region with less than 12 per cent and less than 1 per cent for Maluku and Papua in the east region. In Borneo, SMEs concentrated less than 5 per cent. Accordingly, due to these facts, Indonesian government must admit that there are imbalanced distributions of economic development in each region.

Tambunan (2005) describes the importance of clustering not only for the development of SMEs in the cluster, but also for the development of villages/towns in Indonesia. He gave the example of how the clustering of rattan furniture producers has absorbed an entire village in Tegal Wangi,

West Java, and created several small-scale industrial activities in neighbouring hamlets. Similar evidence can be drawn from the cluster of wood furniture in Jepara, Central Java. The growth of this cluster in the 1980s had transformed the town into a thriving commercial centre with many furniture showrooms and factories, modern hotels, new commercial banks, supermarkets and European restaurants. These examples demonstrate the importance of clustering as a means for the development of SMEs as well as for the regions' social and economic development.

Nevertheless, it has been found that not all the clusters within the district of Java Island were successful. Some of them, particularly those enterprises which were established in rural areas, have found it difficult to thrive because of market competition. This highlights the urgency for both government and universities to get more involved in such rural areas in order to help build up a proper cluster approach.

Clusters in Indonesia can be classified into four types based on the level of their development. The first type is *artisan*, in which the process of clustering is still at an 'infant' stage. The second type is *active*; indicating that it has developed rapidly in terms of skill improvement, technological upgrading and successful penetration of domestic and export markets. Examples of this type of cluster are roof-tile clusters, metal-casting clusters, shuttlecock clusters and shoe clusters. The third type is *dynamic*. This type of cluster is characterised by the decisive role of leading/pioneering firms, usually larger and faster growing firms, to manage a large and differentiated set of relationships between firms and institutions within and outside the clusters. Examples of this type of cluster are clove cigarette clusters in Kudus, tea-processing in Slawi and tourism clusters in Bali. In the case of clove cigarette clusters, their products are able to outperform products from Phillip Morris and British American Tobacco (BAT). Tea-processing clusters led by big companies such as Sosro have grown into market leaders in the Indonesian soft drink market, leaving giant Coca-Cola behind (Tambunan 2005).

Clusters of the fourth type are more advanced, more developed and more complex than the previous types. Examples are three of the well-known cluster agglomerations in Indonesia. The first is in the Yogyakarta-Solo area with its tourism, furniture and interior decoration, metal processing and textile and leather goods, which all mutually benefit each other. The second one is Bali, known as a tourist destination with SMEs which produce traditional handicraft, furniture and interior goods, silver jewellery and paintings. The third one is the Jabotabek area where automotive SMEs are located.

In view of cluster strategy, the Indonesian government assumed that because of scattered geographical conditions, clustering SMEs will be a sensible way to help them engage with the other actors. This strategy is expected to build up the local networks of specialized organizations, whose production processes can be closely linked through the exchange of information, knowledge and ideas. It is also expected that the cluster strategy will strengthen the links between SMEs and LEs from different levels in the industrial chain (i.e. suppliers, customers), with service units (i.e. financial institutions, production-supporting services) and with government bodies, semi-public agencies, universities and research institutes.

BRIDGING THE GAPS AMONGST UNIVERSITY, GOVERNMENT AND INDUSTRY TO SUPPORT SMES IN THE INDONESIAN CONTEXT: AN ANALYSIS FOR IMPROVEMENT

The role of universities in developing national industry, specifically SMEs, is essential as part of the strategy to revitalize Indonesian universities by transferring know-how from research laboratories to industry. The ineffective internal policy, weak funding base and lack of ties with private sectors are the main problems with promoting SMEs in Indonesia through government and university (Hill 2001).

Researchers within universities are required to be the experts in their fields. Consequently, to contribute to technology transfer, research institutions (universities and research centres) have two main functions which include acting as a technological supporter for SMEs and setting up research and development centres.

From the university perspective, the SMEs' development through a cluster approach is important because it is more effective and efficient for other institutions, such as the university itself, government and other international agencies. It provides technical and management support, training and general facilities to a group of firms located in the same region rather than to individual firms scattered in dispersed locations. A cluster approach in a region can also promote the development of other local sectors in that region, and hence promote the economic growth of the region.

The participation of the university in promoting SMEs is certainly helpful in implementing SMEs' cluster development scheme for the regional government. In Indonesia, and more particularly in Java Island, the local universities have played a strong and leading role in the development and promotion of the cluster approach. The universities involved include Institut Teknologi Bandung (ITB) or Bandung Institution of Technology, University of Indonesia and University of Gajah Mada. Their role was to support SMEs through the Pusat Inkubator Bisnis (PIB) or the Centre of Business Incubators. This has led to the development of SMEs particularly in the District of Java with the help of the government, the industry and international supporting development agencies such as Japan International Co-operation Agency, Asian Development Bank and World Bank. The role of the industry was to provide practical marketing advice and help SMEs to expand their local market. The support of the universities was carried out by lecturers, researchers and students in their own field of study and expertise. Research laboratories were part of these activities.

The provision of these services are facilitated and coordinated by the business incubators. These services include research, training, consultancy, planning and design, physical and service implementation, project management supervision and IP management.

Nevertheless, the government should provide a comprehensive development package, including technical upgrading through the provision of common service facilities, export training and investment in the improvement of regional infrastructures such as roads and telephone connections. This package will help the clusters, particularly those in rural areas, to gradually develop the share of their local and international markets. Government is also expected to provide robust policy to strengthen SMEs in non-traditional sectors to develop their technical skills by engaging with large industry, as this demand has risen recently along with the rise of SMEs in non-traditional sectors, particularly in electronic, machining and automotive components.

Ideally, triple helix interactions between the university-industry-government could also help to improve the conditions for innovation in a knowledge-based society. In this context, universities such as ITB, with its special expertise in technology, together with industry, have been trying to create applied technology in the form of devices for local SMEs. ITB has also contributed to help SMEs enhance their technical skills. This service, which is essentially funded by certain industries such as the automotive industry, is aimed at producing spare parts for this industry (Irfan, Manullang and Dou 2004). This joint partnership has been mutually beneficial to all actors involved within this endeavour.

However, universities in Indonesia are still having difficulty in directly transferring technology and knowledge to SMEs. Universities seem to be more active within the indirect mode of technology transfer. This indirect mode consists of creating graduates who will bring knowledge, skills and entrepreneurial attitudes to their subsequent jobs and communities (Hill 2001; Peniwati 2000; Berry, Rodriguez and Sandee 2001).

This can be explained by the major problems in R&D activities that the universities in Indonesia experience. The first difficulty is related to the ineffectiveness of internal policy such as education curriculums at university, high loads of teaching and improper evaluation systems. The second obstacle is the outcome of the low level of funding which affects the capability to hire qualified staff and thus slows down the university in expanding their research-based activities. The third inhibitor is related to the lack of integration with the private sector (Berry, Rodriguez and Sandee 2001).

As a result of the significant dependence of most of the funding of the current R&D activities upon the Indonesian government, R&D programs are not responding to demand from firms and industries but rather are linked to the vision and ideas of politicians and bureaucrats. Therefore, it is essential for researchers to be aware of their roles in engaging with the

SMEs for orienting their laboratory's activities towards the needs of the SMEs and of the local economy.

Technologies developed in universities need to match the needs and the capability of SMEs. For that reason, SMEs must play a more active role in enhancing their capability level in order to better assimilate and use the technology transferred. In order to achieve these objectives, Indonesian universities can use the existing programs where students get training from industry to fulfil their credits for graduation and also the opportunity to be involved with the SMEs in transferring the laboratory's inventions. Such a good opportunity should be taken as a momentum to build beneficial relationships between SMEs and other institutions (Berry, Rodriguez and Sandee 2001).

CONCLUSION

Universities possess much of the formal and organizational technology needed for economic development. This chapter has shown that mutual and joint partnerships have emerged when universities and economic enterprises have effectively collaborated and focused on particular issues of economic and social concern in developing countries.

The Indonesian case study has been presented to describe in detail the role of the university in dealing with SMEs, particularly in promoting the cluster approach to enhance their capability technically and economically. Most official statements emphasize the importance of SMEs as a means of generating employment, achieving greater equality through a more diverse ownership structure in business, promoting rural and regional development, providing a basis for entrepreneurial development and redressing the perceived ethnic imbalance in business ownership.

SMEs in Indonesia are regarded as one of the economic drivers complementing the role and contribution of the LEs. This chapter has argued for close relationships between universities and other institutions, such as industry, the government and international development-supporting agencies, to help SMEs improve their level of capability and competitiveness through a clustering strategy. This chapter has also discussed the importance of universities in developing countries to play an active role in the transfer of their expertise and knowledge to the economic and social actors of their society. Similarly to the triple helix system, this approach places a greater emphasis on the role of the academic sphere within the process of technology transfer. It also plays a crucial role in the development of the capability and competitiveness of SMEs through joint networking with other supporting institutions.

Indonesia has been trying to engage all actors in the implementation of this bottom-up model of innovation, where university, industry (both SMEs and LEs) and government support regional development and prosperity together.

Indonesia is hoping that triple helix relationships will be the key to improving the condition for innovation in a knowledge-based society that guarantees stable interactions and exchange, in particular for the non-traditional sector. The complementarity of the triple helix and cluster approach is being regarded as a suitable and practicable strategy for the development of SMEs spread out and located in diverse areas of Indonesia. The implementation of both strategies requires an active involvement so that they can support each other in order to enhance economic and social development in developing countries.

Promoting the triple helix and cluster approach for SMEs in Indonesia is not a trouble-free and short program. It is complex and involves many actors from different backgrounds. It must be taken into account that a joint program would be of value for all actors involved, based on a consensus among them.

REFERENCES

Audretsch, D., and M. Feldman. 1996. 'Research and development spillovers and the geography of innovation and production'. *American Economic Review* 86:630–640.

Berry, A., E. Rodriguez and H. Sandee. 2001. 'Small and medium enterprises dynamic in Indonesia'. *Bulletin of Indonesian Economic Studies* 37 (3): 363–368.

Carlsson, B., and R. Stankiewicz. 1991. 'On the nature, function and composition of technological systems'. *Journal of Evolutionary Economics* 1 (2): 93–118.

Egelhoff, W., and C. Haklisch, eds. 1991. *Technical alliances in the semiconductor industry: Effects on corporate strategy and R&D*. Troy, NY: Center for Science and Technology Policy, Rensselaer Polytechnic Institute.

Etzkowitz, H., J. Dzisah, M. Ranga and C. Zhou. 2007. 'The triple helix model for innovation: The university-industry-government interaction'. *Asia Pacific Tech Monitor* 24 (1): 14–23.

Etzkowitz, H., and C. Kemelegor. 1997. 'Research centres: The collectivisation of academic science'. *Minerva* 35:3.

Etzkowitz, H., and J. M. Mello. 1994. 'The rise of triple helix cluster: Innovation in Brazilian economic and social development'. *International Journal of Technology Management and Sustainable Development* 2 (3): 159–171.

Gnuschke, J. 2001. *Intellectual property and economic development, Business Perspective, summer/fall.* Memphis: Sparks Bureau of Business and Economics Research (SBBER), University of Memphis.

Harrison, B., M. Kelley and J. Grant. 1996. 'Innovative firm behaviour and local millie: Exploring the intersection of agglomeration, firm effects, industrial organization, and technological change'. *Economic Geography* 72:233–258.

Hill, H. 2001. 'Small and medium enterprises'. *Indonesia Asian Survey* 41 (2): 248–270.

Hughes, T. 1987. *The evolution of large technological systems*. Cambridge. MA: MIT Press.

Irfan, H. K., S. Manullang and H. Dou. 2004. 'Developing competitive technical intelligence'. *Indonesia Technovation* 24:995–999.

Jaffe, A., M. Trajtnberg and R. Henderson. 1993. 'Geographic localization of knowledge spillovers as evidenced by patents citations'. *Quarterly Journal of Economics* 108:577–598.

Leydersdoff, L., and P. Van den Basselaar. 1998. 'Technological development and factor substitution in a non-linear model. *Journal of Social and Evolutionary Systems* 21, 173–192.

Marshall, A. 1920. *Principles of economics*. London: Macmillan.

Peniwati, K. 2000. *Improving the quality of graduate programme management: Applying McKinsey's 7S Framework to strive for fit between the elements of TRI DHARMA PERGURUAN TINGG*. Jakarta: PPM Graduate School of Management.

Porter, M. E. 1998. 'Cluster and the new economic of competition'. *Harvard Business Review* (November–December): 78–90.

———. 2000. *The competitive advantage of nations*. London: Macmillan.

Rosenbloom, R. S., and W. J. Spencer. 1996. *Engines of innovation: US industrial research at the end of an era*. Boston, MA: Harvard Business School.

Shinn, T. 1997. *Instrument hierarchies: Laboratories, industry and divisions of labour*. London: Macmillan.

Soehendro, B. 1998. *Deregulasi Perguruan Tinggi Dalam Memasuki Pasar Bebas*. Jakarta: Nasional Manajemen Stratejik Perguruan Tinggi.

Tambunan, T. 2005. 'Promoting small and medium enterprises with a clustering approach: A policy experience from Indonesia'. *Journal of Small Business Management* 43 (2): 138–154.

Turpin, T., N. Sullivan and A. Deville. 1993. *Crossing innovation boundaries: The formation and maintenance of research links between industry and universities in Australia*. Canberra: AGPS.

10 The Incubator Movement

A Comparative Analysis of Brazil and Mexico

Mariza Almeida, Branca Terra, and Maria Del Pilar Montserrat Perez Hernandez

INTRODUCTION

The use of incubators has become a global phenomenon in terms of the organization of support structures for the creation and development of companies. There are currently some five thousand incubators throughout the world (Knopp 2007). In both Brazil and Mexico, incubators were created along the lines of other countries' organizational models, but their implementation has taken place according to the constraints imposed by local realities. According to Etzkowitz, Mello and Almeida (2005), incubators originally sprang up in the United States as a support structure geared towards aiding individual or groups of entrepreneurs in the commercialization of research via the creation of new companies. In Brazil and Mexico, incubation has been used to address other needs, such as transferring knowledge produced by universities to the private sector, increasing the technology level of already existing companies and creating new businesses or developing business clusters (Etzkowitz 2008; Lalkaka and Bishop 1996, Smillor, Gibson and Dietrich 1990). In both countries, incubators have also been used as an instrument of policy to diminish poverty through professional training and the creation of income-generating opportunities for socially and economically marginalized groups (Pereira 1998; Singer 2002).

Brazil and Mexico are thus strategic *loci* for research into the experience of incubator implantation. The two countries are quite different in this respect. Both began their incubator projects at different times, following different paths, and yet both have produced a similar number of functioning incubators. In 2007, Brazil had 440 incubators in action, including technological, traditional, mixed, cooperative and cultural institutions.[1] Mexico, meanwhile, had four hundred incubators which included high technology, intermediate technology and traditional businesses.[2]

In Brazil, public policies geared towards supporting the creation of incubators were first put in place in 1998, following the demand that arose from the social movement led by the National Advanced Technology Enterprise Promoter Entity (Anprotec). The *Programa Nacional de Apoio a Incubadoras de Empresas do Ministério da Ciência e Tecnologia* (National Program for the Support of Enterprise Incubators of the Science and Technology Ministry) was created in that year by the Brazilian federal government. There were sixty incubators that were already in operation in the country in 1998. This represents a significant growth in number since 1986, when the first two incubators were created. The evolution of the incubation process as an instrument of national policy in Brazil was thus organized in a bottom-up fashion, via the universities and in a context of the rebirth of civil society following the end of the military dictatorship.

The consolidation of the enterprise incubation system in Mexico can be set to span two periods. The first period, which began in the early 1990s, involved mainly academic initiatives with limited reach and impact on the academic structures within which they evolved. This period, referred to as the 'take-off' stage, began with the creation of the country's first incubator in 1990. In 1992, the Incubator of Base Technology Enterprise Program (PIEBT, for their initials in Spanish) was founded by the National Council of Science and Technology (CONACYT, for their initials in Spanish). The PIEBT folded five years later partly because it was found wanting and partly due to the restructuring of the agenda to tackle the country's economic crisis. The incubators were dependent on public funds; once the PIEBT disappeared, most of the incubators created during this period went out of operation (Pérez and Suchil 2005).

The second period, which began around 2001 and continues to this day, is based on public policies which are directed to promote enterprise incubators systematically. This has allowed the proliferation, development and success of incubators based on the 'Mexican model' that had spread to other Latin American countries, such as the Dominican Republic, Peru and Bolivia. This second period, called the 'growth' stage, involves a significant change in policies with respect to the promotion of technological innovation. The new policies recognized the role that small and medium-sized companies play in technological innovation and set in motion a phase where the creation and promotion of innovative enterprises is the strategy's main objective. Several mechanisms and programs have been designed to drive innovative activities within production plants through a combination of federal resources, coordinated with the cooperation of other organizations, such as the Mexican-United States Foundation for Science (FUMEC, for their initials in Spanish; Pérez and Márquez 2006).

The data utilized in the present chapter's bibliographical and documentary analysis is the result of different research projects undertaken by the authors.[3] The research regarding Brazil also includes thirty semi-structured interviews with incubator managers (in technological, mixed, traditional, cooperative and private incubators), Anprotec (Brazilian Association of Science Parks and Business Incubators), directors, representatives of regional incubator networks and directors of science and technology organs at the federal, state and municipal levels. A significant amount of information was also gathered at events organized by Anprotec and the Rio de Janeiro Incubator Network in 2000–2004.

In the case of Mexico, the data was collected from research conducted by National Polytechnic Institute (IPN, for their initials in Spanish), including a technical bibliographic review, a database analysis of incubators and interviews with employees involved in the incubation system as well as incubator managers. The results of this research have been presented at different fora (such as the XI Seminar ALTEC, 1st Ibero-American Congress of Science, Technology and Society, etc.).

Our comparison between incubators in Brazil and Mexico is developed from the analysis of three aspects, which are considered to be relevant for the exhibition of the different trajectories and aspects which make up both national cases. The analysis focuses on the political and institutional contexts which gave rise to the incubator movement, the types of incubators created and the incubation models which were adopted.

The remainder of this chapter is organized in three sections. The section following this introduction discusses the triple helix model. This is followed by a detailed and comprehensive discussion and analysis of the incubator movement both in Brazil and Mexico. The conclusions are given in the final section.

THE TRIPLE HELIX MODEL

One of the main analytical models used in the study of innovation is the triple helix. This is understood to be a spiral process which is developed through negotiations and reflexive interactions created by the intersection of three institutional spheres: the university, government and enterprise. The triple helix model postulates four levels of transformation among the institutions in each sphere: internal transformations in the role of each institution; changes incurred by the influence of one sphere on another in the course of their interactions; the creation of new structures engendered by the intersection of the three helices; and, finally, the recursive effect of these three levels on social institutions such as science itself (Etzkowitz 2008).

Depending upon the dynamics of the relationship between academia, private enterprise and the government, a variable series of configuration may occur in the formation of the triple helix system. Triple Helix I—or the *laissez-faire* model—presumes an institutional separation between the three spheres, with well-defined borders. Triple Helix II—also known as the *etatistic* model—occurs in those cases where the state also encompasses enterprise and the university. Finally, Triple Helix III presumes a superposition of the three spheres with the creation of hybrid institutions or networks. These three configurations generate distinct kinds of interaction and insertion between the component institutional spheres and these, in turn, generate significant theoretical and practical implications (Etzkowitz and Leydesdorff 2000). The foundations of the triple helix thus include room for both 'top-down' and 'bottom-up' initiatives.

Within this model, the locus for research progressively moves from the company's research and development (R&D) laboratories into the university sphere (Etzkowitz 2008). The increase in this sort of activity leads to the creation of an entrepreneurial culture within the university which, in turn, impacts upon both research and teaching. New organizational structures then arise, such as technology transfer offices, spin-off firms, technological parks, incubators and other such mechanisms (Etzkowitz et al. 2000). Within the structures of the triple helix model, company or cooperative incubators are seen as hybrid organizations which internalize the relationship between academia, enterprise and the state, stimulating the creation of an interactive space which connects these spheres (Etzkowitz 2008).

THE INCUBATOR MOVEMENT IN BRAZIL AND MEXICO

The Political, Scientific and Social Context in Brazil and Mexico

Both Brazil and Mexico experienced significant changes which occurred in their respective political environments during the 1980s which, in turn, were reflected in changes in public policy, in general, and in national scientific and technological policy, in particular.

The first incubators were founded in Brazil in the mid-1980s, a decade marked by major changes in the country's political and cultural life. The 1980s saw the return of democracy in Brazil, with direct elections for governors and mayors and increased pressures for direct presidential elections and a definitive end to the military dictatorship. These changes, in turn, created new social actors and marked the return of civil society to the governing sphere. The science and technology policy initiated by the Brazilian government after the Second World War and maintained by the dictatorship was motivated by national security concerns and sought to achieve technological autonomy in sectors related to national defense by undertaking research investment and training in related areas. This meant that, for the first time, research activities were institutionalized in Brazilian public (and certain Catholic) universities as part of a large-scale graduate school–based program. The private sector was generally excluded from this endeavour and did not benefit from the technologies created by the public laboratories and universities. This meant that private firms generally had to acquire mature technologies from overseas sources (Coutinho and Ferraz 1994).

At the end of the 1970s, Brazil experienced a severe economic crisis which combined deep external debt with a high rate of inflation (Dahlman and Frischtak 1993). Macroeconomic instability, lack of resources to finance technological projects and changes in industrial policy led to the government's abandonment of the technological autonomy project. During the final phase of the Brazilian military regime, the dictatorship was gradually phased out and, in 1982, state elections took place with the opposition winning a clear majority. The university community, supporting the return of democracy but hobbled by the military government, participated in the formulation of political platforms during the state elections. At the same time, science and technology policy began to be redefined, in particular with regards to the traditional role played by the universities in private sector endeavours and in the development of action plans.

At the federal level, CNPq (*Conselho Nacional de Pesquisa Científica e Tecnológica*—National Council for Scientific and Technological Research) established programs which focused on technology, seeking to protect the technology developed in national universities and research institutions, as well as identifying demands for technology transfer to the productive sector. In 1984, four months before the end of the military regime, the *Programa de Implantação de Parques de Tecnologia* (Technological Park Implementation Program) was created. Twelve projects were implemented in several different states. An analysis of the Brazilian experience of technological parks by Santos (1987) and Pereira, Emmerick and Pereira (1987), as well as other more recent studies (Gomes 1999) conclude that the program was marked by political-institutional fragility which left it vulnerable to initiatives taken at all levels of governments (municipal, state and federal), resulting in modest results which in no way reached its declared objectives. It was also under-supported and under-funded.

In 1986, the military regime finally came to an end in Brazil, with the return of a full democracy. With the reorganization of civil society on democratic lines, groups and individuals could now express themselves freely without fear of persecution. This led to the foundation of a new set of political forces. This new political environment made it possible for science and technology policies to be implemented based upon the triple helix model, as had already begun to occur in incubators, even though other areas of the society had not yet adopted the same model.

The incubator proposal originally arose as an alternative to the installation of technology parks, as it needed fewer resources and could be relatively easy to set up by state governments. The first two incubators in Brazil were founded in 1986—the first in São Carlos in the state of São Paulo at the initiative of a newly elected state government (Souza and Garcia 1999), and the second in the state of Santa Catarina by the UFSC (*Universidade Federal de Santa Catarina*—Federal University of Santa Catarina) and the state government of Santa Catarina (Cavagnari 1987). Incubators are generally presented as a new instrument for the transfer of technology. As such, the existing social and intellectual structure of Brazilian academia, combined with the difficulties of implementing the military government's innovation policies during the economic crisis and civil society reorganizations of the early 1980s, opened the way for a new economic and social mission involving the training of individuals and organizations.

In the case of Mexico, political power was held by one political party for seventy years; and it was not until 2000 when the democratic transition began with presidential elections. This situation created high expectations among Mexicans, in general, and among members of the country's scientific and technological community, in particular. Public policy was consequently reassessed with the result that science and technology policy had to focus on the needs of the industry and the society. The industrial policy that drives entrepreneurial and industrial policy was also changed, especially the one directed to the small and medium enterprises, Programa Especial de Ciencia y Tecnología—Special Program of Science and Technology (PECYT; Pérez et al. 2006).

The results obtained from different countries based in incubation programs and the problems generated by structural changes in the Mexican economy created the right conditions for new incubation initiatives during the last decade of the twentieth century. These were consolidated in the 2001 Mexican Economic Policy with the enforcement of the Law for the Development of Competitiveness in Micro, Small and Medium Enterprises and, the 2001–2006 Sectoral Program for Enterprise Development and the Fund for the Small and Medium Business (Fondo de Apoyo para la Micro, Pequeña y Mediana Empresa; Pérez et al. 2006). As in the case of many other Latin American countries, the main goal of the Mexican economic policy in the first half of the twentieth century was industrialization. From the technological point of view, this meant that most production systems were imported from overseas. During this period, technological research did not keep up with the needs of industrial development and led to dependence on foreign technology. All this soon crystallized as a structural characteristic of the Mexican industrial economy.

It was only in the 1970s, with the creation of CONACYT, that the endogenous development of science and technology emerged as a relevant objective that was addressed in public policies geared towards improving national competitiveness (Álvarez 1995; Solleiro 1994). Technological research and development activities were mainly funded by public sources during this period. Beginning in the 1980s, the development of research capabilities was focused on the public sector. Research activities were therefore attempting to deal with problems that were of little concern to the dominant industrial system of that time.

With the economic crisis of the 1980s, the subsequent structural adjustment policy of the 1990s and the implementation of the North American Free Trade Agreement (NAFTA) in 1994, the existing productive chains in the Mexican economy ruptured and—in some cases—disappeared altogether. This also caused adjustments in the prevailing technology and innovation policy, which involved, for example, changes in patent laws, which led to the reduction of incentives for local technological development and increased the technological dependency of local industry on foreign companies. This allowed more flexibility for the interaction between university and industry, and also enforced new requirements aimed at improving the competiveness of local firms. Simultaneously, in the 1990s, a new science and technology model emerged which was oriented towards the generation of knowledge that was demanded by economic agents. This model was set in motion, however, with low expectations and achieved limited results. In 2000, this new orientation was made explicit in public policy, with the creation of a plan that integrated several institutional reforms, new laws and programs (such as the PyMES Law) aimed at promoting and supporting innovative firms capable of producing goods with high added value.

In 2004, the Fund for the Small and Medium Business began its operation as a program implemented by the Office of the Secretary of the Economy, and set up support systems for innovative firms. The Fund used different mechanisms, such as 'risk capital', 'angel capital', 'business accelerators' (TechBa) and entrepreneurial programs to achieve its goals, and also set the basis for the National Incubation System (Pérez and Marquez 2006).

A Typology of Incubators

In the science and technology field in Brazil, the emergence of proposals for altering public policy led to the creation of the first incubators. In conjunction with this, the directors of the technology parks (many of whom were associated with or had originated in universities) created Anprotec in 1987, a nationwide movement that was not, however, represented by the academic institutions (Anprotec 2002). Although Anprotec's original focus was on technology parks and poles, the organization began to take a particular interest in incubators, which are seen as a different form of organization with its own identity and needs.

The relationship between Anprotec and the business sector, which is currently administered by Sebrae (*Serviço de Apoio à Micro e Pequena Empresa*—the Support Service for Small and Micro Enterprise), began in 1988 during the elaboration of Brazil's new Federal Constitution[4] with a discussion of articles relevant to small companies. The enterprise incubator concept began to be applied to the creation of new companies in Brazil during the 1990s in traditional sectors of the country's economy. These incubators are called 'traditional incubators'. The first one was founded in 1991 by a private organization, the FIESP (*Federação e Centro das Indústrias do Estado de São Paulo*—São Paulo Federation of Industries of the State of São Paulo), under the auspices of a program which would organize several incubators in various cities of São Paulo state. According to Lalkaka and Bishop (1996, xii, 49–50), this initiative was exceptional in global terms, as the private sector began to invest in incubators only *after* the FIESP program began to show positive results. In this context, then, the creation of traditional incubators represents more than the creation of a new type of incubator in Brazil. It also represents the moment in which an alliance between academia, business and government was first formed in the country in order to meet common developmental objectives. The formal entry of the Sebrae system into the field of incubator support began in 1991 and involved Sebrae activities in conceiving, structuring and gradually consolidating incubator projects (Silva 1997).

A third type of incubator existing in Brazil is the cooperative incubator. The first of this type of incubator, the *Incubadora Tecnológica de Cooperativas Populares* (Technology Incubator for Popular Cooperatives) was founded at Coppe/UFRJ in 1995 in a movement which saw several different civil entities uniting to create a series of actions aimed at reducing poverty: *Ação da Cidadania contra a Fome, a Miséria e pela Vida* (Citizen Action against Hunger, Misery and for Life).[5] In this context, a *Comitê de Entidades Públicas no Combate à Fome e pela Vida* (Committee of Public Entities in the Struggle against Hunger and for Life—COEP) was set up in order to reflect upon and mobilize the social responsibilities of public institutions and organizations.

This initial discussion led to a concrete proposal to incubate projects for economic intervention and the generation of income and work, directed at the socially, economically, culturally and spatially excluded sectors of the Brazilian populace. The model adopted here is similar to the one previously used by Coppe in the incubation of technological companies.[6] Under the auspices of this proposal, the university would receive support from COEP, the *Fundação Banco do Brasil* (Banco do Brasil Foundation) and FINEP (Financiadora de Estudos e Projetos—Project and Studies Financing Agency).

Figure 10.1 shows that incubators started being created in Brazil in 1987. The growth rate of the Brazilian Incubators Movement was slow in the beginning, averaging 30 per cent over 1997 to 2005 and from 2007 to 2008. This growth was a consequence of the interest developed by local and state governments to create traditional incubators in order to increase jobs and generate incomes in the regions.

Figure 10.2 highlights the two phases of the incubation project in Mexico. In the first 'take-off' phase, there were very few incubators. However, in the second or 'growth' phase, the number of incubators increased dramatically after the creation of the Fund for the Small and Medium Business in 2004. The Fund's strategy seeks to further consolidate the process by creating an ever-greater number of linkages within the incubator system and between industry and local environment.

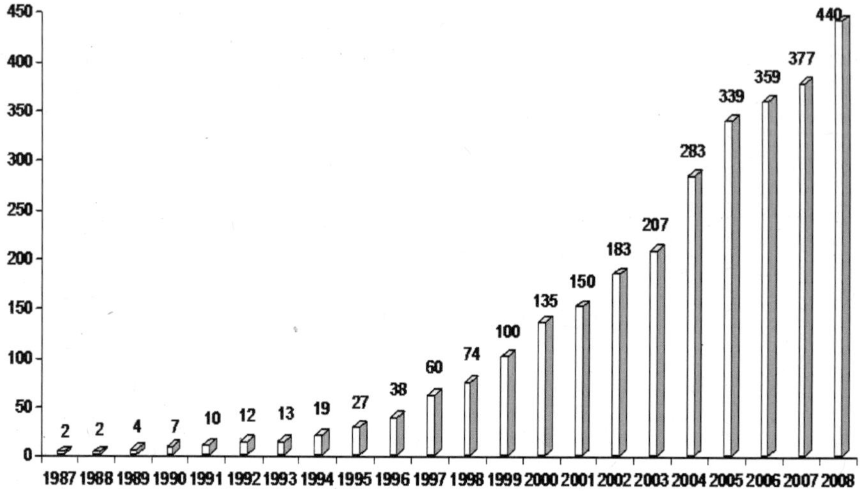

Figure 10.1 Growth of incubators in Brazil.
Source: Based on data collected by authors.

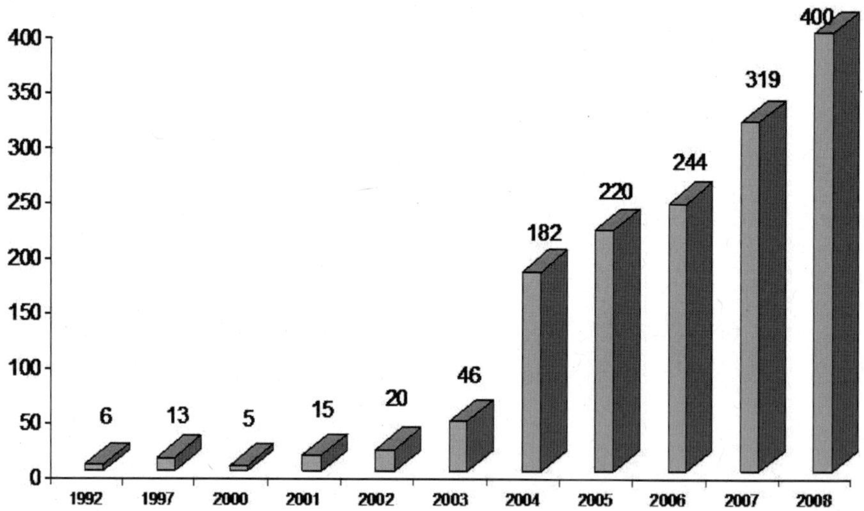

Figure 10.2 Growth of incubators in Mexico.
Source: Based on data collected by authors.

Models of Incubation

Through Anprotec, Brazilian incubators have developed a widespread system of personnel training which covers such diverse themes as the incubation process, models for incubator gestation, indicators and evaluation methods, entrepreneurialism and risk capital, among others. The individual incubators in the system are autonomous and are free to develop their own incubation models. Incubation methodology is elaborated via Anprotec publications and documents which disseminate the best practices collected from various incubators (Etzkowitz, Mello and Almeida 2005).

The redesigned incubating strategy in Mexico emerged in 2001 when the lack of high-technology companies and the relative abundance of micro-, small- and medium-sized companies in the country were recognized. Another element which influenced the current strategy was derived from the results of the Best Practices in Incubation Forum, in which national and international experiences regarding enterprise incubation were exchanged. Several successful models which had promoted the creation of a large number of incubators were recognized at this conference. This was followed by the Fund for the Small and Medium Business defining mechanisms, modalities, schemes and available resources for the creation, support and consolidation of micro-, small- and medium-sized enterprises.

Currently, a large number of diverse incubating models coexist in Mexico. They differ in terms of technological intensity and the sectors which they seek to serve. Seventeen models of traditional incubators have been recognized as have twelve models of intermediate technology incubators, five of the high-technology incubator models, four models for incubating tourism and agricultural business and six for incubating enterprises. The IPN's model (named Robust Incubation Model [RIM]) covers all types of incubation. However, of the four hundred existing incubators in Mexico, 75 per cent currently use the RIM model (Pérez and Súchil 2005).

The focus of Mexican incubation strategy is on creating environments and mechanisms to start-up firms that can achieve self-sustainability. This focus is also aimed at permeating the educational system by means of the entrepreneurial program. Another characteristic of the Mexican incubating process is its national coverage, its confluence with local and regional social sectors and its participation in international cooperative institutions. The main strategy was to give incentives to promote entrepreneurial culture and to commit local agents to the innovation process, creating productive vocations throughout the country.

CONCLUSIONS

Although Mexico started creating incubators after Brazil, it has adopted a centralized public policy under government management. In the seven short years that followed, Mexico has created a number of incubators which are similar to those produced by Brazil over a period of twenty years.

The incubator creation initiatives in both countries arose following changes in science and technology policies and the creation of new organizational structures. Mexico and Brazil currently support several similar types of incubators. However, the incubation models and processes adopted by these countries are different. In addition, what this chapter has sought to show is how the Brazilian incubator movement led to the renewal of the country's technology park project, which was interrupted after its initial implementation in the mid-1980s. In Mexico, on the other hand, the success of the incubation program and its methodologies has resulted in the diffusion of this knowledge throughout several other Latin American countries.

Looking at the political context of science, technology and innovation, we find that Brazil launched a support program only for companies with a wider technological base in 2007. Called the Inovar Project,[7] this program seeks to support the creation of risk capital companies in Brazil. The approval of the 2004 Innovation Law and its implementation in 2005 made it possible for the federal government to channel resources to those companies involved in developing technological innovation, stimulating the creation of Technological Innovation Nuclei. In Mexico, these types of laws were approved and programs were initiated in tandem with the federal program of support for incubators.

Taken as a whole, the Mexican policy is much more interconnected than the Brazilian. Although an incubator support program exists in Brazil, most work along these lines is done by the incubator movement itself, led by Anprotec. The association directors started playing an entrepreneurial role on various fronts, disseminating the concept, convincing universities and institutes to participate directly in the movement and, at a broader level, meeting and persuading representatives of various institutions to support the incubators (Pereira and Bermúdez 1997; Anprotec 2002). In this manner, Anprotec has a decisive influence over the dissemination of this concept, stimulating its growth from the bottom up and developing an incubator movement whose nature is more associative than government policy.[8] This association is quite influential on the national level, intermediating between the institutions which make up the triple helix and the demands of the incubators themselves in an attempt to plan activities which support the association's own developmental policy. For this reason, we can consider the Brazilian incubator movement to be the result of a bottom-up approach, with each individual incubator linking itself to the institutions of the triple helix in order to guarantee its sustainability. This makes the Brazilian experience a typical example of the triple helix model Type III (Almeida 2004). In Mexico, government policy induces the creation of relationships between the helices, defining the goals and results which incubators and their associated enterprises should strive for. As to the Mexican experience, it can be defined as characteristic of the triple helix model Type I.

The success of the Brazilian incubator movement and, in particular, the creation of high-technology enterprises within the university environment, gave new life to the technology park concept, especially after 2003, when these began to receive resources for their implantation in the different institutions which make up the triple helix. This was not a possibility during the 1980s. Furthermore, in Mexico, the success of the government's incubator program, the establishment of incubation models and the results of the enterprises created by these led to the exportation of this model to other

Latin American countries, which later began to establish their own incubators. The types of incubators seen in Brazil (cooperative, cultural, etc.) and Mexico are similar, although they also demonstrate different models and processes of incubation.

NOTES

1. This number was obtained through two sources: Anprotec (2007)—which counted four hundred operating incubators (with at least one company successfully incubated); and PRONINC (2007)—which counted forty cooperative incubators.
2. See http://www.contactopyme.gob.mx/snie/ExportaDirectorioIncubadoras.asp (accessed 30 April 2008).
3. One co-author previously analysed the Evolution of Brazilian Incubator Movement (Almeida 2004; Etzkowitz, Mello and Almeida 2005) and the Cooperative Incubator Movement in Brazil (Almeida 2006, 2007). The second co-author has analysed the Mexican incubation system linked to technological firms (Pérez and Marquez 2006; Pérez and Súchil 2005). The third co-author has been working on incubators in Brazil and United States (Terra 2000).
4. Almeida, interview with Luiz Afonso Bermudéz, president of Anprotec, 26 March 2001.
5. Souza, Herbert, 'Idéias sobre a Conferência da Fome organizada pelo Banco Mundial em novembro de 1993'. http://www.ibase.org.br/paginas/bmundial.html.
6. Testimony of Professor Luiz Pinguelli Rosa, Vice-Director of Coppe/UFRJ, in *Ossos do Ofício cooperativas populares em cena aberta*, Rio de Janeiro, May 1998.
7. Projeto INOVAR.
8. Almeida, interview with Alberto Sampaio Aranha, director of Anprotec and manager of the Gênesis Incubator at PUC/Rio, 9 October 2001.

REFERENCES

Almedia, M. 2006. The incubator as organisational training method. In Ethiopia Triple Helix Conference, 2006, Addis Breba. Transforming Organisation for Knowledge Economy and Enterprise Development, vol. 1, 165–179.

Almedia, M.; Mello J. M. C.; Elizkowitz, H. 2007. Incubators for economic and social development: The Brazilian experience. Paper presented at 6th Triple Helix Conference, 2007, Singapore.

Almeida, M. 2004. 'A Evolução do Movimento de Incubadoras no Brasil'. PhD thesis, COPPE, UFRJ, Brazil.

Álvarez, J. 1995. 'Experiencias de vínculos entre instituciones de educación superior, centros de investigación y desarrollo tecnológico y el sector industrial en México'. In *Aspectos Tecnológicos de la Modernización Industriales de México*, ed. P. Mulás del Pozo, 296–316. Fondo de Cultura Económica. Hexico.

Anprotec. 2002. *15 Anos: Como o Brasil Desenvolveu um dos Mais Importantes Movimentos de Incubadoras e Parques*. Brasília: Anprotec.

Cavagnari, G. 1987. 'A Experiência do Certi'. Paper presented at *Seminário Internacional de Parques Technólogicos, Rio de Janeiro, Annals* 307–312.

Coutinho, L., and J. Ferraz. 1994. *Estudo da Competitividade da Indústria Brasileira.* Rio de Janeiro: Papirus.

Dahlman, C. J., and C. R. Frischtak. 1993. 'National systems supporting technical advance in industry: The Brazilian experience'. In *National innovation systems: A comparative analysis*, ed. R. Nelson, 414–450. New York and Oxford: Oxford University Press.

Etzkowitz, H. 2008. *The triple helix: University-industry-government innovation.* New York and London: Routledge.

Etzkowitz, H., and L. Leydesdorff. 2000. 'The dynamics of innovation: From national systems and mode 2 to a triple helix of university-industry-government relations'. *Research Policy* 29 (2): 109–123.

Etzkowitz, H., J. M. C. Mello and M. Almeida. 2005. 'Towards "Meta-Innovation" in Brazil: The evolution of the incubator and the emergence of a triple helix'. *Research Policy* 34 (4): 411–424.

Etzkowitz, H., A. Webster, C. Gebhart and B. Terra. 2000. 'The future of the university and the university of the future: The evolution of the ivory tower to entrepreneurial paradigm'. *Research Policy* 29 (2): 315–330.

Gomes, E. 1999. 'Polos tecnológicos y promoción del desarrollo: ¿Hecho o artefacto?' *Redes Revista de Estudios Sociales de la Ciencia* 14 (7): 177–216.

Knopp, L. 2007. *2006 state of the business incubation industry.* Athens, OH: National Business Incubation Association.

Lalkaka, R., and J. Bishop. 1996. *Business incubators in economic development: An initial assessment in industrializing countries.* New York: UNIDO.

Pereira, M. G., and L. A. Bermúdez. 1997. 'Parques Tecnológicos e Incubadoras de Empresas em países em Desenvolvimento: Lições do Brasil'. In *A Economia dos Parques Tecnológicos*, ed. M. G. Pereira and P. Fórmica, 147–159. Brasília: Anprotec.

Pereira, M. G., M . C. E. Emmerick and R. S. Pereira. 1987. *A Experiência Brasileira.* Paper presented at *Anais do seminário internacional de parques technólogicos, Rio de Janeiro: NSY WPPE UFRJ Annals*: 108–157.

Pereira, R. S. 1998 'A criac¸˜ao de Cooperativas como instrumento de geracão de trabalho e renda: o caso da Incubadora Tecnológica de Cooperativas Populares da Coppe/UFRJ'. Paper presented at *VIII Seminário Nacional de Parques Tecnológicos e Incubadoras de Empresas,* 265–277. Brasília: ANPROTEC.

Pérez, P., and A. Márquez. 2006. 'Análisis del sistema de incubación de empresas de base tecnológica de México'. Paper presented at the Primer Congreso de Ciencia, Tecnología y Sociedad, Mexico.

Pérez, P., R. Oliver, H. Merrit, A. Márquez and J. L. León. 2006. 'El emprendedor en México: ingenio vs innovación'. Paper presented at the Primer Congreso de Ciencia, Tecnología y Sociedad, Mexico.

Pérez, P., and O. Súchil. 2005. 'Caracterización del Sistema de Incubación de Empresas de Base Tecnológica del Instituto Politécnico Nacional'. Paper presented at the X Seminario Ibero-Latinoamericano de Gestión Tecnológica, Brazil.

Santos, S. A. 1987. 'A Experiência Brasileira: O caso de São Paulo'. Paper presented at *Seminário Internacional de Parques Technólogicos, Rio de Janeiro. Annals* 73–107.

Silva, M. L. 1997. 'Incubadoras de Empresas: atuação do Sistema Sebrae', in VII Anais do seminário nacional de parques technólogicos e incubadoras de empresas, Salvador, Anais, Brasília: 12–21.

Singer, P. 2002. 'A recente ressurreição da economia solidária no Brasil'. In *Produzir para viver: os caminhos da produção não capitalista*, ed. B. S. Santos, 81–129. Rio de Janeiro: Civilização Brasileira.

Smillor, R. W., D. V. Gibson and G. B. Dietrich. 1990. University spin-out companies technology stand-ups from UT-Austin. *Journal of Business Venturing* 5:63–76.

Solleiro, J. 1994. 'Gestión de la vinculación universidad-sector productivo'. In *Estrategias, planificación y gestión de ciencia y tecnología*, ed. Eduardo Martínez. Caracas, 403–429. CEPAL-ILPES/UNESCO/UNU/CYTED-D, Editorial Nueva Sociedad.

Souza, M. C., and R. Garcia. 1999. 'Sistemas Locais de Inovação em São Paulo'. In *Globalização & Inovação Localizada*, ed. J. E. Cassiolato and H. M. M. Lastres, 300–344. Rio de Janeiro: IBICT.

Terra, Branca. 2001. A Transferência at Tecnologia am Universidades papa a inovaqáo Tecnológica. Rio de Janeiro: Qualitmark.

11 The Challenges Arising in the Evolution of the Triple Helix Institutional System

The Case of Malaysia

Azley Abd Razak and Mohammed Saad

INTRODUCTION AND BACKGROUND OF STUDY

Malaysia's first formal science and technology policy was introduced in 1986 (National Science and Technology Policy II 2000–2010 [NSTPII] 2000). One of the major programs of this policy was the provision of greater opportunities for interactions between industry and public research institutions. Its aim was to increase economic development at a national level in tandem with support for Malaysia's industrialization program. Since its inception, the government has been proactive in seeking industry collaboration and cooperating in enhancing several economic and industrialization programs. For instance, a greater focus on research and development (R&D) projects that involve industry has been reflective of this new policy implementation (NSTPII 2000). The policy was originally formulated by the Malaysian National Scientific and Research Council, membership of which consists of both government and industry representatives (NSTPII 2000).

Due to macro environmental changes, namely developments in the economic, political and socio-cultural landscapes, the policies of the Malaysian government on innovation have subsequently evolved and developed. This change is perceived to be more flexible in meeting the needs of all key stakeholders in national industry growth, especially the three key tripartite partners: the government, industry and university research bodies. In fact, one of the main objectives of the Eighth Malaysia Plan (2001) was to encourage the introduction of the network form of collaboration between universities, government and industries (Eighth Malaysia Plan 2001). One of the government's responses to these needs was to start encouraging more R&D collaboration between government agencies, universities and private sector industrial bodies. This reaction by the government was supported by the view that more active participation by industry in R&D collaboration (with the government and academia) would improve the state's innovation and economic growth agenda for an increasingly turbulent and unpredictable global marketplace (Dodgson 2000; Martin and Scott 2000).

Indeed, this led to recognition of the role that universities could play in shaping and enhancing existing government and industry policies and programs (Abd Razak and Saad 2007). Malaysian universities have since been actively encouraged to participate more in R&D activities and to be entrepreneurial through knowledge transfer partnerships, for example, between industry and government bodies. In fact, some of the local public universities have set up, or are in the process of setting up, commercial arms of their own (namely, private holding companies). This idea of the 'entrepreneurial university', as stated by Etzkowitz and Leydesdorff (1997, 1999) and later reiterated by Leydesdorff (2000), is a unique feature of the triple helix model as a dynamic process and evolutionary model for enhancing innovation. The Malaysian government has also awarded research university status to four top national universities designated under the Ninth Malaysia Plan, to be the country's first full-fledged research universities (Abd Razak and Saad 2007).[1]

This study, which examines the challenges arising in the evolution of the triple helix institutional system in the context of the Malaysian socio-economic environment, comprises four key sections. The first section provides an overview of the existing literature on the triple helix model. The second discusses the evolutionary process followed by the triple helix system in general and more specifically in Malaysia. The third section investigates the issues and challenges facing Malaysian institutions in the course of their evolution as key actors within the implementation of the triple helix model. A fourth and final section concludes with a brief note on the way forward for the development of the triple helix system in Malaysia and other developing countries.

KEY FEATURES OF THE TRIPLE HELIX MODEL

Etzkowitz and Leydesdorff (1997, 1) first described the triple helix model as a spiral model of innovation able to capture multiple reciprocal linkages at different stages of the capitalization of knowledge. Etzkowitz and Leydesdorff (1997) argued that the triple helix model will be the key strategy of the national or multinational innovation agenda in the twenty-first century. The model is presented as the only feasible solution for intranational cooperation between the key drivers for national economic growth. This view was later echoed by Godin and Gingras (2000), who, using the success of the Canadian model, suggested that in the past two decades the Canadian government had profusely focused on the need to develop, promote and indeed integrate ties between universities and businesses in conjunction with national level policy. However, according to Saad (2004), the triple helix model is generally viewed as lacking strong theoretical and empirical bases. For example, Saad (2004) and Saad and Zawdie (2005) have questioned the very relevance of the role played by institutions within the triple helix system in knowledge generation and subsequently in the growth of national and regional innovation. Furthermore, they point to various key issues and challenges relating to successful implementation of innovation strategies based on the triple helix model in developing countries.

The dynamic nature of the triple helix model allows a flexibility and integrative function which becomes embedded in the interfaces of the interlocking helical relationships. For example, industry gains some of the values derived from universities through a sharing of research knowledge. This further encourages research links and collaboration between private firms and universities in order to deliver goods or services to the general public and society, thereby attaining mutual, common, long-term strategic goals of socio-economic growth (Etzkowitz and Leydesdorff 1997). Industries also develop an academic dimension and begin to appreciate the value of research, sharing and disseminating knowledge within their own cultures through, for instance, the provision of higher skill and knowledge levels. In addition to performing their traditional functions, the stakeholders of the triple helix also engage in enhancing the roles of the other spheres by practicing values derived from the helical partners (Etzkowitz and Leydesdorff 2001). This conjoint dynamic behavioural process model also means the long-term objectives become synergistic to each other and complement the partners. This partner coordination and cooperation is considered central to the success of the triple helix model.

Another main argument for the triple helix model is the emphasis on the importance of academia in the capitalization of knowledge. According to Etzkowitz and Leydesdorff (1997), in the existing dynamic macro environment, universities are changing their mission, establishing new relationships with industry and becoming more entrepreneurial. Etzkowitz et al. (2000, 326) define this new type of university as the 'amalgam of teaching and research, applied and basic, entrepreneurial and scholastic interests'. However, to fulfil this new role, universities are expected to obtain some business skills to effectively carry out activities, such as commercializing their research, setting up their own start-ups and having a concrete business plan (Abd Razak and Saad 2007). As highlighted by many authors, such as Leydesdorff and Etzkowitz (1998), Martin (2000), Harman (2002) and Calvert and Patel (2003), universities are increasingly undertaking entrepreneurial tasks and creating opportunities through business development.

On the whole, and as emphasized by Etzkowitz and Leydesdorff (1997), the triple helix model is about the interactions and relationships between university, government and industry. However, the organizing principle and its nature and form are inevitably different from one country, and indeed from one context, to another. For instance, a developing country like Malaysia might require a reduced role for the government and a greater role for other institutional bodies such as local public universities, whereas in others, a more active and enhanced role from the government is needed (Abd Razak and Saad 2007). In brief, according to Etzkowitz (2003), the triple helix model comprises three basic elements:

1. A prominent role for the university in innovation, on a par with industry and government in a knowledge-based society. In Malaysia, the government is striving to encourage universities to be actively involved in research and commercialization (*Sunday Star* 2007).[2]
2. A movement towards collaborative relationships among the three major institutional spheres, in which the innovation policy is an outcome of interactions among the spheres rather than a prescription from the government or an internal development within an industry. Several authors have attempted to highlight the complexity of collaborative relationships in the context of developing countries, such as Saad and Zawdie (2005) for Algeria; Sutz (2000) for Latin America; and Malairaja (2003) for Malaysia.
3. In addition to fulfilling their traditional functions, each institutional sphere also 'takes the role of the other' (Etzkowitz 2003, 309).

EVOLUTION OF THE TRIPLE HELIX INSTITUTIONAL SYSTEM

According to Etzkowitz (2003, 302), the transition of the triple helix model starts 'from two opposing positions: a statist model of government controlling academia and industry, and a laissez-faire model, with industry, academia, and government separate and apart from each other, interacting only modestly across strong boundaries'. The next stage in the evolutionary process is the hybrid form of triple helix in which each institutional sphere keeps its own distinctive characteristics and at the same time also assumes the role of the others. The signpost in the evolutionary process underlying the triple helix system is graphically depicted in Figure 11.1.

The major element for a statist triple helix element is that the government plays the major role 'driving' academia and industry while at the same time planning and controlling and managing activities aimed at encouraging innovation. Meanwhile, industry is regarded as the national champion whereas the university's role is reduced mainly to teaching and academic research (Etzkowitz 2003). The benefit of this model is that industry and universities will receive strong support and guidance from the government. However, with this model, neither government nor industry will be able to

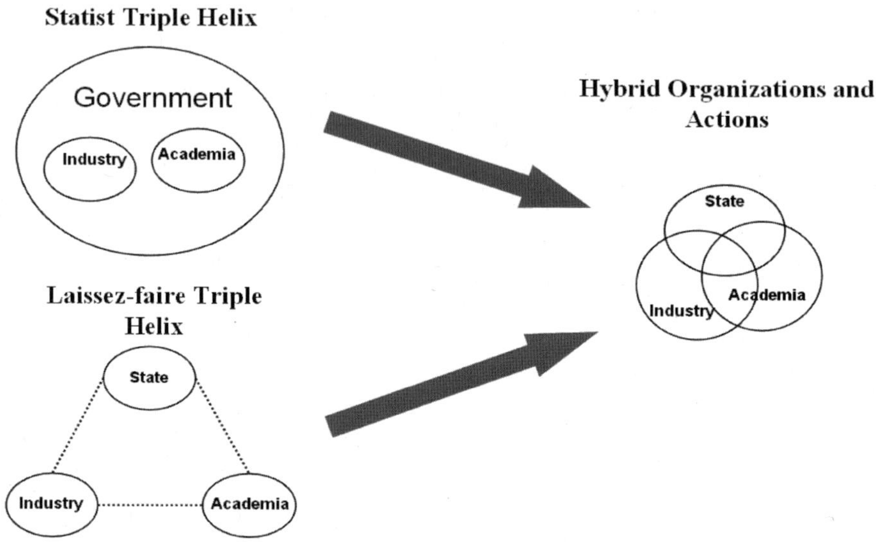

Figure 11.1 Evolution of the triple helix system.
Source: Etzkowitz (2003).

exploit the potential knowledge-generating activities within universities as both teaching and research tend to be far removed from industry needs and universities do not have any incentive to engage in the commercialization of research (Etzkowitz 2003).

As for the laissez-faire triple helix model, governments, universities and industry operate independently as separate institutional spheres (Etzkowitz 2003). The firms are expected to operate completely apart from each other in competitive relationships and are only linked through market forces. The government's role is limited to dealing with only those problems that can be regarded as market failures (Etzkowitz 2003). In this model, industry is the driving force, with the other two spirals acting as ancillary supporting structures. In the laissez-faire triple helix model, the individualistic mentality is more prominent and creates a type of 'heroic' entrepreneur. The advantage of this model is that industry will be able to grow without any undue interventions by the government. However, the main downside is related to the difficulty for the three institutional spheres to interact in a way that would maximize the synergy in the relationship, and therefore benefit from value inherent in each sphere.

A third transition is the hybrid triple helix model where each institutional sphere keeps its own distinctive characteristics while at the same time being able to assume the role of the other (Etzkowitz 2003). All three spheres within this model will be able to interact and collaborate with one another actively in order to promote strong innovation activities. All of them will gain values from each other which can help them achieve common long-term strategic goals. However, the problem with the hybrid triple helix is that its emergence as an institutional system could be elusive, as it involves a complex process based on high levels of commitment, understanding and trust between all three spheres (Abd Razak and Saad 2007; Saad, Zawdie and Malairaja 2008). The next section discusses the application of the triple helix model to the Malaysian context.

EVOLUTION OF THE TRIPLE HELIX INSTITUTIONAL SYSTEM IN MALAYSIA

The Malaysian government started to recognize in its second National Science and Technology Policy the increasing importance of the collaboration between universities, government agencies and private industrial partners (NSTPII 2000; Malairaja and Zawdie 2008). This is regarded as an attempt by the government to shift the national economy from a labour-intensive manufacturing base to a knowledge economy; therefore, in concordance with the demands of the turbulent and increasingly competitive environment of the twenty-first century. The government has, for instance, developed several means to promote strong links among universities, industry and government through funded programs such as the intensification of research in priority areas (IRPA), which is central to the completion of the Eighth Malaysian Plan (2001).

The Malaysian government started to also acknowledge the importance for industries and universities to be less dependent (on the government) and to become more active in promoting innovation. For example, the percentage of innovating firms in the manufacturing sector (35 per cent) has exceeded that of Portugal (26 per cent) and Spain (29 per cent; see Malaysian Science and Technology Information Centre [MASTIC] 2004). However, the rate of innovation in other European countries such as Ireland (74 per cent), Denmark (71 per cent) and Germany (69 per cent) was significantly higher than that of Malaysia. The survey conducted by MASTIC for the period 2000–2001 also indicates that, in general, large-sized firms are more innovative than the smaller firms, where they account for 25 per cent of the total number of innovating firms. However, the number of smaller firms active in innovation is, as illustrated in Figures 11.2 and 11.3, considerably higher than the corresponding figure of 1.8 per cent for the 1997–1999 period (MASTIC 2004). There is also clear evidence from Figures 11.2 and 11.3 that the percentage of innovating entrepreneurs (sole proprietorship) has increased dramatically from 0.5 per cent to 12.9 per cent.

As already discussed and highlighted in Chapters 7, 8 and 12, the triple helix model is essentially based on a prominent and more entrepreneurial role

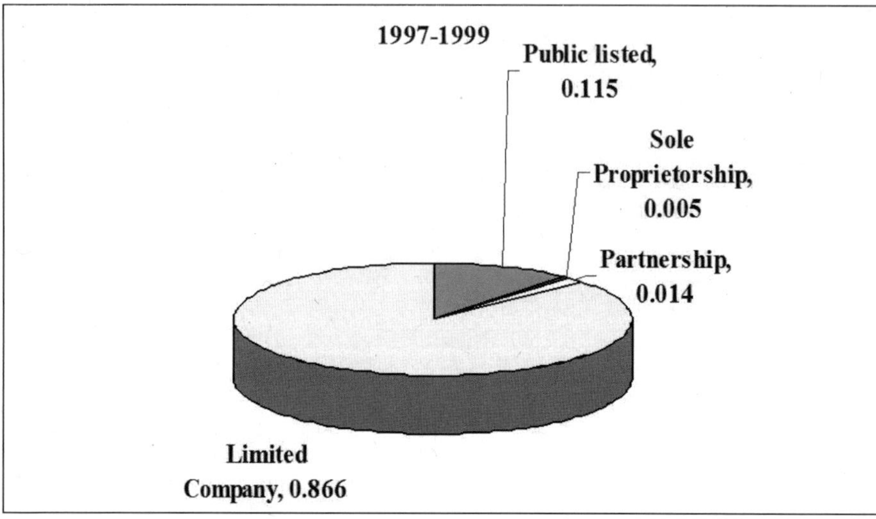

Figure 11.2 Ownership structure of innovating firms (1997–1999).
Source: MASTIC (2004).

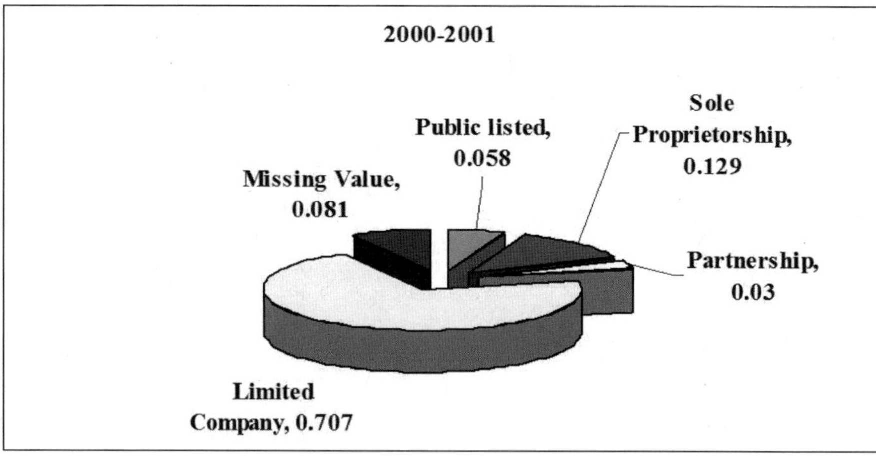

Figure 11.3 Ownership structure of innovating firms (2000–2001).
Source: MASTIC (2004).

for the university in the creation and diffusion of knowledge and innovation. There is, in the context of Malaysia, evidence of entrepreneurial behaviour as some of the local public universities have set up or are in the process of setting up commercial arms (a private holding company). An example of this is the establishment of USains Holdings (the university's commercial arm) by the University Sains Malaysia (Malairaja 2003). Furthermore, the Malaysian government has recently conferred research university status to its top four local universities which are designated under the Ninth Malaysia Plan to be the country's first full-fledged research universities.

Studies by Abd Razak and Saad (2007) and Malairaja and Zawdie (2008) have provided some further evidence suggesting that universities in Malaysia are being perceived by their partners as being of good quality, credible and capable researchers. The majority of the respondents in the interviews in this study stated that universities in Malaysia are determined to launch active relationships with the industrial sector. However, according to Abd Razak and Saad (2007), these relationships are more related to educational development, consultancy and training. The university is seen by respondents, from both the government and industry, as a key institution that can provide consultations and advice (mostly based on technical expertise). In fact, the government also uses universities to implement its policies (Abd Razak and Saad 2007). For example, the government is using the universities to implement its policies in increasing the number of researchers, IT workers and researchers for the country to achieve its Vision 2020[3] (MASTIC 2004). Table 11.1 provides a summary of how Malaysian universities are perceived by their partners. Further explanations regarding the table are given in the research methodology section.

Table 11.1 Current Perceptions of Malaysian Universities

Role of universities	**Status of universities**	**Credibility and capabilities**	**Procedures and structures**	**Culture of partnership within universities**	**Type of relationship with government and industry**
• Clear evidence of the desire to see the universities being more entrepreneurial. • Some universities are more entrepreneurial than others. • Establishment of the research universities. • A clear definition of *research universities* is required.	• Does not really reflect the quality of teaching and research institution. • The effectiveness of the relationship does not depend on the status of the universities.	• The local universities have capable researchers. • The main issue is about the flexibility and the attitude of the universities towards change. • Still lacking in capability and experience in strengthening the relationship between universities and industries.	• There is still some room for improvement about the efficiency of procedures within universities. • Need to reduce red tape.	• High competitiveness between universities and researchers that limited the culture of partnerships. • Significantly depending on the individual staff.	• Act as a consultant for both government and industries, especially in the area based on technical expertise. • Universities are places for government to implement the policies. • More active relationship in commercialization of experts such as education, consultancy, training and development.

Source: Abd Razak and Saad (2007).

RESEARCH METHODOLOGY

This research is achieved through the use of a qualitative case study approach which, according to Yin (1994), is an empirical inquiry of a specific contemporary phenomenon within its real-life context. Such an approach is useful where the boundaries between the phenomenon and the context are not clearly evident (Yin 1994), as featured in the Malaysian context. For the purpose of this study, semi-structured interviews were used to elicit information from managers.[4] The interview questions were semi-structured as they can include specific and closed questions in order to obtain specific information or confirm facts about open-question answers. They were also aimed at encouraging the interviewees to provide an extensive or developmental answer which may be used to obtain deep and rich data (Eisenhardt 1989).

The samples for the interviews were taken from the three spheres of the triple helix: the government (government ministries and agencies), universities (researchers, deputy vice-chancellors and staff from research management centres) and industries (managers and executives). All of the eighteen respondents constituting the sample were carefully selected to represent the position of stakeholders in the evolving triple helix institutional culture in Malaysia. The university respondents were from seven different universities of which four are categorized as research universities. From the government sphere, there were seven respondents in total, of which five were from different government agencies[5] and the other two from the Ministry of Science, Technology and Innovation (MOSTI). There were four respondents from industry: two from local and two from international multinational corporations (MNCs).

The data was analysed using the thematic analysis approach. It was later transcribed, from which several patterns were identified. According to Aronson (1994), from interviews, the ideas and patterns can be better understood by using thematic analysis. Therefore, it is relevant to this study to use thematic analysis as it is focusing on a developing country and aimed at better understanding the role of each actor (in the triple helix model) and the relationships between them. According to Taylor and Bogdan (1984) and Leininger (1985), these patterns can be later identified as the main themes. The next step was to catalogue the sub-themes or, according to Ritchie and Lewis (2003), create a thematic index. This has led to a clear identification of a pattern expanding from this process which was followed by the development of a valid argument justifying the choices of themes (Aronson 1994). This approach is based on greater links between findings from the literature review and the data collected. It leads to facilitating of the analysis and bringing a greater degree of validity to the study (Ritchie and Lewis 2003).

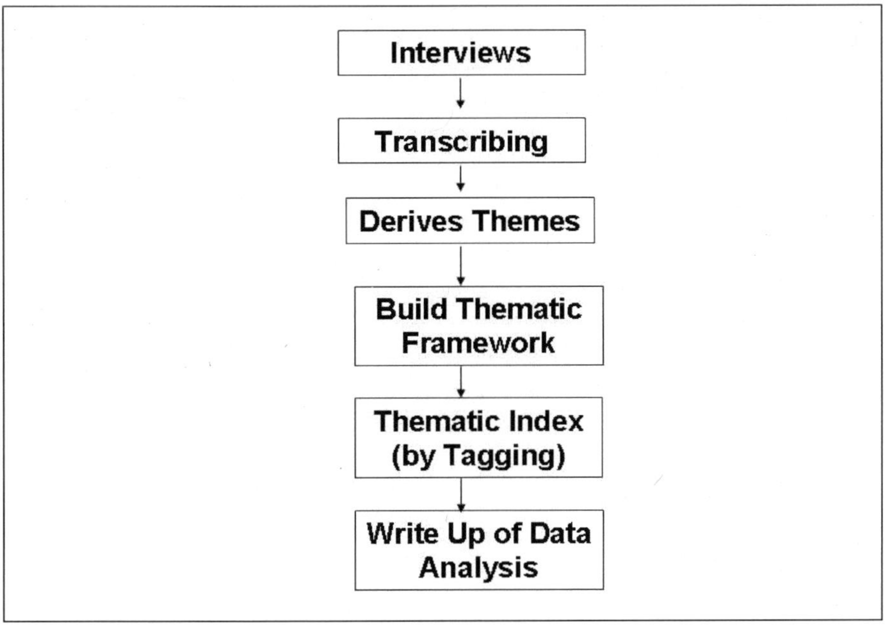

Figure 11.4 The process of data analysis.
Source: Derived from Ritchie and Lewis's (2003) approach.

LINKS WITHIN TRIPLE HELIX INSTITUTIONAL SYSTEM IN MALAYSIA

To evaluate the type and quality of links existing within the triple helix institutional system in Malaysia, the discussion is divided into two subthemes which investigate the current situation and views on effectiveness of the links in Malaysia.

All the respondents suggested, and agreed, that links between the three key spheres exist in Malaysia. According to a senior general manager of a commercial arm of one of the local universities, there are several government agencies such as the Invest Penang and PSDC[6] striving and working towards improving the relationship between university and industry. The type of collaboration (in this particular case, for an MNC) as highlighted by a senior manager of manufacturing and training is more inclined to develop programs on training and educating staff at universities. He elaborated that the company has given a monthly salary for lecturers to conduct training programs especially related to fibre optics as the industry is very new. For Sharma, Garg and Wani (2004), it is the university's role to prepare and make available trained manpower with the right type and level of knowledge, attitude and skill base needed by the industry. Furthermore, effective utilization of resources and interactions between industry and university is crucial for the improvement of their respective efficiency, productivity and quality.

In the case of the biotechnology industry, the responses were quite encouraging. However, the respondents found it difficult to define the level of effectiveness and the types of collaboration achieved between the university, industry and government in Malaysia. As mentioned by a general manager of a government agency, this difficulty is related to a lack of studies on the triple helix relationships in Malaysia.

Having acknowledged that the relationships between university-industry-government exist, all respondents believe that there is still significant room for improvement and it is not yet at a satisfactory level, as suggested by the deputy vice-chancellor of a local university: 'It is moving, but we need a more speedy progress to deliver this especially with RM9[7] activities'. This view, as illustrated in the following quote, was also echoed by a senior government officer: 'We need the improvement, from our Seventh and Eighth Malaysian Plan, therefore, in our Ninth Malaysian Plan, hopefully it will be more about commercialization, demand and supply'.

The current nature of the links is essentially related to the technical training, education and professional development, especially with MNCs. This type of relationship is also identified by a senior manager of an MNC who claimed that they have hired some university lecturers even though they do not have hands-on experience on knowledge transfer, training and developing new products.

The preceding discussion highlights the following three major findings regarding the implementation of triple helix relationships between key institutions in Malaysia:

1. Links between the university, industry and government exist; however, their level and effectiveness remain uncertain.
2. There is still significant room for improvement although for certain industries like Biotech the links are blossoming.
3. The current nature of the links is more about the technical training, education and professional development, especially with MNCs.

CHALLENGES ARISING IN THE EVOLUTION OF THE TRIPLE HELIX INSTITUTIONAL SYSTEM IN MALAYSIA

The analysis of the interviews identified the following seven main groups of challenges which influence the introduction of the triple helix approach within Malaysian universities. The findings of this study are consistent with several other studies. Saad (2004), for instance, highlighted the need for effective governance by the developing countries. This is certainly true for Malaysia, especially with the administration of universities. The issues are summarized in the following with supporting verbatim evidence from respondents.

Technological and Human Resources Factors

Malaysian universities are lagging behind in terms of expertise in the development of new and advanced technologies, such as fibre-optic technologies. As emphasized by a senior manager of an MNC, most Malaysian universities are not well prepared and do not even have the resources and equipment to develop new and innovative technologies. This was also noted by Schramm (2004), who argues that research, especially in life sciences, computing and engineering, has not been successfully commercialized largely due to bureaucracy and lack of applied skills and resources in universities.

All the respondents also highlighted the lack of human resources in science and technology development as an ongoing issue for Malaysia. A resolution from the Malaysian Science and Technology Convention (MASTEC) identified, in 2003, several issues related to human resources development in the country. These issues, which include the low supply in skills development, inappropriate policy instruments to encourage in-house employee training and the inadequate supply of lecturers and instructors in skills development, were perceived to be the main causes for the lack of human resources in science and technology. The number of researchers, despite an increase from 15.5 to 18.0 researchers per 10,000 between 2000 and 2002, remains relatively low when compared to international standards (MASTIC 2004).

Policies, Procedures and Processes within the University

Universities are faced with the dilemma and debate about the balance between teaching and research responsibilities. According to a fellow researcher working for an MNC, universities are having problems managing resources and time between research and teaching.

Universities also face the issue of inflexibility of staffing policies that prevent them from hiring the right staff for the right jobs. One of the main difficulties is related to the lack of experts in technology transfer, which is crucial for the relationship to ensue with the industrial sector. Furthermore, according to a senior manager of a university's commercial arm, these factors have contributed to the quality and speed of work of the university, as the staff is only allowed 40 per cent of its time for consultancy. It takes academic staff two years to do what the full-time staff can do in one year.

Commercialization Issues

Commercialization has always been identified as a key challenge for the effective establishment of university-industry-government relationships. A lack of entrepreneurs and of support from industry (for funding) seems to be central to poor commercialization. As highlighted by the director of a government agency, 'a critical mass of experienced and skilled entrepreneurs is needed both within the industry and university spheres'. Entrepreneurs wanting to commercialize their product to medium-sized companies may perhaps be interested in purchasing the foreign technology rather than relying on local universities to develop it. More funding and support (from industry) is needed. However, the existing regulations and practices related to the funding and loan applications remain very rigid and need to be changed and improved.

The preceding factors are mainly related to the issue of links between academic and industrial actors, as suggested by Danell and Perrson (2003), for whom a region needs a critical mass of interacting academic and industrial actors in order to develop technologically and prosper in economic terms.

Relationship between the Main Stakeholders and Spheres

On the whole, there are issues of misperceptions related to the nature of the actual relationship and its parameters and boundaries. The industry perceives university staff as being 'too theoretical' (senior manager of MNC) and not having 'sufficient practical knowledge about real-life situations' (senior government officer of a government agency). According to a senior government officer, this lack of exposure and knowledge about industry has affected the nature of their relationship and the type of collaboration with their industrial and academic partners. Furthermore, universities appear to be unable to prioritize the strategic research areas which correspond to the industry's interests. It is evident that the engineering and science faculties are receiving more demand for collaboration from industry (through consultancy and research projects) as opposed to the social science faculties. However, it can be risky and inappropriate for this type of relationship to emerge only in certain faculties and not in others.

The perception of universities by the industry also affects the relationships between the university and its main partners. For instance, small and medium enterprises (SMEs) often perceive universities as 'ivory towers'. A significant number of industry managers still consider overseas researchers are much more capable than locals. There is also a feeling that no formal procedures are embedded in order to harness a continuity that will promote long-term relationships between universities, government and industry. The relationship is very much based on, as one respondent put it, a 'need-by-need basis'.

Meanwhile, several respondents from industry highlight that it is industry that should create the initiatives, and that they require help and guidelines from the government about handling the academia–industry relationship.

Work Cultures

The apparent difference in work cultures between the universities, industry and the government was raised by several respondents as an important issue in managing the interrelationships in the tripartite model of the helix model. One of the main issues in work cultures is the response time (timeliness) especially between the universities and industry. The CEO of a state investment agency highlighted that timeliness is the biggest barrier in the relationship. According to a senior manager of one MNC, it is simply because universities and their staff are not ready, or do not fully comprehend the work culture of industry.

This difference between cultures is also raised by a senior manager of a university's commercial arm, for whom 'industry wants things very quickly at their pace of time, however, we (the university) always insist that we need to take into consideration our teaching responsibilities, which is the major duty for Malaysian academics, whether we like it or not.'

Intellectual Property (IP)

There is also evidence that the Malaysian government is planning to set up a broad intellectual property (IP) policy. However, some concerns about the clarity of this policy and its guidelines have been expressed by representatives from key stakeholders from the university, government and industry spheres. Current IP policies and practices are seen as significantly affecting the nature of relationships between the universities, industry and the government. This view is clearly echoed by the following quote from one senior government official:

> IP is the most important barrier. In Malaysia we don't have a clear IP guideline. There is like fifty-fifty partnership. However, for both the industry and universities, they are still confused with the IP requirements. It is because the IP awareness in Malaysia is still low. We basically need clearer guidelines.

Government Policies

It was noted by several authors, such as Gibbons (1998), that government policy is one of the important factors contributing to the change of roles in universities. This is quite evident in Malaysia, with the introduction of government initiatives to establish research universities. However, the inflexibility, inconsistency, rigidity, vagueness and lack of direction in government policies are examples of the main challenges stressed by respondents from all parties. One senior government officer suggested that the country needs clear direction and the ministries should communicate with one another to make sure there is a focus in terms of direction and policies.

The issues related to the government policies are summarized by the senior government officer:

> There are so many policies such as education policy and science technology policy. For example, now we have fewer than sixteen scientists per ten thousand people. It is our aim to achieve twenty scientists per ten thousand. However, we are not clear on what do we have to do in order to achieve this aim. That is, as a result of lack in clarity [as] every ministry is doing its own things.

CONCLUSION

From the findings and discussion in this chapter, it is obvious that the Malaysian government continues to behave as the dominant sphere in the development of knowledge and innovation. Both academia and industry remain dependent upon the government in terms of overall coordination, objectives, planning and resources. However, the respondents from universities cited in the study raised concerns regarding clarity from government sources. This, as discussed earlier, corresponds to a statist triple helix model. However, the universities and industry are still operating separately and they are only linked through the market need which is a key feature of a laissez-faire triple helix (Etzkowitz 2003).

Although all the institutions have made an effort to move the triple helix relationship into the next transition, there are still key issues and challenges that need to be overcome. For example, efforts should be made to address issues such as commercialization, IP policy and differences in work cultures. It is clear that the situation has been improving for the past ten years with the introduction of policies aimed, to a certain extent, at strengthening the role of universities and their links with their immediate external environment. However, there is no doubt that the government, industry and universities need to work closer with one another. All respondents agree that there is still a long way to go and a strong commitment is clearly required from all parties for the successful implementation of the triple helix institutional system in Malaysia. This study has provided seven key areas which can form a platform for interrelational development between the partners for the Malaysian context and therefore provide a public policy agenda for the successful implementation of the triple helix model.

NOTES

1. Each university received RM50 million (U.S.$15 million) for research, development and commercialization activities (*Sunday Star* 2007).
2. *Sunday Star* is the Sunday edition for the Malaysian daily newspaper *The Star.*

3. Vision 2020 is a program dedicated to Malaysia's aspiration to be a fully developed country by the year 2020.
4. According to Saunders et al. (2000), managers in developing countries are more likely to agree to be interviewed than to complete questionnaires.
5. The five agencies are:Invest Penang (a state investment agency), Malaysian Technology Development Corporation Sdn Bhd (MTDC), Small and Medium Industries Development Corporation (SMIDEC), Malaysian Agriculture Research and Development Institute (MARDI) and Kulim High Tech Corporation (owned by a state government).
6. Penang Skills and Development Centre.
7. Ninth Malaysian Plan.

REFERENCES

Abd Razak, A., and M. Saad. 2007. 'The role of universities in the evolution of the triple helix culture of innovation network: The case of Malaysia'. *International Journal of Technology Management and Sustainable Development* 6 (3): 211–225.

Aronson, J. 1994. 'A pragmatic view of thematic analysis'. *Qualitative Report* 2 (1). http://www.nova.edu/ssss/QR/BackIssues/QR2–1/aronson.html (accessed 1 June 2007).

Calvert, J., and P. Patel. 2003. 'University–industry research collaborations in the UK: Bibliometric trends'. *Science and Public Policy* 30 (2): 85–96.

Danell, R., and O. Persson. 2003. 'Regional R&D activities and interactions in the Swedish triple helix'. *Scientometrics* 58 (2): 205–218.

Dodgson, M. 2000. *The management of technological innovation: An international and strategic approach*. Oxford: Oxford University Press.

Eighth Malaysia Plan. 2001. '2001–2005 economic planning unit, prime minister's department, Malaysia'. http://www.epu.jpm.my (accessed 20 August 2008).

Eisenhardt, K. M. 1989. 'Building theories from case studies research'. *Academy of Management Review* 14 (4): 532–550.

———. 2003. 'Innovation in innovation: The triple helix of university-industry-government relation'. *Social Science Information* 42 (3): 293–338.

———. 1997. *Universities in the global knowledge economy: The triple helix of university-industry-government relations*. London: Cassell Academic.

———. 1999. 'The future location of research and technology transfer'. *Journal of Technology Transfer* 24 (2–3): 111–123.

———. 2001. 'The transformation of university-industry-government relations'. *Electronic Journal of Sociology* 5 (4). http://www.sociology.org/archive.html (accessed 1 May 2007).

Etzkowitz, H., Webster, A., Gebjardt, C., Cantisano Terra, B. R. (2000). 'The future of the university and the university of the future: evolution of Ivory Tower to entrepreneurial paradigm: Research Policy 29(2): 313–330.

Gibbons, M. 1998. *Higher education relevance in the 21st century*. London: ACU.

Godin, B., and Y. Gingras. 2000. 'The place of universities in the knowledge production'. *Research Policy* 29 (2): 273–275.

Harman, Grant. 2002. 'Australian university–industry research links: Researcher involvement, outputs, personal benefits and "withholding" behaviour'. *Prometheus* 20 (2): 143–158.

Leininger, M. M. 1985. 'Ethnography and ethnonursing: Models and modes of qualitative data analysis'. In *Qualitative research methods in nursing*, ed. M. M. Leininger, 33–72. Orlando, FL: Grune and Stratton.

Leydesdorff, Loet. 2000. 'The triple helix: An evolutionary model of innovations'. *Research Policy* 29 (2): 243–256.
Leydesdorff, L., and H. Etzkowitz. 1998. 'The triple helix as a model for innovation studies'. *Science and Public Policy* 25 (3): 195–203.
Malairaja, C. 2003. 'Learning from the Silicon Valley and implications for technological leapfrogging—the experience of Malaysia'. *International Journal of Technology Management and Sustainable Development* 2 (2): 73–95.
Malairaja, C., and G. Zawdie. 2008. 'Science parks and university–industry collaboration in Malaysia'. *Technology Analysis and Strategic Management* 20 (6): 727–739.
Malaysian Science and Technology Information Centre. 2004. *Malaysian science and technology indicators 2004 report*. Ministry of Science, Technology and Innovation (MOSTI) KL.
Martin, Michaela, ed. 2000. *The management of university–industry relations: Five institutional case studies from Africa, Europe, Latin America and the Pacific Region*. Paris: International Institute for Educational Planning, UNESCO
Martin, S., and J. T. Scott. 2000. 'The nature of innovation market failure and the design of public support for private innovation'. *Research Policy* 29 (4–5): 437–447.
National Science and Technology Policy II 2000–2010. 2000. 'Building competitiveness in a knowledge-driven economy'. Unpublished manuscript, Academy of Sciences, Malaysia.
Ritchie, J., and J. Lewis. 2003. *Qualitative research practice: A guide for social science students and researchers*. Thousand Oaks, CA: Sage.
Saad, M. 2004. 'Issues and challenges arising from the application of innovation strategies based on the triple helix culture: Experience of the incubation system in Algeria'. *International Journal of Technology Management and Sustainable Development* 3 (1): 17–35.
Saad, M., and G. Zawdie. 2005. 'From technology transfer to the emergence of a triple helix culture: The experience of Algeria in innovation and technological capability development'. *Technology Analysis and Strategic Management* 17 (1): 1–15.
Saad, M., G. Zawdie and C. Malairaja. 2008. 'Triple helix strategy for universities in developing countries: The experiences in Malaysia and Algeria'. *Science and Public Policy* 35 (6): 431–443.
Schramm, C. J. 2004. 'Accelerating technology transfer and commercialisation'. Paper presented at IP Commercialisation and Research Spinouts Conference, Boston, MA, 4 November.
Sharma, S. K., T. K. Garg and V. P. Wani. 2004. 'Effective industry/institute interaction for developing entrepreneurial vision amongst engineers for the sustainable development of SMEs in India'. *International Journal of Technology Transfer and Commercialisation* 3 (1): 38–55.
Sunday Star. 2007. 'Boost for research'. 19 August, 4.
Sutz, J. 2000. 'The university-industry-government relations in Latin America'. *Research Policy* 29 (2): 279–290.
Taylor, S. J., and R. Bogdan. 1984. *Introduction to qualitative research methods: The search for meanings*. New York: John Wiley and Sons.
Yin, R. 1994. *Case study research, design and method*. London: Sage.

12 Innovation in a Developing Country Context

Developing an Integrated Framework through India's Experience

Surja Datta and Mohammed Saad

INTRODUCTION

A recent World Bank report starts with the following claim: 'India is increasingly becoming a top global innovator for high tech products and services' (Dutz 2007, 15). The paradox of India which is embedded in this claim eludes the author of the report. In the same report, it is pointed out that India is a country where approximately 90 per cent of Indian workers are still employed in the informal sector. Spread of higher education is still severely limited amongst the working population and 70 per cent of the workforce is still employed in the agricultural sector. The report rightly sees the preceding as key challenges to India's ambition of sustaining and improving this excellence in IT and IT-based services but the puzzle is how India has reached such a position *despite* these challenges.

Traditional explanations have focused on reforms that took place in the early 1990s and also on the cheap availability of skilled labour in India. Such explanations do have some merit but it is the argument of this chapter that they do not get to the heart of the matter. The rapid transformation of India from a primarily agrarian economy to a post-industrial economy based on services and one that has enjoyed considerable success in exporting knowledge services to developed nations cannot be solely explained on the basis of economic policy changes or the low cost of human resources. This chapter suggests that of the many factors that have contributed to India's success, two stand out. The reason for focusing on these two factors is not only because they are central to any explanation of India's success, but also because they have hitherto not been highlighted in traditional explanations of this success.

One of these factors has to do with the nature of post-industrial innovation that India has excelled in; the other relates to the particular social structure created by the trans-national circulation of skilled labour, in which certain categories of service firms in India find themselves embedded.

The case of India is an interesting one from the innovation perspective for two main reasons. Firstly, as already alluded to, the rise of India as a top innovator of knowledge-based services in the face of considerable challenges is a phenomenon that is worthy of detailed investigation. Secondly, national success in innovative activities have been primarily explained through the concept of national innovation systems (Lundvall 1992; Nelson 1993) or the triple helix thesis (Etzkowitz and Leydesdorff 2000), which has focused on the national institutions of industry, government and university. The case of India provides an opportunity to evaluate these theories in the context of a developing nation that has excelled in service innovation.

The chapter is structured around three main sections. In the first section, a discussion which explores the relation between innovation and social capital is carried out. This section reviews existing literature on social capital and its role in innovation and attempts a new synthesis, the outcome of which is an explanatory framework that sees innovation as recombination of various forms of capital. In the second section, a critical evaluation of theories of national innovation and the triple helix thesis is carried out, where it is suggested that such theories do not adequately represent the reality of innovation in developing countries. The missing link, it is argued in this section, is the concept of transnational social circulation, which can be seen as a form of social capital arising out of the circular movement of skilled labour across national boundaries. In the last section, through the case study based on India's experience in provision of informational services, it is illustrated how transnational social circulation has played a major role in fostering innovation in the country.

FORMS OF CAPITAL AND ITS RELATION WITH INNOVATION

Innovation has been thought of in different ways. Most academics agree that it was Schumpeter (1934) who first brought the idea of innovation as an engine for economic growth to the forefront (Fagerberg, Mowery and Nelson 2005). For Schumpeter, innovation is essentially a novel recombination of existing resources brought about by the leadership and vision of the entrepreneur. By introducing new combinations of existing resources, entrepreneurs earn excess rents from the market, which is ultimately eroded by imitators who copy the new way of doing things. This conception of innovation is quite broad; but this is necessary as otherwise the scope of investigation is unnecessarily restricted. As Lundvall (1992) points out, innovation can happen in products as well as processes. Innovation can consist of the form of introducing novel products into the market but also in a new way of organizing activities. So innovation is ubiquitous—it's happening everywhere and all the time.

Innovation did not start with the Industrial Revolution but it is reasonable to argue that the pace of innovation did rise significantly during this period. As Schumpeter pointed out, it was in the industrial age when financial capital took centre stage in the innovation process. Access to financial capital often became a prerequisite to be in the innovation game, and the people who dealt with financial capital became entrepreneurs in their own right (Schumpeter 1934).

Financial or economic capital is only one of the different forms of capital that innovators can possess. Adapting from Bourdieu (1985) and others (Schumpeter 1934; Lin 2001; Becker 1962; DiMaggio 1997; Burt 1992), capital can be conceptualized as human, social, cultural and economic. These capitals are fungible; that is, it is often possible to convert one form of capital to another, but it can be useful to see them as distinct from one another and understand their roles in the innovation process. This section attempts a classification of different forms innovation and capital for this purpose.

FORMS OF CAPITAL

Concepts of what constitutes capital have been changing since Adam Smith. Early economists thought of labour as a factor of production, much like land, rent and material. The productivity gain came mainly from the division of labour which allowed the labour to specialize. A capitalist was seen as a person having capital that allowed him to hire and purchase factors of production including labour. The concept of capital in this era was essentially economical. Marx, more so than the economists before him, highlighted the role of economic capital and suggested that ownership of economic capital can lead to exploitation of labour, which increased profits for the capitalists at the expense of the labour. This monolithic conception of capital started changing in the 1960s with some economists realizing that human resource itself can be seen as capital. Knowledge and skills of the labourer allow the bearer of the knowledge to transform himself into a capitalist (Becker 1962; Schultz 1960). The concept of capital underwent further expansion in the 1970s and 1980s, when a host of academics started actively discussing social capital, the idea that social relations impart value to the bearer of these relations (Coleman 1990; Bourdieu 1985; Lin 2001; Burt 1992; Granovetter 1973). Around the same time, sometimes interchangeably with the idea of social capital, the concept of cultural capital came into prominence. The assertion here is that values, norms and beliefs embedded in particular cultures can be put to productive uses by people embedded in such cultures (DiMaggio 1982). This concept of cultural capital has its origin in Max Weber's seminal work *The Protestant Ethic and the Spirit of the Capitalism*, which has been subsequently elaborated, modified and expanded by scholars such as Bourdieu (1985) and DiMaggio (1982).

Economic Capital

Many equate capital with economic capital and although this is, as already highlighted, an important form of capital, it is by no means the only one which matters in economic development in general, and in innovation in particular.

Economic capital for a firm or an entrepreneur can be financial or resources that can be easily converted into money. Soto (2000) has shown that resources owned by individuals can be unevenly converted into money across different countries. Individual property taken for granted as a form of economic capital in the West may turn out to be not so readily convertible into capital in other parts of the world. Formal representation of a resource that is recognized by the members of a social system (which may be a nation) converts the resource from a dead asset to live capital. Thus we can see how firms and entrepreneurs across nations will have differential access to economic capital that is not only dependent on the resources they possess but also on what kind of representational system exists in the country for those resources.

Human Capital

Human capital can be thought of as the value embodied in skills and cognition of individuals. Skills and knowledge can be acquired while performing activities, that is, learning-by-doing, and they can also be acquired through education, which can be achieved by attending formal organizations such as schools and universities but also from the family environment in which an individual is brought up. Economists have recognized for some time that human capital can be an engine for economic growth and have been forcefully arguing for higher investments by government in education (Schultz 1960; Becker 1962).

Social Capital

Social capital has been conceptualized in different ways. The individualistic conception of social capital is advocated by scholars such as Burt (1992) and Lin (2001), who suggest that social capital is the value that resides in social relations shared by individuals with one another. Here the focus is on social ties that individuals may have and on the position of the individual in specific social structures because they can profit from such relations and positions. The other conception of social capital is more collective, where it is suggested that social capital can be seen as an indicator of trust and community spirit within societies. This view is usually accompanied with an associated suggestion that the more social capital of this kind there is, the better it is for society (Putnam 2000; Fukuyama 1999). For the purposes of this research, the individualistic conception is more useful as it looks at capital as something that is valuable to entities, who may be firms or entrepreneurs. The concept of social capital is further elaborated later in the chapter as it is central to the themes of triple helix and national systems of innovation.

Cultural Capital

The concepts of cultural and social capital have sometimes been used interchangeably to mean similar things. Weber (1978) was amongst the first to put culture at the forefront of economic analysis. He made the useful distinction between 'class'—a person's position in the social structure—and 'status'—a person's social esteem in a social group. Whereas 'class', according to Weber, is an impersonal concept, 'status' is intrinsically linked with a person's participation in a defined group. Scholars have found it hard to pin down the concept of cultural capital (DiMaggio and Mohr 1985), but perhaps the best description of cultural capital again goes back to Weber who characterised it as a 'style of life . . . expected from all those who wish to belong to the circle' (1978, 932).

It is perhaps stating the obvious that all forms of capital are social constructs. The form of capital that seems most objective and quantifiable—which is economic capital—is at the heart of it a social reality and has an ontology that is very different to that of gravity,[1] for example. Human capital, that is, knowledge and skills of individuals, is also context specific to different social systems. Social and cultural capital are sometimes considered as more abstract forms of capital and are found even more difficult to measure compared to human and economic capital. The social ontology of different forms of capital, however, does not make them less 'real'. These different forms of capital affect the innovation process in a fundamental way, which will be discussed in the next section.

MIX OF CAPITAL IN INNOVATIONS

The different forms of capital—economic, human, social and cultural—are influenced by each other and at the same time they are fungible; that is, one form of capital can often be converted into another form. From the perspective of the triple helix model, one of the assertions of this chapter is that innovation inevitably needs a mix of these different forms of capital. Different eras have required different configurations of capital. As Schumpeter (1934) pointed out, in the industrial era, economic capital and those who dealt in it came to be particularly relevant for innovation, but even those innovations that require substantial economic capital to come into being need other forms of capital, such as human, social and cultural. For Schumpeter, the centrality of economic capital to the innovation process arises from the fact that resources are already engaged in productive activities. A recombination of these resources requires an incentive to disengage them from their present activities and redeploy them in new ones. Economic capital is thus useful to create incentives for re-combinatory activities.[2]

For other innovations, the economic capital may be a very minor part of the capital mix; other forms such as social or cultural being dominant. As Rogers (1995) showed, a simple innovation such as boiling water to prevent spread of disease in a village in Peru did not require great economic capital to succeed. However, the program initiated by the government of Peru failed to make substantial impact as the social capital of the change agents (government representatives) was not sufficient and the cultural capital of the villagers made them associate boiled water with ill health. In more recent times, one has seen innovations such as Wikipedia, Linux and Apache. Wikipedia, the online encyclopaedia, has many more entries than the traditional *Encyclopaedia Britannica*, with arguably similar quality levels (Giles 2005); however, what is striking is that Wikipedia is entirely produced by the voluntary contributions of people across the globe. The main ingredients in innovation of Wikipedia are human, social and cultural capital. Similar is the case of Apache, a Web server software, and Linux, an operating system which competes with Microsoft. In Table 12.1, a list of major innovations is provided along with the configuration of capital mix that is associated with each innovation.

Two things need to be highlighted from Table 12.1. Firstly, as it has often been pointed out (Lundvall 1992; Fagerberg, Mowery and Nelson 2005), innovation is often incremental and cumulative. What appears as radical and discontinuous is often built upon an innovation that preceded it. Wikipedia would not have been made possible without the Internet. Moreover, the configuration of capital mix differs across the evolution of innovations. Whereas economic capital played a significant role in the innovation of the Internet, it played only a peripheral role in the innovation of Wikipedia. Weber has argued that capitalism which is an innovation in its own right has much to owe to the cultural capital that the 'Protestant Ethic' instilled in its followers.

Table 12.1 Examples of Relationships between Innovations and Types of Capital

Innovation	*Forms of Capital*			
	Economic	*Human*	*Social*	*Cultural*
Internet	Initial investment by U.S. Defence	Knowledge of scientists involved with ARPANET	ARPANET and other networks of scientists constitute the social capital	Culture of open sharing of knowledge contributed to the success
Wikipedia	Economic capital invested by producers is minimal	Knowledge of volunteers is a key ingredient	In the beginning, the contributors and moderators formed a social group which was co-located	Culture of the founder members was of open sharing of knowledge and also of open access
Global procurement of knowledge-intensive services	Access to financial capital is relatively less important	Access to human resources with relevant knowledge is central to the process	Social ties, both direct and indirect, are a key resource to establishing initial trust amongst exchange partners	Cultural capital of the entrepreneurs often plays a significant role

Source: Adapted from Benkler (2006), Von Hippel (2005), Datta and Saad (2008) and Tuomi (2002).

In this section, it has been shown how different innovations require different configurations of capital mix. In the next section, innovation is understood not from the perspective of individual entrepreneurs or firms but from a system perspective. Both the micro and macro perspective is important as innovation does happen in a vacuum—it is often incremental, and the environment in which an innovation takes place is often a key enabler. It is thus essential to understand how different institutions, national and otherwise, influence the innovation process.

INNOVATION FROM A SYSTEM PERSPECTIVE

The system perspective on innovation suggests that innovation is intrinsically linked with new knowledge creation and learning and such learning and knowledge creation takes place within specific social systems (Lundvall 1992). As some social systems are more conducive to knowledge creation than others, it is important to understand the key features of such social systems, as such an understanding can lead to policy-making that encourages innovations in societies (Nelson 1993; Lundvall 1992). The challenge, however, is to decide *which* social systems to study. How do we draw boundaries around different social systems that invariably intersect and overlap with one another? One way to get out of this conundrum is to draw boundaries across social system along national lines. National actors such as government, industry, university and research laboratories arguably make up discrete social systems and can be studied for this purpose.

This approach can be justified on two counts. Firstly, the economic performance of nations has differed quite considerably from one to the other, and, as innovation is a key engine for economic growth, it makes sense to study social systems based on national boundaries. Secondly, one of the main purposes of studying social systems from the innovation perspective is to make policy decisions which will affect the way actors within the systems interact with each other with the objective of improving the innovation capacity of the system. As national governments are usually in a position to make and implement such policy decisions, it is useful to limit the inquiry to national systems. It must be recognized, however, that demarking social systems in such a fashion is not without its problems. In today's globalized world, actors who are involved in an innovation often transcend national boundaries. Users, producers and suppliers interact across national boundaries and it is only in the most insulated nations that innovation or, more likely, the lack of it can be explained solely through national actors. Scholars who advocate the study of the national innovation system agree that demarking social systems on national boundaries has its limitations but maintain that it is fruitful to study social systems in such a fashion (Edquist 2005; Lundvall 1992; Nelson 1993). Actors who are essentially national in character, for example, governments and universities, have been central to the innovations witnessed in certain nations. The role of the government in Japan in successfully boosting the innovation capacity of the nation has been frequently cited in this context (Lundvall 1992; Freeman 1992).

It would be fair to say that a national innovation system perspective is less of a theory with definite propositions and more of a broad framework that lends itself well to the consideration of innovations happening in society and the carrying out of empirical research after formulating specific research questions. In the programmatic book *National Systems of Innovation*, Lundvall states the following:

> we may make a distinction between a system of innovation in the narrow sense and a system of innovation in the broad sense. The narrow definition would include organisations and institutions involved in searching and exploring—such as R&D departments, technological institutions and universities. The broad definition . . . includes all parts and aspects of the economic structure and the institutional set up affecting learning as well as searching and exploring—the production system, the marketing system and the system of finance present themselves as sub systems in which learning takes place. (1992, 12)

It is useful to elaborate the concept of institutions here as it is often interchangeably used to denote organizations and the 'rules of the game'. According to North (1990), institutions are the rules of the game, which include both formal rules such as legal rules set out by the body politic and intellectual property rights amongst others, and informal rules such as the norms of behaviour. Lundvall (1992) also views institutions from this perspective. Here organizations, such as the government, firms and universities, are actors who interact with each other within the constraints of formal and informal institutions. One way innovation can be encouraged is to make policies that facilitate interaction between organizational actors such as governments, universities and the industry. The key to innovative society is continuous interactive learning amongst organizational actors and institutions that can go a long way either to facilitate or hinder such a process. It is important to note that the relation between institutions and innovation goes both ways; that is, institutions influence innovation activities and are themselves influenced by innovation.

The triple helix thesis also looks at national organizational actors such as universities, industry and government and strives to understand how their interrelationships affect the innovation capacity of nations (Etzkowitz and Leydesdorff 1997). The model is built on the idea that universities take on a leading role in directing national and/or regional economic development, through academic entrepreneurial activities (Etzkowitz and Leydesdorff 1997). According to Etzkowitz and Leydesdorff (1997, 1), the triple helix model is a 'spiral model of innovation, which is able to capture multiple reciprocal linkages at different stages of the capitalization of the knowledge'. The triple helix thesis highlights the role of the university in the innovation process.

THE MICRO AND MACRO INNOVATION PERSPECTIVE IN A DEVELOPING COUNTRY CONTEXT

In the first part of this section, a more individual perspective of innovation is presented. It is discussed how the entrepreneur's (or the firm's) capital, which is a prerequisite for innovation, can be seen as a mix of economic, human, social and cultural capital. In the second part, a system perspective of innovation will be discussed that focuses on how national institutions have been found effective in promoting innovation. In this section, the linkage between the two perspectives is developed. It should be clear from the preceding discussion that a holistic explanation of innovations which takes place in a society would need to incorporate both perspectives. Innovations can depend to a large extent on the vision of the entrepreneur or on the research and development (R&D) activities of firms (Schumpeter 1934). The capital mix of the firm or of the entrepreneur is essential for innovation to take place. On the other hand, innovations are also dependent, to a large extent, on the institutional environment that organizations find themselves embedded in. As discussed in the preceding section, one of the main problems of the system perspective on innovation is the drawing of boundaries around social systems that are to be studied. The nation-state provides a convenient solution for this problem but this has its limitations, as discussed earlier. Theories of National Innovation Systems (NSI) and the triple helix thesis have also been studied primarily in a developed country context. NSI was originally developed and essentially based on the experiences of a small number of Scandinavian countries (Lundvall 1992) whereas the triple helix thesis draws heavily from experiences in Anglo-Saxon countries (Leydesdorff and Etzkowitz 2000). The over-reliance on the experiences of developed countries in generating a system theory of innovation has an obvious limitation. It is harder to identify causal relations when studying a system which has already been successful in innovation activities. Adopting a system perspective in a developing country context may be more useful in identifying key variables that lead either to success or lack of success in innovation.

One key variable that is likely to be different in a developing and developed country context is related to the movement of skilled labour. Developing nations often witness large-scale emigration of skilled labour. Human capital that is embodied in skilled labour is thus disengaged from national systems and attached to new social systems (in countries the skilled labour emigrate to). The large-scale emigration of skilled labour from developing nations has periodically given rise to fears of 'Brain Drain'; the idea that the vital intellectual resources of the country are lost due to such emigration.

However, there have been relatively very few studies that have systematically examined the effects of large-scale emigration on economic development. Those who have looked at the issue (Khadria 2002) find it more complex than a simple, reductive 'Brain Drain' problem. Whereas the initial loss of human capital is real enough, there is often gain from the Diaspora that is difficult to account for due to a variety of reasons. In the following section, a conceptual framework is developed that can explain one form of gain that may arise due to movement of skilled labour across nations.

THE MISSING HELIX: TRANSNATIONAL SOCIAL CIRCULATION

The emigration of skilled labour from developing nations has been a cause of concern for policymakers and academics for a considerable period of time. The reasons for these concerns are not hard to fathom. The productivity of nations is often dependent to a large extent on the quality of skilled labour they have available. Engineers, doctors, scientists, amongst other skilled labour, are central to delivering critical services to the population at large and when such labour leaves the country in large numbers the concerns become quite obvious. Many developing nations have faced the issue of large-scale emigration of skilled labour. Somewhat ironically, developing countries that have invested more in higher education seem to be in a less advantageous position as these countries produce cohorts of skilled people through their education system but have not always been able to match the skills of these people with relevant jobs and remunerations. The process has been termed the 'Brain Drain', signifying the loss of intellectual capital that results from emigration of skilled labour. Lately, some scholars have recognized that 'Brain Drain' is only part of the story. Whereas emigration of skilled labour undoubtedly leads to short-term loss of intellectual capital for the nations, what can be gained through this process is the international experience, networks and social contacts of the mobile labour force. Some countries, such as Taiwan and India, have successfully tapped into these and have reaped great rewards. Saxenian (2002) calls this process the 'Brain Circulation'. The focus of research which has acknowledged 'Brain Circulation' is often the intellectual capital that countries regain as a result of the circular movement of skilled labour.

It is suggested here that the term 'Transnational Social Circulation' aptly captures this process. What is gained is not only the intellectual capital of the Diaspora but the linkage of transnational social networks of the ego (the skilled labour). Moreover, as the ego usually is a knowledge worker, what gets created due to the linking are knowledge networks that span different nations. In the following section, this process of creation of knowledge networks is explored in greater depth.

Granovetter (1973) showed how different kinds of social ties serve different purposes. Whereas the importance of social relations and the relative position of an actor in the social structure in gaining financial and human capital has long been recognized (Bourdieu 1985; Blau 1964), Granovetter pointed out that *weak* social ties are particularly adept at bringing fresh information to the network and the ego (focal actor). Burt (1992) later expanded the idea and introduced the concept of *structural hole*, that is, the relationship between two closed networks. In other words, structural holes denote the lack of social relations between two densely connected social networks. People within each sphere share strong ties but there are no actors within the two spheres who transcend the individual spheres and share social relations with people in both the spheres. Burt (1992) goes on to say that people who do transcend such close networks and link both the spheres together perform a valuable activity which benefits the networks as well as the actor who helps in the linking process. The networks benefit as new information is circulated within the networks due to the bridging process (Kilduff and Tsai 2007). For example, the bridging of networks may bring information that may allow organizations to identify potential exchange partners and initiate new business ventures. Such bridging may also help in generating initial trust that is often so essential to fructify business relationships (Huxham and Vangen 2003). Actors who bridge such networks also gain, as they are often in a position to benefit from being the carrier of fresh information to a network. Burt (1992) suggests that this form of information brokering can be seen as a form of entrepreneurial activity and innovation. Whereas it is certainly not the case that all innovations are forms of information brokering, it has been highlighted before that bringing established ideas to a new context can be legitimately seen as a form of innovation (Rogers 1995).

The suggestion here is that many entrepreneurs of successful outsourcing firms from India excelled in this bridging activity; that is, they linked disparate networks of people, thus creating value for the networks as well as for themselves. The social capital of the entrepreneurs of successful outsourcing firms, seen from this perspective, has an important role to play in the success of these firms. What emerges through the transnational social circulation of skilled labour is a form of social capital that is shared between an ego and an ego's social network. Transnational social circulation can thus be defined as a linking of erstwhile disconnected networks of people located in two distinct geographical spaces (different nations) being linked together due to circular movement of skilled labour. The question that can legitimately arise here is why a circular movement of people is essential for this process. Can such networks arise due to a one-way movement of skilled labour? A significant body of literature has been written about Diaspora networks and also about the fact that countries have found it very hard to effectively tap into their Diaspora networks (Kuznetsov 2006). Whereas *émigrés* themselves have used Diaspora networks very effectively for personal benefits, such networks have usually remained out of the grasp of people back home. The situation changes when these emigrants return (either in a physical or in a virtual sense) to their country of origin, especially when they are still in a productive stage of their careers (as opposed to returning in retirement). In this process, the professional and social networks of the ego (the skilled professional) in both countries get enmeshed, creating a form of a transnational social network in which the ego acts as the bridge.

There is, of course, the question of why transnational circulation happens in certain contexts and does not materialize in others. In the case of India, for example, although emigration of skilled labour has been witnessed since the 1960s, circulation of such labour is a fairly recent phenomenon, beginning in a significant way from the mid-1990s onwards. In other developing countries, all too often, the movement remains one-way and not circular. An in-depth analysis of the reasons of transnational social circulation is beyond the scope of this chapter but one can speculate that it has much to do with the improvement of standard of living and economic opportunities that act as incentives for the expatriates to return home.

Transnational social circulation thus can be seen as the helix that is often missed in the analysis of innovation in a developing country context. Along with the other helices manifested in relationships between universities, government and industry, the social capital generated out of transnational social circulation can be seen to significantly contribute to innovative activities in the society.

To conclude this section, the suggestion that has been put forward is that in a developing country context, it is important to understand the dynamics of large-scale emigration of skilled labour on innovation capacity of the nation. In the first instance, when such labour emigrates, there is undoubtedly a loss of human capital which can lead to the detriment of the innovation capacity of the nation. However, if such labour returns to the home nation after a period of time, it may lead not only to regaining of the human capital (which may be enhanced through new experiences and knowledge), but also to a gain in social capital, in the form of the social networks that the ego has developed while studying or working abroad. Whereas intranational circulation of personnel within national entities such as industry, university and government can facilitate knowledge exchange and innovation (Etzkowitz and Dzisah 2008), such a process can be strengthened and augmented by transnational social circulation as has been discussed here.

THE CASE OF INDIA

In recent years, India has emerged as a hub of service exports, often excelling in the provision of innovative knowledge-based products (Dutz 2007). As suggested before, India's success constitutes a puzzle, which is how a developing nation with a high rate of illiteracy which originates historically from a lack of adequate investment in primary, secondary and higher education has transformed itself into a powerhouse of service exports specializing in knowledge-based activities. Conventional explanations have focused on the economic reforms initiated in the country in the early 1990s and/or the telecommunication reforms that started in the mid-1980s and accelerated from the mid-1990s onwards.[3] However, a more holistic explanation of India's success must look not only at these macro changes but also at a micro level. How have firms that have been at the forefront of this information revolution succeeded in exporting services to the competitive markets of North America and Western Europe? As a developing nation that has found a way to excel in knowledge innovations, India is, in many ways, an ideal case study for exploring and illustrating the conceptual assertions made earlier in the chapter. To reiterate the main assertions here: firstly, it has been noted that different innovations need different configurations of capital mix. Whereas industrial innovations often require access to economic capital, the main ingredients of post-industrial innovation are often social, cultural and human capital. Secondly, it has been suggested that in the context of developing nations, interaction amongst national actors often will not sufficiently account for the innovation activities that take place. It is important to understand the dynamics of the large-scale emigration of skilled labour to study the growth or lack of innovations in such societies. Thirdly, it has been argued that the transnational social circulation of skilled labour can explain, in part, some of the innovations that have been witnessed in certain developing countries' contexts.

It is the contention of the chapter that India's success is based on the twin pillars of post-industrial innovations and transnational social circulation of its skilled labour. The case study, which is designed to illustrate this suggestion, is based on the analysis of forty successful outsourcing firms from India. This analysis is combined with a meta-analysis of the evolution of India's service industry and the factors that have been responsible for the large-scale emigration of skilled labour from India from the mid-1960s to the late 1990s. In the following section, an outline of the structure of the case study is presented.

Evolution of the Service Industry in India

One needs to go back at least to the time of independence of India, which happened in 1947, to get a comprehensive perspective of the changes that led to the emergence of India as a service powerhouse. One can argue that the most influential legacy of the British occupation of India has been its 'language'. English is now spoken by around four hundred million Indians, which is more than the combined population of the U.S. and the UK put together (which are the other two major English-speaking nations). The use of English as the official language of communication in India was a conscious decision by the political leaders in India, who in 1949, while drafting the constitution of India, settled on the language as the official medium of communication (Guha 2007). The motivation for the decision was pragmatic, as India is a land of multiple regional languages which are quite different from each other (and not dialects of one main language). There was, therefore, a need for a common language with which at least most of the political elites were familiar to enable political transactions to take place post-independence.

Initially, the decision was to keep English as the official language of communication for fifteen years and then replace it with the national language of Hindi, but when the time came in 1965, English was allowed to continue as the main official language—again mainly for pragmatic reasons. English by then was even more widely used within the populace in general and, especially, amongst the political elites. Other regional languages were by now being used for official communication, especially in regional areas, but crucially at the central level the use of English as the official medium of communication continued (Guha 2007).

Although the use of English as an official language was mainly a matter of showing pragmatism in post-independence India, its significance as a driver of economic growth in the late twentieth and early twenty-first century is now being fully appreciated. Because English is the unofficial language of business in modern times, the proficiency of educated Indians in the language has undoubtedly been a major asset in India's drive towards globalization.

Apart from adapting English in a uniquely Indian way to enable political and business exchanges in post-independence India, the other major initiative that the political leadership undertook in the late 1940s was the drafting of the constitution of India. It was a major exercise that ushered in dramatic reforms. Amongst the many reforms it initiated, a few stand out; providing equal voting rights for men and women (dramatic as this was late 1940s, when North America and many parts of Europe were only waking up to this issue), granting special privileges to low castes, mandating multiparty democracy and underlying the secular nature of the country. The constitution of India set out the framework under which the political game is to be played in the years to come and the fact that democracy is thriving in India, sixty years thereafter, can be taken, at least in part, as an indicator of the robustness of the framework. That a developing country like India has successfully practiced democracy for over sixty years makes it an even more interesting case study. It gives an opportunity to study how the different institutional actors like university, industry and government have evolved in a developing country context in which democracy has been practiced.

In the initial years of post-independence India, the focus of the government was on 'self-sufficiency' in terms of technology and food. The motivation for this can be partly attributed to the recent history of imperial rule in India and the need felt by the political leaders at the time to minimize dependence on external trade to ensure security and stability. At the same time, the consensus amongst the economists and political leaders was that India needed to industrialize rapidly to catch up with the developed nations of the time. This period was also characterised by intensive collaboration amongst the economists and government policymakers (Byres 1998), which led to the formulation of 'five-year' plans that were developed as road maps of India's economic development.

The dual need of self-sufficiency and industrialization required a large contingent of home-grown engineers and scientists who could carry the policy decisions forward. With this in mind, the Indian government invested quite heavily in the higher education sector,[4] more specifically, they set up the Indian Institute of Technologies (IITs), higher education engineering institutes of repute for producing high-quality engineers.

Although the initiative of producing a large home-grown breed of engineers turned out to be quite successful, the policy of self-sufficiency led to a relatively closed economy with very high tariff barriers in place for inhibiting external trade. As a result of this, economic growth suffered. From 1960s to 1990s, India's share in world trade shrunk. Economic growth, though positive, was sluggish throughout, earning itself the not so complimentary sobriquet of 'Hindu rate of growth', which implied a slow but steady rate of growth. Partly as a result of these two issues, stagnant economic growth and a highly educated and skilled labour force, India witnessed a large migration of skilled labour from 1970 until the late 1990s, prompting some commentators to write about 'Brain Drain'—the fear that most of the skilled people will migrate from the country resulting in a skills shortage within the country.

Where did the skilled labour migrate to? Khadria (2002) points out that there are no comprehensive national estimates of the proportion of the output of Indian higher education institutions migrating to developed nations for work or study. However, according to his own estimates, the rate of migration can be as high as 22 to 33 per cent of total graduates from IITs.

This estimate is further corroborated by immigration data in the U.S. which indicates substantial numbers of Indian immigrants in professional and managerial occupations in the domestic labour market. The U.S. has been undoubtedly the chosen destination for skilled Indian professionals since the 1960s. Before 1960, Britain was the preferred destination, a trend which is surely connected to the historical links that Britain and India shared. What led to the shift in the choice of destination for employment and higher education for the highly skilled Indians in the 1960s? This question is outside the scope of this research but one can speculate that the rapid economic growth and existence of world-class engineering institutes in the U.S. are amongst the motivating factors. What matters to this research is the fact that large numbers of educated Indians have migrated in increasing numbers to the U.S. in search of employment or higher education since the 1960s.

A review of several reports that focused on foreign immigration to the U.S. can be summarized as follows:

- Eighty-seven per cent of the Indians working for Silicon Valley high-technology industries arrived after 1970 (Saxenian 2002).
- In 1999, in the U.S., the leading countries of origin for H1B Visas (visa issued for foreign professionals) were: India 55,047; UK 6,665; China 5,779 (Khadria 2002).
- In a revealing survey that was carried out on foreign professionals of Chinese and Indian origins, Dossani (2002) finds that only 3.3 per cent of Indian professionals came before 1970, 9.5 per cent between 1970 and 1979, 29.3 per cent between 1980 and 1989 and 44 per cent between 1990 and 1999.
- Investigating the effects of immigration on trade in Silicon Valley, Saxenian (2002) finds that Chinese and Indian engineers were running 29 per cent of Silicon Valley's technology businesses by the end of the 1990s. These companies collectively accounted for more that $19.5 billion in sales and 72,839 jobs.
- In the same study, Saxenian finds a correlation between immigration from a country to California and California's trade with that country. He points out that with every 1 per cent increase in number of first-generation immigrants from a given country, California's exports to that country go up by nearly 0.5 per cent.
- It is difficult to say with certainty where and how the labour moved, but from related studies, few patterns can be established. It is clear that educated Indians started to migrate in large number from 1970 onwards. It is also clear that the U.S. became the preferred choice of migration amongst educated Indians from 1960 onwards. A significant number of Indian professionals landed jobs in the technology sector, which was concentrated in the Silicon Valley in California. India's trade with the state of California rose along with the rise of migration of Indian professionals to that state.

India's gross domestic product (GDP) showed steady growth during the period 1950–1980 of around 3.7 per cent, which accelerated to around 5.8 per cent during 1980–1990 and maintained this relatively high rate of growth during 1990–2000 (Gordon and Gupta 2003). However, the sharp acceleration witnessed in the 1990s requires further explanation, especially as it can be seen that the rate of GDP growth stayed pretty much the same during 1980–1990 and 1990–2000. Gordon and Gupta (2003) point to the rise in business services and service exports as an explanatory factor of this. They argue that something fundamental happened in terms of change in the telecommunication infrastructure and economic policy which accelerated the growth of services during this period.

Indeed, there is much to support this line of argument. The government initiated major economic reforms in 1991, although minor reforms started taking place from 1980 onwards (Panagariya 2004), which certainly boosted the trade of services, in terms of allowing the easy set up of businesses, less bureaucratic red tape, allowing to raise foreign capital from overseas, freeing of capital good imports and so on. There was also lot of investments that went into upgrading the telecommunication infrastructure of the country (Kapur 2002). These technical explanations—economic reforms and innovation in information and telecommunication technologies—are certainly important elements in the overall explanation but it is equally important to recognize that there are other elements that are essentially linguistic, cultural, institutional, political and social.

The over-reliance on technical explanations for the growth of offshore outsourcing perhaps stems from the need to provide a clean causal link between reforms and technological innovations, on one hand, and its effect in terms of higher export of services from India on the other. What is missing is the contingent and relatively unplanned nature of this growth. Of course, the British rule in India was not a planned event (not even by the British themselves; see Roy 2002) and thus the absorption of English language by political elites in India was an historical contingency. Even after independence, India could have opted for a non-English regional language for its political and business transactions but chose English. What was also certainly not planned was the outward migration of educated Indians during 1960–1990s, which was generally considered to have a negative effect on the economy by the government and economists of the time. During the late 1990s and in the early years of the twenty-first century, the fear of 'Brain Drain' has been transformed into a positive idea of 'Brain Circulation', where it is recognized that the social ties that skilled migrants carry with them while going to work overseas often result in new business opportunities between the country of emigration and the country where the labour force emigrate to (Saxenian 2002). The contingent nature of the growth of offshore outsourcing in India underlies the importance of context specificity of such a phenomenon and should caution against trying to replicate such success in a different context by imitating the economic and technological reforms.

However, there is plenty of incidental evidence that the success in offshore outsourcing in India owes a large part to the 'social web' that underpins such transactions. Such social structures which have not yet been systematically investigated are examined in the following section of this chapter.

Social Capital of Successful Indian Outsourcing Firms

From the preceding discussion, it can be suggested that the skilled Indians who had emigrated for higher education and work during the period

1960–1990 have had a role to play in India's outsourcing success. The question is how to explore this proposition empirically in the context of social capital and the university-industry-government network in offshore outsourcing.

For this purpose, the first task is to identify successful Indian firms specializing in outsourcing. There have been several surveys that have tried to rank the top global outsourcing firms. International Association of Outsourcing Professionals (IAOP) produces a yearly ranking of the top one hundred outsourcing firms worldwide. A similar ranking is produced by GlobalServices 100 and another, The Black Book of Outsourcing, ranks the top fifty outsourcing providers. The rankings are derived from a survey administered by these agencies, participation in which is entirely voluntary. As many of these firms are privately held entities, financial information related to their activities is not freely available in the public domain. The surveys thus use secondary proxies such as number of employees, number of service centres and countries served to assess the success of participating firms. Quality certification of the service provider is also an important component of the evaluation process. There are, of course, firms that feature in all three surveys but others feature in only one or two of the surveys. This variance is attributable to the voluntary nature of the survey as well as to the different methodologies that are used. The appropriateness of these methodologies is open to question, but for this study, all the firms that appear on these surveys constitute the population of successful outsourcing firms worldwide.

The next step is to identify the firms amongst this population that are based in India, which can be done readily through a search of their website. The Indian-based firms can be roughly classified into three categories for the purpose of this study. In the first category are firms that have been well established in other industry sectors before branching out in the outsourcing industry. In the second category are firms that specialized in the outsourcing sector from inception. They are usually started by first-generation Indian-born entrepreneurs. The third category consists of firms that are similar in all other respects to the second category but are incorporated outside India. These firms have an Indian-born 'founder' entrepreneur and have their service centres based in India. However, the headquarters are incorporated overseas, due to reasons that may have more to do with the political dynamics of the outsourcing decision in countries such as the U.S.

For the purpose of this study, which is to identify whether network ties play a role in the success of outsourcing firms in the offshore context, firms in the second and third categories are included whereas the firms in the first category are excluded. Success of incipient firms has been linked to the social capital of the founder (Lin 2001). If the social ties of the founder, which in part represents his/her social capital, can be directly associated with success in offshore outsourcing sector, it can lend credence to the suggestion that these ties are essential for offshore transactions to take place. From the three surveys mentioned earlier, forty firms were identified that fit either in the second or third category.

Obviously for this purpose, the 'founder' entrepreneur of these firms needs to be identified. Most of the firms in the surveys have an account of their organizational history on their websites which includes a profile of their founder. In some cases where this could not be found, the details were gathered by doing some Internet research. In other cases, where such data is unpublished in the public domain, firms were contacted through email and/or telephone and the details were obtained.

What is of interest to us are the 'social ties' of the 'founder' entrepreneur that existed prior to setting up of their businesses. Following from Watts's (2004) suggestion, that social ties are made on a substrate of social identities, the focus here is to identify the affiliation network of the founders. Affiliation networks and social identity are closely related concepts. Social identity would arise out of membership to different affiliations.

As over 70 per cent of the offshore outsourcing transactions take place out of the U.S., a main point of interest would be to identify if the founders share any major U.S. affiliation. If they do, it would follow that they also would have been endowed with social ties arising out of these affiliations and these ties would be invaluable in generating the initial trust, which, it is suggested here, is of vital importance in offshore outsourcing.

What can be considered major U.S. affiliations? Affiliations that are characterised by emotional intensity, reciprocity and frequent face-to-face interactions can engender strong ties (Granovetter 1973). For Indian-born immigrants to the U.S. such ties can arise out of educational and/or organizational affiliations. This is not to say that social ties cannot be developed through other affiliations, but education and jobs are the two main reasons why Indians migrate to North America. For this study, if the founders have pursued education in the U.S. for a substantial period of time (for example, time required to acquire a bachelor's or master's degree) and/or have worked for more than two years in the U.S. prior to setting up their own businesses in India, he or she is considered to have a U.S. affiliation.

Another interesting affiliation to look into would be that with Europe, especially with the UK as India and Britain share many similar institutions including the language. Despite sharing many institutional similarities as a result of the legacy of British rule in India, relative to North America, offshore outsourcing has been slow to take off in Britain.

It was decided that the affiliation network of the founders would be investigated with the objective to specifically look for affiliation with the U.S. and Europe. At the same time an inductive approach was adopted to find whether there existed other patterns of affiliations amongst these founder entrepreneurs.

Information on the founders was retrieved from secondary sources such as websites of the companies they founded, personal home pages, annual reports, SEC filings, social network websites and other archival data.

After collecting the data and doing a search for patterns, it was decided to include two additional categories of affiliation as it was found that a significant number of the founders had these affiliations.

These are affiliation to the main engineering and business management higher education institutions in India—IIT and IIM (premier Indian Technology and Management Institutes). Both of these educational institutes enjoy a reputation of being amongst the top-tier premier engineering and management institutions worldwide.

From the analysis, it emerged that many of the firms adopted a 'boot-strapping strategy', which means that the firms often relied on their own internal revenues for growth rather than depend on external funding. In general, firms who depended on a boot-strapping strategy for growth would have a low debt-equity ratio, so data was gathered on this aspect also whenever it was available.

The data was collected in the following format.

- name of the firm
- name of the founder(s)
- U.S. affiliation
- Europe affiliation
- IIT affiliation
- IIM affiliation
- growth Strategy of the firm—boot-strapping/external borrowings
- debt-equity ratio of the firm

Apart from this, a lot of related qualitative and quantitative data was also collected. Quantitative data such as sales turnover and number of employees was collected whenever such data was available. Qualitative data such as educational and work background of the entrepreneurs, especially of the period before they started their firms, was also collected.

From the analysis, it emerges that the majority of the founders share U.S. affiliation prior to setting up business in India. Approximately 75 per cent shared U.S. affiliation whereas only 13 per cent of the entrepreneurs share European affiliation. The majority of the founders are engineering graduates either from India, in most cases from IIT, or from the U.S. A significant number of them have prior working experience in the U.S. and in some cases their firm's present clients include their previous employers.

It was also found that the majority of the firms (around 70 per cent) depended on a boot-strapping strategy for expansion of the firm. Arguably, this has been made possible due to the nature of services in which these firms specialize in. As has been discussed earlier, in knowledge-intensive businesses often the ingredients that are key to success are social capital and intellectual capital of the entrepreneurs, precisely the forms of capital in which the firms in the study were initially richly endowed with.

NOTES

1. For an interesting discussion on the ontology of money, see Searle (1995).
2. For a detailed discussion on the relation between economic capital and innovation, see O'Sullivan (2005).
3. For a detailed discussion on the effects on these reforms on the Indian economy, see Panagariya (2008).
4. The disproportionate investment in higher education was done, some argue, at the expense of primary education, for which India has suffered long-term negative consequences (see Sen 1999).

CONCLUSION

This chapter has suggested that India's success in service exports has been partly dependent on the dynamics of post-industrial innovation and transnational social circulation of skilled labour. Through the case study, the main ideas of the research have been highlighted. It can be said in conclusion that although the success of service firms in India has been dependent on factors that are highly contingent on the specific social context they found themselves embedded in, some insights can be drawn to generalize the findings. First, many developing countries face the same issue of large-scale emigration of skilled labour. India's experience shows that a circular movement of such labour can bring real benefits to developing nations as social networks across large geographical spaces gets interconnected due to such movement. It has been shown in the chapter how the triple helix thesis can be augmented by taking into account the effects of large-scale emigration of skilled labour on the innovation capacity of societies. The dynamics of post-industrial innovation has also meant it is possible to be a knowledge entrepreneur at a very low cost, with the key forms of capital that the entrepreneurs need access to being social, human and cultural capital. For entrepreneurs in developing nations this provides a viable model to get onto the innovation fast track.

REFERENCES

Becker, G. S. 1962. 'Investment in human capital: A theoretical analysis'. *Journal of Political Economy* 70 (5): 9–49.
Benkler, Y. 2006. *The wealth of networks: How social production transforms markets and freedom*. New Haven, CT: Yale University Press.
Blau, P. M. 1964. *Exchange and power in social life*. New York: John Wiley and Sons.
Bourdieu, P. 1985. 'The forms of capital'. In *Handbook of theory and research for sociology of education*, ed. J. G. Richardson, 241–258. New York: Greenwood.

Burt, R. 1992. *Structural holes: The social structure of competition*. Boston, MA: Harvard University Press.

Byres, T. 1998. *From ivory tower to the belly of the beast: The academia, the state, and economic debate in post-independence India, the Indian economy: Major debates since independence*. Delhi: Oxford University Press.

Coleman, J. S. 1990. *Foundation of social theory*. Cambridge, MA: Harvard University Press.

Datta, S., and M. Saad. 2008. 'Social capital and university-industry-government networks in offshore outsourcing—the case of India'. *Technology Analysis and Strategic Management* 20 (6): 741–754.

DiMaggio, P. 1982. 'Cultural capital and school success: The impact of status culture participation on the grades of U.S. high school students'. *American Sociological Review* 47 (2): 189–201.

———. 1997. 'Culture and cognition'. *Annual Review of Sociology* 23:263–287.

DiMaggio, P., and J. Mohr. 1985. 'Cultural capital, educational attainment, and marital selection'. *American Journal of Sociology* 90 (6): 1231–1261.

Dossani, R. 2002. *Chinese and Indian engineers and their networks in Silicon Valley*. Stanford, CA: Asia/Pacific Research Center.

Dutz, M. A., ed. 2007. *Unleashing India's innovation: Towards sustainable and inclusive growth*. Washington, DC: World Bank.

Edquist, C. 2005. 'Systems of innovation: Perspective and challenges'. In *The Oxford handbook of innovation*, ed. J. Fagerberg, D. C. Mowery and R. R. Nelson, 181–208. Oxford: Oxford University Press.

Etzkowitz, H., and J. Dzisah. 2008. 'Rethinking development: Circulation in the triple helix'. *Technology Analysis and Strategic Management* 20 (6): 653–666.

Etzkowitz, H., and L. Leydesdorff. 2000. 'The dynamics of innovation: From national systems and mode 2 to triple helix of university-industry-government relations'. *Research Policy* 29 (2): 109–123.

Fagerberg, J., D. C. Mowery and R. Nelson. 2005. *The Oxford handbook of innovation*. Oxford: Oxford University Press.

Freeman, Chris. 1992. 'Formal scientific and technical institutions'. In *National systems of innovation: Towards a theory of innovation and interactive learning*, ed. B. A. Lundvall, 169–187. London: Pinter.

Fukuyama, F. 1999. 'Social capital and civil society'. Paper presented at the IMF Conference on Second Generation Reforms, Washington, DC.

Giles, J. 2005. 'Special report: Internet encyclopaedias go head to head'. *Nature* 438 (December): 900–901.

Gordon, J., and P. Gupta. 2003. 'Understanding India's service revolution'. Paper presented at the IMF-NCAER Conference, A Tale of Two Giants: India's and China's Experience with Reform, 12–14 November.

Granovetter, M. 1973. 'The strength of weak ties'. *American Journal of Sociology* 78 (6): 1360–1380.

Guha, R. 2007. *India after Gandhi: The history of the world's largest democracy*. London: Macmillan.

Huxham, C., and S. Vangen. 2003. 'Nurturing collaborative relations: Building trust in interorganizational collaboration'. *Journal of Applied Behavioral Science* 39 (1): 5–31.

Kapur, D. 2002. 'The causes and consequences of India's IT boom'. *India Review* 1 (2): 91–110.

Khadria, B. 2002. *Skilled labour migration from developing countries: Study on India*. Geneva: International Labour Office.

Kilduff, M., and W. Tsai. 2007. *Social networks and organizations*. London: Sage Publications.

Kuznetsov, Y., ed. 2006. *Diaspora networks and international migration of skills: How countries can draw on their talent abroad.* Washington, DC: World Bank.

Lin, N. 2001. *Social capital: A theory of social structure and action.* New York: Cambridge University Press.

Lundvall, B. 1992. *National systems of innovation: Towards a theory of innovation and interactive learning.* London: Pinter.

Nelson, R. 1993. *The national innovation system: A comparative analysis.* New York: Oxford University Press.

North, Douglass C. 1990. *Institutions, institutional change and economic performance.* Cambridge: Cambridge University Press.

O'Sullivan, M. 2005. 'Finance and innovation'. In *The Oxford handbook of innovation*, ed. J. Fagerberg, D. C. Mowery and R. R. Nelson, 240–265. Oxford: Oxford University Press.

Ranagriya, A. 2008. India: The Emerging Giant, USA: Oxford University Press.

Putnam, R. D. 2000. *Bowling alone: The collapse and revival of American community.* New York: Simon and Schuster.

Rogers, E. M. 1995. *Diffusion of innovations.* New York: Free Press.

Roy, T. 2002. 'Economic history and modern India: Redefining the link'. *Journal of Economic Perspectives* 16 (3): 109–130.

Saxenian, A. 2002. 'Silicon Valley's new immigrant high-growth entrepreneurs'. *Economic Development Quarterly* 16 (1): 20–31.

Schultz, T. 1960. 'Capital formation by education'. *Journal of Political Economy* 68:571–583.

Schumpeter, J. 1934. *The theory of economic development.* Cambridge, MA: Harvard University Press.

Searle, J. 1995. *The construction of social reality.* London: Allen Lane.

Sen, Amartya. 1999. *Development as freedom.* London: Oxford University Press.

Soto, Hernando De. 2000. *The mystery of capital: Why capitalism triumphs in the West and fails everywhere else.* New York: Basic Books.

Tuomi, I. 2002. *Networks of innovation: Change and meaning in the age of Internet.* Oxford: Oxford University Press.

Von Hippel, E. 2005. *Democratizing innovation.* Cambridge, MA: MIT Press.

Watts, Duncan J. 2004. *Six degrees: The science of a connected age.* London: Vintage.

Weber, M. 1978. *Economy and Society.* G. Rowth & C. Wittich, eds, Berkeley: University of California Press.

13 Science Parks and University–Industry Collaboration in Malaysia

The Malaysian Experience

Chandra Malairaja and Girma Zawdie

INTRODUCTION

Science parks are increasingly used in many countries as a popular strategy for promoting innovation, especially among small and medium-sized enterprises (SMEs). Co-location of firms in a science park, it is argued, encourages networking and collaboration, which can foster innovation. Science parks thus provide an environment that assist firms to develop their technological and innovative capabilities. In particular, science parks provide a resource network that enables firms located in them to collaborate with universities and research institutions. This chapter investigates the role of science parks in enhancing university–industry collaboration and puts the case for science parks to the test with reference to the experience of Malaysia.

UNIVERSITY–INDUSTRY LINKS

Universities are an important source of knowledge and provide the basis for innovation in emerging industries (Feldman and Kelly 2002). Innovation in this context refers to the creation and successful introduction of new products and processes in the marketplace. As such, university–industry links have emerged as key components of the national innovation system.[1] Universities no longer confine themselves to their traditional role in teaching and research, but are increasingly engaging in entrepreneurial and business activities known as 'third stream activities'. This phenomenon is attributed to the pressure exerted on the universities to commercialize their research findings to generate revenue to cover some of the operating costs including research costs. The success of many academics as entrepreneurs in various technology fields has also contributed to this trend. The direct involvement of universities with industry can be seen in activities such as research funding, training partnerships and technical service contracts. Apart from these, industry also sponsors research centres and researchers and offers sponsorship or endowment of chairs (Laursen and Salter 2003; Siegel, Westhead and Wright 2003).

However, too much emphasis on university research for industrial application may undermine the academic value of research in universities. The role of universities, it is argued, is in the creation of knowledge and talent, which are crucial for the growth of a knowledge-based economy. But Howitt (2000) argues, for instance, that attempts to weaken university–industry links would be counterproductive in view of the increasing mutual interdependence between universities and industry brought about by the biotechnology revolution.

Universities have set up Technology Transfer Offices (TTOs) to facilitate the transfer of technology from the university labs to markets, and have also established mechanisms to create academic spin-offs. The emergence of the entrepreneurial university phenomenon is the driving force behind the triple helix university-industry-government interactions (Etzkowitz et al. 2000).[2] Associated with this phenomenon is the practice of setting up science parks by universities to achieve the aforementioned objectives, as is the case of the UK, where the majority of science parks are owned by universities. However, despite these developments, universities throughout the world have to overcome many obstacles that stand in the commercialization route. For example, Schramm (2004) points out that a great deal of research in life sciences and in computing and engineering has not been successfully brought into the marketplace for practical use mainly due to bureaucracy and lack of applied skills and resources in universities.

Although mechanisms have been put in place to facilitate university–industry collaboration, there are still many problems that need to be overcome. For example, problems exist between academics and industrialists in determining the type and suitability of research for industry needs. On one hand, academics cherish freedom in their research endeavour and do not want to be dictated by the whims of people from industry. On the other hand, industrialists complain that university research does not to a large extent meet the needs of industry. Johannesen and Rasmussen (2000) point out the existence of a mental and cultural gap between the universities and businesses, with the former preoccupied with 'blue sky research', whereas the latter's priority is to achieve success in the marketplace.

THE SCIENCE PARKS STRATEGY

The two most often-quoted definitions of a science park are those given by United Kingdom Science Parks Association (UKSPA) and the International Associations of Science Parks (IASP). Both these definitions share three main elements in terms of (a) knowledge and technology transfer through interaction between universities, research and development (R&D) institutions and enterprises; (b) growth of knowledge-based and innovative companies; and (c) creation of supportive environment. The importance of creating a supportive environment for firms, especially technology-based

Table 13.1 SME Features versus the Knowledge Economy

Strengths:	*Weaknesses:*
Enthusiasm	Financial shortage
Flexibility	Information shortage
Threats:	*Opportunities*:
Globalization	Globalization
Speed of changes	New business niches
Local Mentality	International mentality

Source: International Association of Science Parks (IASP).

SMEs, is important as these enterprises possess specific features (see Table 13.1) that would make them potential vehicles for the growth and development of knowledge economy.

Science parks are set up to facilitate commercialization of technologies, stimulate development of technology-based SMEs and promote regional development.[3] The differences between science parks and industrial parks are shown in Table 13.2. The rationale for building a science park is that it helps create a resource network (see Table 13.3) for firms located in them (Westhead and Storey 1994). Firms located in an innovation cluster benefit from knowledge spill-over from nearby universities and research institutes (Jaffe 1989, Vedovello 1997). This is apparent from the fact that all the science parks in the United Kingdom are located in or near universities (Siegel et al. 2003, HM Treasury 2003, Westhead and Wright 2003). These parks also help create spin-offs, stimulate higher research productivity and promote entrepreneurship. Science parks have had their appeal felt significantly in rapidly industrializing developing countries where they are employed as 'essential elements for a structured indigenous capacity in support of national technology diffusion approach' (UNIDO 2004, 6).

Table 13.2 Science Parks versus Industrial Parks

Science Parks	***Industrial Parks***
High-tech firms with R&D capability	**Manufacturing firms**
Technical manpower	Low-wage labour
New, innovative products	Mature products
Complete product cycle	Manufacturing only

Source: Ho (1999).

Table 13.3 Science Park as a Resource Network

Type of Resource	*Description*
University related	University links, access to university resources, university education, academics and graduates as skilled manpower
Science park facilities	Business advisory services, venture capital, flexibility of premises, car parking, administrative facilities, science park management
Cluster effects	Image, reputation and credibility of location and collective learning

Source: Based on Lowegren (2001).

Science parks are built to foster enhanced university partnership leading to greater utilization of university research results. These parks are perceived as effective interfaces between university research and industrial application that can also assist in the creation of spin-offs and enable academics to become entrepreneurs. Specifically, science park firms get involved in joint research projects with faculty and students, participate in seminars and workshops that provide a forum for the exchange of best practice technical information, recruit university graduates and make use of university resources (library and laboratory facilities). Firms that conduct R&D are capable of producing new products and processes for commercialization, although R&D may not necessarily result in innovation.

According to Westhead and Storey (1994), the survival rate of science park firms having university links is higher than off-park firms, which do not have such links. Based on a study of firms in three technology parks in the Basque region of Portugal, Lilian and Inaki (2003) found firms that experience the highest growth rates to be those that establish formal links with R&D-intensive universities. But despite these potential benefits, there is also evidence of a lack of collaboration between science park firms and universities. For example, according to a survey conducted by the Manchester Science Park (MSP) in 2003, 46 per cent of the companies in the MSP have no links at all with HEIs (see Harper 2003). Dierdonck, Debackere and Engelen (1990) also found lack of collaboration between Belgian science park firms and local universities.

The world's first science park was the Stanford University Research Park in Palo Alto, California, which played a major role in the evolution of Silicon Valley as the most successful high-technology cluster.[4] This prompted policymakers elsewhere to emulate the Silicon Valley model, leading to the establishment of science parks ranging from a size of less than sixty acres to parks covering thousands of acres[5] supporting firms including SMEs, involved in a variety of technology sectors. Currently, there are about seven hundred science parks in the world, also known as research parks (in the U.S.), technopoles (in France), innovation centres (in Germany) and technology parks in many developing countries (Sanz 2002 and 2003).

THE MALAYSIAN EXPERIENCE

Towards Knowledge-Based Economy

Malaysia has emerged as a major global producer and exporter of technologically sophisticated high-value-added products. This achievement, which is normally associated with a mature industrial economy (UNCTAD 2003), however, begs the question of innovation and technological progress insofar as Malaysia falls short of the level of technological competence to reflect industrial 'maturity'.[6] Malaysia's economic performance over the last couple of decades has nonetheless been impressive. Foreign direct investment (FDI) played a major role in the transformation of the agriculture-based Malaysian economy into one which is largely based on high-technology manufacturing activities. However, the extent of technological spillover from FDI-driven growth has not been all that encouraging in Malaysia (Tidd and Brocklehurst 1999), as it has been in Taiwan, Korea and Singapore (Wong 1999).

The emergence of China as a more favourable destination for FDI than Malaysia has also posed a further challenge to the Malaysian economy. Malaysia can no longer rely on low-wage cost advantage and the provision of incentive schemes to attract FDI. Moreover, the emphasis on a knowledge-based economy and increasing complexity of technologies require Malaysian firms to acquire new knowledge and competencies.

Malaysia has adopted best practice networking strategies to assist firms to meet these needs and enhance their technological capabilities. These strategies include high-tech cluster development based on the Silicon Valley model, technology incubators, science parks, venture capital funds and other instruments under technology policy within the national innovation system (NIS) framework. Malaysia is, however, a latecomer to the science park phenomenon. Although the first science park was set up in 1988, it was not a science park in the true sense of the term because it was housed in a small building with hardly any science-based activities in it. The first proper science park was in fact launched in 1996.

Malaysia's bid for transition to a knowledge economy would, however, greatly depend on the ability of the country to harness the creative and innovative capacity of its people through the provision of a policy environment amenable to the development of entrepreneurship and the growth of investment in R&D. It is significant that Malaysia's competitiveness ranking had dropped from sixteen in 2004 to twenty-eight in 2005 in the face of the country's strong export performance, including high-tech exports, high FDI inflow and low inflation. The implication is that Malaysia had lagged behind technologically in relation to other countries during this period. The situation has consequently prompted the government to adopt the NIS framework to review existing science and technology (S&T) policy and evolving strategies to strengthen the country's S&T capabilities through support to local firms, especially SMEs.

The Growth of Universities

In 1962, there was only one university in Malaysia. At present, there are eighteen public universities, twenty-seven private universities and university colleges and 559 private colleges (see Table 13.4). The growth in the number of universities is attributed to increasing demand for tertiary education and the government's liberalization policies aimed at making Malaysia a centre of excellence for higher education. The government introduced the Action Plan for Industrial Technology Development (APITD) in 1990, which provides for the setting up of private universities by local private enterprises and allowing foreign universities to set up branch campuses in Malaysia and establish human resource development funds.

Universities in Malaysia are expected to play a greater role in creating the manpower to meet the needs of a growing knowledge-based economy. This is reflected in the entrepreneurial orientation of many local universities as well as in greater university–industry links. However, intense debate has emerged as to the quality of higher education provided by local universities. The trigger for this debate is the increasing level of unemployment among local university graduates. Reports suggest that there are about eighty thousand unemployed graduates in the country (*New Straits Times* 2005). This appears to be something of a paradox, considering the fact that there are almost two million foreign workers in Malaysia. In effect, the situation reflects the extent of mismatch in the demand for and supply of skills in Malaysia and the mismatch between university education and training and the needs of industry. In view of this, universities have been urged to offer courses that are relevant to the needs of industry.

Universities in Malaysia have established various mechanisms to facilitate links with industry, especially in the area of technology transfer and commercialization of research results. One noticeable effort is the setting up of investment arms or business units to market expertise, skills and technology that could be utilized by private enterprises. A good example is Usains Holdings Sendirian Berhad (Usains) formed in 1997 by the Universiti Sains Malaysia (USM) in Penang. Two additional units to complement and work together with Usains were established by USM in April 2002, and are known as the Engineering Innovation and

Table 13.4 Tertiary Educational Institutions in Malaysia (2000 and 2005)

Institution	*2000*	*2005*
Public		
University	11	11
University College	0	6
Polytechnic	11	20
Community	0	34
Total	**22**	**71**
Private		
University	5	11
University College	0	11
Polytechnic	3	5
Community	632	532
Total	**640**	**559**
Total	**662**	**630**

Technology Development (EITD) unit and the Medical Innovation and Technology Development Unit (MITD).

A general survey conducted by the EITD unit in USM to gauge its perspective about university–industry collaboration yielded the following findings:[7]

- insufficient innovative products for commercialization
- lack of research results worthy of commercialization
- culture of conducting quality research for commercialization still not widespread
- lack of conviction and commitment amongst academic staff to participate in research

According to Ali (2003), the constraints faced in university–industry collaboration include:

- dominance of foreign investments in the critical sectors of manufacturing
- lack of really effective funding by government and industry
- lack of highly capable scientists who can lead in terms of knowledge frontiers
- the nascent venture capital industry
- lack of innovative entrepreneurship

- focus of universities towards teaching, thus creating a divergence of objectives between university and industry

Science Parks

The government has been supporting local technological development by establishing science and technology parks, technology incubators and the Multimedia Super Corridor (MSC).[8] These physical development initiatives involve offering affordable office spaces and access to modern facilities (e.g. telecommunications) and linkages to researchers, networking and venture capital funding. The science park strategy has been adopted by the government to stimulate innovation amongst SMEs and to enhance university–industry collaboration. To date, five science parks have been set up throughout the country by the federal and state governments (see Table 13.5). These are Technology Park Malaysia (TPM) in Bukit Jalil in Kuala Lumpur; Kulim Hi-Tech Park (KHTP) in the northern state of Kedah; Selangor Science Park (SSP); UPM-MTDC Incubation Centre in the state of Selangor, located in University Putra Malaysia (UPM); and Technovation Park based at the UTM Campus in Skudai in the state of Johore. The proximity of these parks to universities and research institutes is intended to enhance prospects for the development of technology-based companies through university–industry collaboration. As Porter (2003) argues, clusters cannot be started from scratch.

Table 13.5 Science Parks and Technology Incubators in Malaysia

Name/Location of Science Park	*Size (acre)*	*Year of est.*	*No. of firms in park*	*Technology Focus*
Technology Park Malaysia Kuala Lumpur	750	1995	120	ICT, biotechnology
Kulim Hi-Tech Park Kulim, Kedah	630	1996	33	High-tech manufacturing
Selangor Science Park	478.4	2001	n.a	
Technovation Park UTM Campus, Skudai, Johor	130	1995	21	High-tech activities
UPM-MTDC Technology Incubator Centre One	18	1997	32	IT and multimedia
UKM-MTDC Smart Technology Centre	6	1999	10	Biotechnology, pharmaceuticals
UTM–MTDC Technology Innovation Centre One	n.a	1999	n.a	Advanced electronics, advanced manufacturing
MSC Central Incubator	n.a	2000	35	IT and multimedia

Apart from science parks, technology incubation centres have also been set up within local universities to nurture the growth of small high-tech companies and start-ups. These include three such centres set up by Malaysian Technology Development Corporation (MTDC) and the MSC central incubator set up by Multimedia Development Corporation. The biggest incubator is the UPM-MTDC Technology Incubator Centre One, based at University Putra Malaysia, which caters to thirty-two companies. The MSC central incubator is home to twenty start-ups.

METHODOLOGY AND SAMPLE

The present study used a matched sample research design, which involves a comparative analysis of firms located on science parks and a comparable group of firms sited off-park to see whether there are significant differences between these two groups of firms in university–industry collaboration. This approach has been used extensively in previous studies investigating similar research questions (Westhead and Storey 1994; Ferguson 199;, Lindleof and Lofsten 2003; Westhead 1997).

Of the aforementioned five science parks, only TPM and KHTP seemed suitable for study because they have been in operation for more than five years whereas the other three are either recently set up or do not contain sufficient member of companies. However, KHTP was subsequently excluded from the study because almost all the companies located in this park are multinational companies and locally owned large enterprises involved in high-tech manufacturing. Thus, inclusion of KHTP was not considered to be of much use to the study where the focus is on shedding light on the impact of science parks on Malaysian SMEs involved in technology-based businesses.

The focus of analysis in this study is on high-tech SMEs in Malaysia. For the purpose of the study, an SME is defined as a business entity employing not more than 150 employees and having a turnover of not more than M$25 million (£3.7million). Only companies located in TPM were chosen for the sample study of science park firms in Malaysia. TPM, located near universities and research institutes, has been in business for more than five years, and has since its establishment shown substantial development in terms of physical expansion and growth of tenant companies. It has to date 120 tenant companies.

The sample on which the study is based covers a total of 101 technology-based companies in TPM, which is more than a reasonable representation of all such companies located in science parks in Malaysia. This is because TPM is the largest science park in terms of number of tenants. Even the combined number of tenants of the other four science parks in operation at the time of study does not exceed that of TPM.

The sample for off-park firms was drawn from the databases of SMEs maintained by Malaysian Science and Technology Information Centre (MASTIC); Multimedia Development Corporation (MDC); Federation of Malaysian Manufacturers (FMM); and Small and Medium Industries Development Corporation (SMIDEC). These databases were found useful as they contain the business address registers of companies that fall within the definition of SMEs used by MITI and meet the criteria of a technology-based business set by MESDAQ.

Cooperation and collaboration of a firm with local universities can be construed as a throughput in the innovation process (Kemp et al. 2003; Klomp 2001). The variable in the study is the type of links that exist between firms (science park and off-park firms) and local universities. The study identified six types of links as shown in Table 13.2. This variable was assigned binary values of one (1) for an indication of 'yes' and zero (0) for an indication of 'no'. Thus the total value obtained by science park firms having collaboration with local universities is the same as the number of firms indicating such collaboration. Apart from an overall score of one (1) for a 'yes' response and zero (0) for a 'no' response in the case of collaboration with university, further scores were given to each of the six types of university links developed for the study. Respondents could respond to more than one of these categories. These six links were given scores of 1 to 5. Informal contacts with academics and universities as customers were given a score of one (1); two (2) for sponsorship of research trials/projects; three (3) for employment of academics on a part-time consultancy basis; and four (4) for access to equipment and other facilities. The highest score of five (5) was given to contacts involving joint/collaborative research. These scores were then analysed for mean and standard deviations.

RESULTS

Multiple responses were received on the types of links science park firms and off-park firms forge with local universities. The number and the percentage of the firms involved in each of these relationships are shown in Table 13.6. Overall, 59.1 per cent of science park firms have links with universities compared to 50 per cent of the off-park firms. In terms of joint research collaboration, only seven (31.81 per cent) science park firms have such collaboration compared to nine (30 per cent) off-park firms.

The six categories of university links were given scores of 1 to 5. These scores were totalled and analysed for mean and standard deviations and the results are presented in Table 13.7. Based on t-test of the means of 4.5 for science park firms and 4.4 for off-park firms, it was found that there is no statistically significant difference between the two groups of firms with regard to links with universities. The analysis was extended to determine how both groups of firms perform with regard to joint research collaboration with universities. Here, too, no statistically significant difference was found between science park and off-park firms. A further test using the chi-square method on both groups of firms with and without joint research collaboration showed no evidence that

Table 13.6 Number and Types of University Links between Science Park and Off-Park Firms

Type of Link with Universities	*Science Park Firms* No.	%	*Off-Park Firms* No.	%	*All Firms* No.	%
Informal contact with academics	4	18.18	11	36.67	15	28.85
Employment of academics on a part-time consultancy basis	5	22.28	9	30.00	14	26.92
Conducting joint/ collaborative research	7	31.81	9	30.00	16	30.77
Sponsorship of research trials/ projects	4	18.18	5	16.67	9	17.31
Access to equipment and other facilities	7	22.28	8	26.67	15	28.85
University/HEI as a customer	9	40.91	6	20.00	18	34.42
Total no. of links	36					
Total no. of firms	22		30		52	
No. of firms having links	13		15		28	
Percentage of firms	59.1%		50.0%		53.85	

Note: Chi-square tests on conducting joint/collaborative research:
x^2=0.020 significance level = 0.4442 no statistical significance at 0.05 level of significance.

Table 13.7 Types of University Links between Science Park and Off-Park Firms (Means and Standard Deviations)

Type of link with Universities	*Science Park Firms*		*Off-Park Firms*		*All Firms*	
	Mean	*SD*	*Mean*	*SD*	*Mean*	*SD*
Informal contact with academics	0.1818	0.3948	0.3667	0.4901	2.885	0.4575
Employment of academics on a part-time consultancy basis	0.6818	1.2868	0.9	1.3983	0.8076	1.3437
Conducting joint/ collaborative research	1.5909	2.3837	1.5	2.3305	1.5387	2.3302
Sponsorship of research trials/projects	0.3637	0.7895	0.3333	0.7581	0.3462	0.7640
Access to equipment and other facilities	1.2727	1.9069	1.667	1.7991	1.1538	1.8297
University/HEI as a customer	0.4091	0.5032	0.2	0.4068	0.2885	0.4577
Mean	4.5		4.3667		4.4231	
Standard Deviation	5.8125		5.4360		5.5425	

Note: t = 0.08 d.f = 50, no statistical difference at the 0.05 level of significance (one-tailed test).

would give credence to the veracity of the hypothesis. Therefore, the hypothesis that science park firms are in a better position than off-park firms to establish links with local universities is duly rejected.

DISCUSSION

The results of the study show that the level of interactions between science park and off-park firms and local universities is reasonably robust. Overall, science park firms have more links with universities, although the difference is not shown to be statistically significant. Most universities in Malaysia are funded by the government and provide a lot of opportunities for local businesses in the area of procurement of technology, equipment and services. Local universities do require the products and services of local software and IT firms, thus becoming major customers for local technology-based businesses. Informal contacts still feature high in the study and this is consistent with the observation made by World Bank (2004, 125) which describes the university–industry relationship in Malaysia as 'arm's-length' type, 'where a team of university's professors or research scientists is called in on an as-needed basis'.

One notable observation in the findings is the reasonably high percentage of science park and off-park firms having joint/collaborative research with local universities. In fact, the percentage of science park firms (31.81 per cent) engaged in joint/collaborative research is higher than that found in other studies. For example, in the study by Felsenstein (1994), the percentage is only 13 per cent. There are possibly three reasons for the higher percentage shown in the present study. The first reason relates to the condition imposed on firms receiving government R&D grants. Two of the science park firms (9 per cent) and eight of the off-park firms (23 per cent) are recipients of grants provided under the Industrial R&D Grant Scheme (IGS) by the Ministry of Science, Technology and Environment (MOSTE).[9] An important requirement of this grant is the need for the grant recipient firm to collaborate with researchers in universities or research institutes on the particular research project funded by the grant. However, this condition is not strictly imposed in the case of grants under the Multimedia R&D Grant Scheme (MGS) given by MDC to local ICT and multimedia companies.[10] This partly explains why most of the ICT companies in both the science park and off-park samples do not engage in research collaboration with local universities, although other forms of links are also in evidence. The overall finding is also consistent with the study by Mohnen and Hoaraeu (2002), which found that firms receiving government support for innovation engage in outright collaboration with universities and government laboratories.[11]

The second reason is about the real need of firms to use scientific know-how offered by universities in specific technologies, such as biotechnology and nanotechnology, as observed by Veugelers and Cassiman (2005). In fact, the biotechnology firms in both sample groups have links with local universities. The third reason is the deliberate attempts by both science park management and the technology transfer officials in the universities to link park companies with various university facilities. It should be noted that the CEO of the park is a university professor appointed on secondment by the local university to head the park management. Whereas some companies do not need such links for reasons stated earlier, others do find it essential to forge links with local and foreign universities. For example, TPM Academy, a subsidiary company of the TPM Corporation (park owner), has forged smart partnerships with a number of foreign universities to facilitate exchange of expertise and intelligence in the field of biotechnology, engineering, ICT and business.[12]

However, had it not been for the research grants provided by the government, the level of interaction between the firms and local universities would have been much lower than what it is at present, if not non-existent. In fact, according to a survey by the World Bank (2004), less than 10 per cent of Malaysian firms collaborate with local universities and research institutes. One of the main reasons is the perception amongst most SMEs that university research is not relevant to their needs—a fact acknowledged in the Eighth Malaysian Plan document (see GoM 2001). The World Bank (2004) also cites lack of relevance of services offered by research and technology institutions (including universities) to the firms as the primary reason for lack of collaboration between university and industry in Malaysia. Most of the firms did not view links to university as an important factor influencing their decision to locate in TPM. Another possible reason is that the universities prefer to deal with more established companies for reasons of bigger research grants, prestige and absorptive capacity as shown in a study by Mohnen and Hoaraeu (2002). For example, the Multimedia University in Cyberjaya collaborates with major foreign companies but not with local companies located in the MSC, including those located on TPM.[13]

CONCLUSIONS

The findings of the current study indicate that overall, local R&D institutions in Malaysia have been ineffective in assisting local firms to upgrade their technological capabilities.[14] There is, therefore, a need to strengthen university–industry collaboration to enhance commercialization of research results.[15] This may require policy initiatives to remove the constraints that impede the development of the 'triple helix culture' (Saad and Zawdie 2005). Officials from local universities need to visit science parks to explain the type of research and other facilities available to tenant firms. Industry leaders including those from foreign firms in Malaysia and especially those with expertise in the area of research and technology development should be invited to sit on university boards. TTOs in universities also need to recruit individuals with expertise not only in patent law, licensing or technical expertise, but also those with marketing skills and entrepreneurial experience (Siegel, Westhead and Wright 2003). Besides commercialization of research results, universities can organize programs aimed at assisting science park firms to internationalize their operations. Lessons can be learned from the program run by the University of Oulu to assist science parks firms involved in the software sector in Finland to expand their businesses into the U.S. market.[16]

NOTES

1. A national innovation system is defined as 'a set of distinct institutions which jointly and individually contribute to the development and diffusion of new technologies and which provide the framework within which governments form and implement policies to influence the innovation process. As such it is a system of *interconnected institutions* to create, store and transfer the knowledge, skills and artefacts, which define new technologies' (Metcalfe 1995, 462–463). See also Nelson (1993).
2. See Etzkowitz et al. (2000) for an account of the expanding role of universities from teaching and research to economic and social development.
3. See definition of a science park provided by IASP at http://www.iasp.ws/ (accessed 15 November 2004).
4. Whereas many have acknowledged the role played by the park in the early growth of Silicon Valley, there are some who argue that proximity to university is not a sufficient condition for a successful cluster. See Dosi et al. (2005) for elaboration.
5. The largest science park in the United States is the Research Triangle Park covering more than seven thousand acres. The Aberdeen Science and Technology Park in Scotland is less than sixty acres. See Malairaja (2003).
6. Such maturity would entail possessing manufacturing capabilities, existence of large clusters of high-tech activities with a well-developed local supplier and subcontracting system, technically trained workforce and high level of industrial R&D (UNCTAD 2003, 52). See also Narayanan and Wah (2000) for an account of Malaysia's technological maturity without a strong research base.
7. See Sadullah (2002).
8. The biggest project ever undertaken is the MSC, launched in 1996 to promote and develop information and communication technologies (ICTs) and multimedia activities. The MSC is based on the Silicon Valley model to propel the country into a knowledge-based economy through technological leap-frogging. See Malairaja (2003) and Malairaja and Zaidie (2004).
9. The Ministry is now known as the Ministry of Science, Technology and Innovation (MOSTI).
10. One of the criteria for MGS grants includes linkages to local research institutes and leading companies.
11. Mohnen and Hoaraeu (2002) conducted the study of firms in France, Germany, Ireland and Spain using the information contained in the second Community Innovation Survey (CIS). Large firms and those with patents are the ones that mainly forge such links.
12. TPM Academy was established in 1998 as a subsidiary of TPM and become fully operational in February 2000. As a government company it is owned by the Ministry of Finance Incorporated. It operates under MOSTI.
13. According to Mohan et al. (2004), the Multimedia University has collaboration with thirty-seven companies and twenty-nine universities from all over the world. The collaboration takes the form of scholarships, research grants, setting up of laboratory facilities and sponsorships (equipment, visiting staff).

14. A similar observation has been made by Capannelli (1997).
15. Lessons could be learned from Finland where networks have been successfully developed to promote university–industry links and R&D cooperation (Vestergaard 2003). Similarly Mitra (2000) explains how the Graduates Enterprise programs of the University of Adelaide not only address the issues related to university–industry links, but also promote entrepreneurship.
16. For more details, see Bengtsson and Lowegren (2000) about the success of the Global Software program conducted by the University of Oulu in Finland. The focus of the program is on business and management practices and not on technology.

REFERENCES

Ali, A. 2003. 'Engaging economic development through the commercialisation of research: The Malaysian experience'. Paper presented at the General Conference of the Association of Commonwealth Universities, Belfast, 31 August–4 September.

Bengtsson, L. and Lo Wegren, M. 2001. 'Internationalization in Nordic and Baltic Science Parics. Fianl Report on Nordisk Industrifond. Ideon Science Park Lund.

Capannelli, A. 1997. 'Industry-wide Relocation and Technology Transfer by Japanese Electronic firms a sutdy on buyer-supplier relations in Malayasia' Unpublished PhD dissertation, Tokyo: Hitotsubashi University.

Dierdonck, R. V., K. Debackere and B. Engelen. 1990. 'University–industry relationships: How does the Belgian academic communityss feel about it?' *Research Policy* 19:551–566.

Dosi, A., P. Llerena, and M. S. Labini. 2005. "Evaluating and Comparing the innovation performance of the United States and European Union," Expert report prepared for the Trendchart Policy Workshop 2005. Etzkowitz, H., A. Webster, C. Gebhardt and B. R. C. Terra. 2000. 'The future of the university and the university of the future: Evolution of ivory to entrepreneurial paradigm'. *Research Policy* 29:313–330.

Feldman, M., and M. R. Kelley. 2002. 'How state governments augment the capabilities of technology-pioneering firms'. *Growth and Change* 33 (2): 173–195.

Felsenstein, D. 1994. 'University-related science parks—"seedbeds" or enclaves of innovation?' *Technovation* 14 (2): 93–110.

Ferguson, R. 1995. 'What's in a location?: Science parks and the support of new technology-based firms'. PhD diss., Swedish University of Agricultural Sciences, Uppsala, Sweden. http://www.ekon.slu.se/¬richardf/thesis/spthesis04.html (accessed 15 January 2003).

GoM. 2001. *Eighth Malaysian plan: 2001–2005*. Kuala Lumpur: Economic Planning Unit, Prime Minister's Department.

Harper, J. C. 2003. *Improving the links between tenant companies and higher education institutions: Exploring emerging scenarios for the Manchester science park*. Manchester: PREST, University of Manchester.

HM Treasury. 2003. *Lambert review of business–university collaboration, final report*. Norwich: HSMO Licensing Division.

Ho, I. T. 1999. 'Silicon Valley'. http://www.asis.standford.edu/events/fall99/slides/noslide1.ppt (accessed 29 September 2004).

Howitt, P. 2000. 'The economics of science and the future of universities'. Paper presented at the 16th Timlin Lecture, 16 February, University of Saskatchewan, Saskatoon. http://www.econ.brown.edu/fac/Oeter_Howitt/Publication/Timlin.pdf (accessed 22 May 2004).

Jaffe, A. B. 1989. 'Real effects of academic research'. *American Economic Review* 79:957–970.

Johannesen, J. S., and P. E. Rasmussen. 2000. 'SYLAB—a novel approach for promoting innovation and employment'. Conference Paper, 2000 International Conference on Business Innovation and Technology Innovation, Shanghai, China, 18–19 April.

Kemp R., Folkeringa, J. De Jong and E. Wubben. 2003. 'Innovation and firm performance: Scientific analysis of entrepreneurship and SME's Research Report H200207 & Zoetermeer, The Netherlands: EIM.

Laursen, K., and A. Salter. 2003. 'Search low and high: What types of firms use universities as a source of innovation?' DRUID Working Paper, No. 03–16, Danish Research Unit for Industrial Dynamics, Copenhagen, Denmark.

Lilian, M., and P. Inaki. 2003. 'Technology parks and firm growth'. Paper presented at the XX IASP World Conference on Science and Technology Parks, Lisbon, Portugal, 1–4 June.

Lindelof, P., and H. Lofsten. 2003. 'Science-park location and new technology-based firms in Sweden—implications fro strategy and performance'. *Small Business Economics* 20 (3): 245–258.

Lowegren, M. 2001. 'Advantages of a science park location: Case studies from the Ideon Science Park'. Paper presented at the 17th Colloquium of European Group for Organisation Studies, Lyon, France, 5–7 July.

Malairaja, C. 2003. 'Learning from the Silicon Valley and implications for technological leapfrogging: The experience of Malaysia'. *International Journal of Technology Management and Sustainable Development* 2 (2): 73–95.

Malairaja, C., and G. Zawdie. 2004. 'The "black-box" syndrome in technology transfer and the challenge of innovation in developing countries: The case of international joint ventures in Malaysia'. *International Journal of Technology Management and Sustainable Development* 3 (3): 233–251.

Metcalfe, J. S. 1995. 'The economic foundations of technology policy: Equilibrium and evolutionary perspectives'. In *Handbook of the economics of innovation and technological change*, ed. P. Stoneman, 410–512. Oxford and Cambridge, MA: Blackwell Publishers.

Mitra, J. 2000. 'Nuturing and sustaining entrepreneurship: University, science park, business and government partnership in Austrailia' *Industry and Higher Education* 14(3): 183–190.

Mohan, A. V., A. A. Omar and K. A. Aziz. 2004. 'ICT clusters as a way to materialize a National System of Innovation: Malaysia's multimedia super corridor flagships: The Electronic Journal of Information Systems in Deveoping Countries (EJISDC) 16(5): 1–18.

Mohnen, P., and C. Hoaraeu. 2002. 'What type of enterprise forges close links with universities and government labs? Evidence from CIS2'. MERIT Research Memorandum 2002–008.

Narayanan, S. and L. Y. Wah. 2000. 'Technoloyical maturity and development without research: The challenge for Malaysian manufacturing: Development and change 31, pp. 435–457.

Nelson, R. R. 1993. *National innovation systems: A comparative analysis*. New York and Oxford: Oxford University Press.

New Straits Times. 2005. 'Opinion: The unemployable Malaysian graduate, by Abdul Razak Ahmad'. http://www.nst.com.my/Current_News/NST/Sunday/Columns/NST (accessed 20 March 2005).

Porter, M. 2003. 'The economic performance of regions'. *Regional Studies* 37 (6–7): 549–578.

Saad, M., and G. Zawdie. 2005. 'From technology transfer to the emergence of triple helix culture, the experience of Algeria in innovation and technology capability development'. *Special Issue: The Management of Innovation Revisited: The Strategy, Control and Culture of Innovation, Technology Analysis and Strategic Management* 17 (1): 1–15.

Sadullah, A. F. 2002. 'Commercialisation of research results: Issues and challenges: USM engineering and campus Perspective'. *USM Frontiers* 1 (1): 10–11.

Sanz, L. 2002. *The role of science and technology parks in economic development*. Beirut: IASP. http://www.iasp.ws (accessed 25 August 2003).

———. 2003. 'Science parks: Future perspectives'. Paper presented at the EURADA Conference, Brussels, November.

Schramm, C. J. 2004. 'Accelerating technology transfer and commercialisation'. Paper presented at IP Commercialisation and Research Spinouts Conference, Boston, Massachusetts, 4 November.

Siegel, D. S., P. Westhead and M. Wright. 2003. 'Assessing the impact of university science parks on research productivity: Exploratory firm-level evidence from the United Kingdom'. *International Journal of Industrial Organisation* 21 (9): 1357–1369.

Tidd, J., and M. Brocklehurst. 1999. 'Routes to technological learning and development: An assessment of Malaysia's innovation policy and performance'. *Technological Forecasting and Social Change* 62 (3): 239–257.

UNCTAD. 2003. *Investment and technology policies for competitiveness: Review of successful country experiences*. New York and Geneva: United Nations.

UNIDO. 2004. *IPT's integrated institutional capacity building services and programmes for technology centres and parks*. Vienna: United Nations Industrial Development Organization.

Vedovello, C. 1997. 'Science parks and university–industry interaction: Geographical proximity between the agents as a driving force'. *Technovation* 17 (9): 491–502.

Vestergaard, J. 2003. 'Promoting university interaction with business and community–a comparative study of UK, Sweden and Finland: Research report. Copenhagen: Copenhagen Business School.

Veugelers, R., and B. Cassiman. 2005. 'R&D cooperation between firms and universities. Some empirical evidence from Belgium manufacturing'. *International Journal of Industrial Organisation* 23:355–379.

Westhead, P. 1997. 'R&D inputs and outputs of technology-based firms located on and off science parks'. *R&D Management* 27 (1): 45–62.

Westhead, P., and D. J. Storey. 1994. *An assessment of firms located on and off science parks in the United Kingdom*. London: HSMO.

Wong, P. K. 1999. 'National innovation systems for rapid technological catch-up: An analytical framework and a comparative analysis of Korea, Taiwan and Singapore'. Paper presented at the DRIUD Summer Conference on National Innovation Systems, Industrial Dynamics and Innovation Policy, Rebild, Denmark, 9–12 June.

World Bank. 2004. *World development report 2005*. Washington, DC: World Bank.

Part III

Shift towards a Triple Helix Policy Paradigm and Implications for Developing Countries

14 Creative Reconstruction

A Triple Helix–Based Innovation Strategy in Central and Eastern Europe Countries

Marina Ranga and Henry Etzkowitz

INTRODUCTION

The question has oft been asked in Central and Eastern Europe (CEE) countries[1]: 'What marks the end of transition from the planned economy to the free market?' For some, the answer was the accession to the European Union (EU), in 2004 and 2007, whereas others[2] saw the end only when functional fiscal institutions and social welfare measures were in place. However, irrespective of time horizons, transition no longer means a return to a nineteenth-century liberal society model or even a mid-twentieth-century welfare state, moderating the deficits of capitalism. Transition now takes the meaning of a broader movement towards the twenty-first-century objective of transformation from an industrial to a knowledge-based society that is a common objective for all countries around the world. From this perspective, the CEE countries' transition is part of a broader global movement of 'creative reconstruction'.

The 'creative reconstruction' process thus complements Schumpeter's thesis of 'creative destruction', the demise of mature techno-economic paradigms (Schumpeter 1942), by integrating technological and organizational elements into new configurations. This integration has the potential to replace the separation instituted by the socialist regime to ensure political control. Overcoming separation of institutional spheres is a common problem not only in CEE, but also in the rest of the EU and elsewhere in the world. During the early post-war period, the U.S. (followed by Western Europe) pursued an 'Endless Frontier' thesis in which scientific research was expected to drive the economy. The U.S. gradually developed linkages between sites of advanced research and their use, going against ideological principles of separation, supported by both liberals and conservatives, albeit for different reasons.[3] Perhaps ironically, some of the same principles that the U.S. has left behind in practice have been introduced in CEE by American advisors in the form of a neo-liberal vision that was contradicted by U.S. practice.[4] Despite vast differences in economic, political and academic systems, there is a common global focus on the emerging role of the triple helix in generating knowledge-based economic growth.

In this chapter, we ask whether 'creative reconstruction' processes based on the triple helix model can supersede sub-optimal path dependencies and set in motion the transition to a knowledge-based society supported by indigenous innovation capacities. We start with an analysis of reform drivers of national research and development (R&D) and innovation systems in the CEE countries in the next section, followed in the third section by a few key elements of the triple helix model and the stages of an innovation-based strategy defined by this model. In the fourth section through the seventh section we examine changes in university, industry and government associated with each of the stages of the triple helix–based innovation strategy. In the final section we conclude with an assessment of the viability of a triple helix innovation strategy and a discussion of several areas of intervention to accelerate the process.

NATIONAL INNOVATION SYSTEMS AS DRIVERS OF TRANSITION

The collapse of communism in the CEE countries in the late 1980s revealed the elements of fragmented and dysfunctional national R&D systems, heavily government led and with weak links between institutional spheres, despite ideological principles of integration and science-based progress as a core strategy for the 'scientific technological revolution'. Subsequent attempts to rebuild national R&D and innovation systems in the CEE countries from the 1990s until the present have been part of a broader process of political and economic integration into the EU, which was driven by two main factors:

1. legislation, policies and programs associated with accession to the EU on the political level
2. foreign direct investment (FDI) on the economic level.

Overall, the interplay of these factors can be considered to have been successful in accelerating CEE countries' catching up process: the adoption of the *acquis communautaire* is largely completed and the transition from planned to free market economies is well advanced in many of these countries.[5] Although the time to catch up[6] to EU average innovative performance varies among the new member states of CEE the economy of these countries is improving and the development gap is narrowing. CEE countries have significantly outpaced their Western European counterparts in terms of average gross domestic product (GDP) growth rate (e.g. Latvia with 11.9 per cent in 2006 and Romania with 9 per cent in 2008, close to China, the world leader at 9 per cent on average for the past twenty-five years). Much of this massive growth was the effect of government commitments to stable monetary policy, export-oriented trade policies, low flat-tax rates and the use of relatively cheap labour, as well as a cyclical low in the Western European economies at the turn of the century.

In contrast, the reform of national R&D and innovation systems has been very slow and has ambiguously evolved between restructuring and erosion of potentially viable R&D and technology capacities (Radosevic 1998; Müller and Etzkowitz 2000). Therefore, R&D and innovation had only marginal effects on the economic revival of CEE countries, which was primarily the result of the two factors mentioned earlier, the political influence of the EU, on the one hand, and the economic impact of FDI on the other. Under the influence of these two factors, two complementary strategies for reforming national R&D and innovation systems have been pursued, which are largely similar to the corporate 'make-or-buy' strategic choice between internal (in-house) production of goods or external acquisition from outside suppliers:

1. An FDI-based renewal of the technology base, skills, markets, management techniques, etc. ('buy'), focused on technology imports and acquisitions, and the expectation of knowledge spillovers from large corporations establishing local branch plants to take advantage of a highly skilled workforce and cheap labour costs. This short-term strategy, however, is a 'mixed blessing', encompassing both positive and negative effects of FDI that are discussed in detail in the following. In addition, it provides only a temporary competitive advantage, as FDI flows may be redirected to other countries with similar highly skilled workforces when the labour costs increase in the recipient countries. Moreover, it may have the unintended consequence of depleting the local potential for R&D human resources, at least temporarily, because they cannot find relevant employment and leave the region or indeed the country.[7] By accessing technology from abroad rather than developing it indigenously, the FDI-based renewal of technology can also be considered a continuation of the socialist strategy of 'reverse engineering' of Western technologies, as opposed to developing technology from a domestic science and technology base. Knowledge gains can be relevant in both cases, but they require lower, yet not negligible, technical skills, especially for reverse engineering, than in the case of in-house development of technology.
2. A strategy of restructuring and enhancing local R&D potential ('make'), supported through various policy instruments of national and regional governments, as well as EU funding, especially through the Structural Funds. This medium- to long-term strategy, although more beneficial for the local R&D and innovation base, has been less significant than the former, given the very limited funding made available by the government. For example, government expenditures for

R&D (GBAORD) for all the CEE countries that recently joined the EU (2004 and 2007 enlargements) were below the EU average, with Poland, Slovakia, Romania and Latvia at the lowest end of the scale, with shares of under 0.3 per cent. All the new Member States were not only below the EU average, but also allocated less than EUR 100 per inhabitant to GBAORD. In Romania, the figure was actually less than EUR 10 per person (Eurostat 2008).

Most reform and restructuring efforts in the CEE countries have been guided by the theoretical framework of national/regional innovation systems, which highlights various innovation actors and the interactions between them at the sectoral, regional and national levels (Freeman 1987; Lundvall 1992; Nelson 1993; Nelson and Rosenberg 1993; Lundvall et al. 2002; Edquist 1997; Edquist and Hommen 1999). In contrast, the triple helix model of university-industry-government relations has been much less visible and had significantly less impact in these countries, possibly because of some structural characteristics that impeded the diffusion and use of the model, such as universities being primarily teaching institutions and having relatively low levels of academic research and entrepreneurial activities (patenting, licensing, formation of academic spin-offs and incubation, etc.), weak R&D capacity of domestic firms and internal R&D demand, low public and private R&D expenditure, virtual absence of seed capital, relatively weak horizontal policy coordination, etc.

Many of these aspects have changed dramatically, especially over the last decade, which raises the question of whether the recent transformations within and among university, industry and government institutional spheres can now provide a good basis for the application of a triple helix–based innovation strategy in the CEE countries.

KEY ELEMENTS OF THE TRIPLE HELIX MODEL

The triple helix thesis is that the potential for future economic development in a knowledge-based society lies in a more prominent role for the university and the hybridization of elements from university, industry and government to generate organizational innovation. The model comprises three basic elements.

1. There is a more prominent role for the university in innovation, on a par with industry and government in a knowledge-based society.
2. There is a movement toward collaborative relationships among the three major institutional spheres, in which innovation policy is increasingly an outcome of interactions among the spheres rather than a prescription from government or an internal development within industry.

3. In addition to fulfilling their traditional functions, each institutional sphere also 'takes the role of the other', operating on a vertical axis of their new role as well as on the horizontal axis of their traditional function.

Triple helix formats comprise several configurations of the key actors, ranging from a 'statist' mode, in which the state encompasses and controls academia and industry, and directs the relations between them, to a 'laissez-faire' mode, where government, university and industry are separate and operate apart from each other, with interaction typically arranged by intermediary organizations across strong boundaries. Also, an 'interactive' or 'balanced' mode consists of relatively autonomous yet overlapping institutional spheres, each taking the role of the other, with hybrid organizations emerging at the interfaces (Etzkowitz 2008) (see Figure 14.1).

A new role for universities in society is emerging worldwide, beyond the traditional 'servicing' or networking role that arose in the aftermath of World War II and even predated it in certain areas like agricultural sciences. This new role has arisen due to changes in the technological foundations of economic growth and the development of science-based technologies, a more aggressive commercial stance among universities and a transformation of normative patterns among academic scientists. Universities are combining their research and teaching capabilities in new formats. They not only train students in various scientific disciplines, but also train and encourage them to become entrepreneurs and take up new roles as firm founders, especially in advanced areas of science and technology. Moreover, universities are extending their teaching capabilities from educating individuals to shaping organizations in entrepreneurial education and incubation programs.

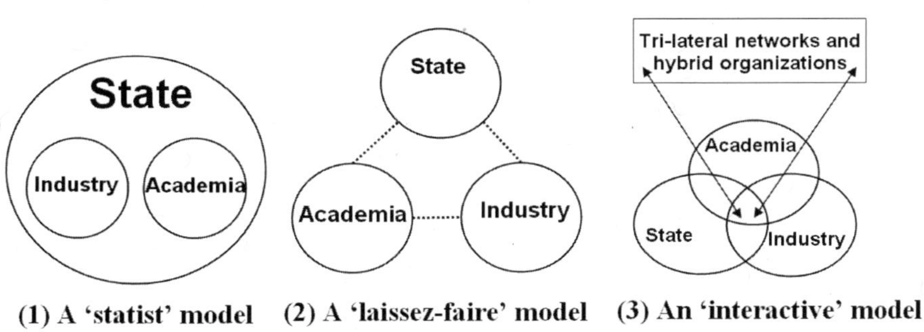

Figure 14.1 Triple helix configurations.
Source: Etzkowitz (2008).

The academic universe is transformed through the advancement of entrepreneurial dynamics to the forefront of university activities. The emergence of the 'the entrepreneurial university' is the result of two interrelated dynamics.

1. *The 'first academic revolution'*—the inclusion of research, next to education, into the university mission. Research as an inextricable part of the teaching process was expanded to an increasing number of disciplines, from humanities to sciences. Practical implications were discerned in some of these research results and steps were taken to put them to use.
2. *The 'second academic revolution'*—the contribution of university to socio-economic development, by producing potentially commercializable research results.

The closing gap between basic and applied research and the emergence of basic research objectives from industrial objectives accelerated the convergence between two types of universities: on the one hand, the 'ivory tower' university focused on the development of academic disciplines with the exclusion of consulting, contract research and other industrial activities, and, on the other hand, universities specializing in applied research for agriculture and industry, the 'land grant' foundations in the U.S. and polytechnic institutions in Europe. The expansion of university missions

Table 14.1 Expansion of University Mission

Teaching	*Research*	*Entrepreneurial*
Preservation and dissemination of knowledge	*'First Academic Revolution'*	*'Second Academic Revolution'*
One mission: teaching (twelfth century to the mid-nineteenth century)	Two missions: teaching and research (mid-nineteenth century to present) The new research mission generated conflict of interest controversies. Tensions largely solved by creating the conditions for the research university as a new academic mode.	Three missions: teaching, research, contribution to socio-economic development (various starting points in time, with the earliest forms at MIT and Stanford in the early twentieth century) The new mission of contribution to socio-economic development also generates conflict of interest controversies. Tensions largely solved by creating specific rules and entities specific to the third mission, and legitimizing the entrepreneurial university as a new academic mode.

Source: Based on Etzkowitz (2003a).

and the essential elements of the first and second academic revolutions are presented in Table 14.1.

Etzkowitz (2003a) describes the development of a triple helix innovation strategy as a sequence of four main stages.

1. First, *internal transformations in each helix that strengthen research and innovation capacity*, such as the development of lateral ties among companies through strategic alliances or changes in the resource base of university systems.
2. Second, the *influence of one helix upon another* and the emergence of 'hybrid' organizations at the interfaces between institutional spheres, such as university technology transfer offices, business incubators and science parks.
3. Third, the *creation of a new overlay of trilateral networks and organizations* among the three helices, which can generate new ideas and formats for high-tech development, particularly at the level of regional industrial clusters.
4. Fourth, the *recursive effect of these exchanges among institutional spheres*, both on the spirals from which they emerged and on the larger society. One such effect is on science itself, as a result of internal changes within academia, strengthened and diffused by government policy (Gibbons et al. 1994). Another is on the firm, which increasingly operates as a distributed network, drawing resources from across the triple helix as well as from various nations and regions. Yet another is the more fundamental shift from an industrial to a knowledge-based society.

In the next sections we examine a number of changes arising in the triple helix actors in the CEE in relation to each of the stages mentioned earlier, in order to assess the feasibility of a triple helix–based innovation strategy.

INTERNAL TRANSFORMATIONS IN THE TRIPLE HELIX INSTITUTIONAL SPHERES IN CEE COUNTRIES: THE LONG JOURNEY OF TRANSITION

University

Most of the CEE universities are still at the stage of the 'first academic revolution', with predominantly teaching functions and less involvement in research, following the Soviet pattern of basic research being performed primarily in academies of science and applied research in branch institutes. However, important changes have occurred both in the teaching and in the research function of the university that we briefly discuss in the following.

The Teaching Function of the University

The reform of the higher education sector in the CEE focused primarily on the *teaching function* and was a slow and difficult process due to the heterogeneity of national systems of higher education (Scott 2002). Two notable developments characterise the reform of the university teaching function.

1. An *unprecedented shift from elite to mass higher education*, due to a rapidly expanding demand for higher education as a way to reduce unemployment pressure. For instance, in 2002 unemployment rates for tertiary educated people in the CEE countries were similar to those in the EU, both standing at 3 per cent, with small deviations from this average in Bulgaria with 6 per cent (Eurostat 2003). Access to higher education improved significantly over the 1989–1999 decade, and the trend continued in the 2000s.[8] Over the 1998–2002 period, the highest gross enrolment rates (GER) in Europe were recorded in Romania (61 per cent), Poland (60 per cent), the Baltics (40–57 per cent), Czech Republic, Slovakia and Hungary (32–46 per cent), significantly higher than the EU25 average of 16 per cent (Kwiek 2006b).
2. The *emergence of private higher education providers*, particularly in Poland and Romania, where the private higher education enrolments reached almost 30 per cent of total enrolments in 1999–2002 (Kwiek 2006a). One of the ironies of transition is that some former teachers of Marxism-Leninism in state universities morphed into entrepreneurship educators in the new private universities when their former academic home departments were disbanded.

The government played a key role in the reform process, with more or less success in different countries (e.g. Eisemon et al. 1995). In addition, some EU initiatives such as the TEMPUS program (Trans-European Mobility Scheme for University Studies) made a notable contribution, particularly in curriculum modernization, teacher training, university management and mobility among higher education institutions, both from the EU and the partner countries.

Increasing access to higher education resulted in large numbers of graduates, up to ten times higher than in the EU15 (Table 14.2). From a disciplinary perspective, the absolute number of graduates in engineering, manufacturing and the construction sectors was higher than in the sciences, but the growth rate in sciences in the CEE countries was considerably higher, possibly due to the tendency of universities to move towards theoretical areas that do not require expensive scientific equipment (Radosevic and Kriaucioniene 2006). These developments are also reflected in high values of such indicators as science and engineering graduates, tertiary education and particularly youth education, which exceed the EU25 average for almost all CEE countries (Figure 14.2).

Table 14.2 Graduation from Tertiary Education and Annual Average Growth Rates (AAGR) in Total and Selected Fields of Study

	Total graduates		*In science*		*In engineering, manufacturing and construction*	
	AAGR 1998–2001	*Per 1,000 population aged 20–29*	*AAGR 1998–2001*	*Per 1,000 population aged 20–29*	*AAGR 1998–2001*	*Per 1,000 population aged 20–29*
EU15 [(1)]	2.5	40.4	5.4	4.5	1.9	5.9
Accession countries (ACC)	19.5	55.3	42.3	2.3	7.3	4.9
Czech Rep.	12.5	25.9		2.7		3.0
Estonia	10.0	39.9	28.1	2.4	31.9	4.8
Latvia	24.8	62.6	32.9	3.2	-3.0	4.4
Lithuania	12.7	51.5	18.0	2.5	13.2	10.6
Hungary [(2)]	15.0	36.1	-16.6	0.9	-0.9	3.6
Poland	23.6	74.3	83.8	2.6	9.8	5.1
Slovenia	7.1	40.0	-1.8	1.5	2.3	6.6
Slovakia	14.6	29.4	23.6	2.6	21.6	5.0
Bulgaria	4.3	43.7	10.8	1.8	10.7	6.6
Romania	4.0	22.7	2.0	1.3	4.9	4.2

(1) EU15 and ACC are estimated.
(2) Annual average growth rate (AAGR) calculated for 1998–2000.
Source: Eurostat (2003).

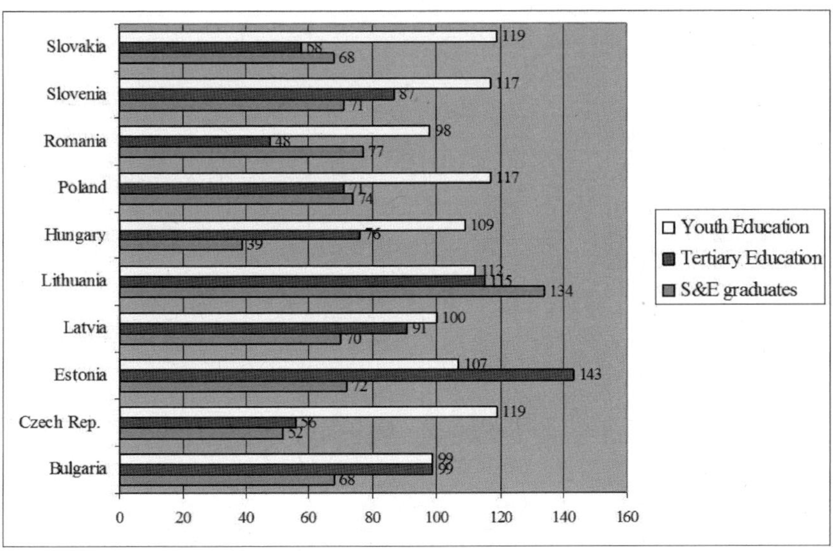

Figure 14.2 Innovation performance indicators in CEE countries (% relative to EU25), 2005.

Source: European Trend Chart on Innovation (2005). European Innovation Scoreboard 2005. Comparative Analysis of Innovation Performance. Available at: www.proinno-europe.eu/extranet/admin/uploaded.../EIS_2005.pdf. Last accessed: 7 September 2010.

Larger numbers of higher education graduates also translated into larger numbers of people working in science and technology (S&T) areas, at similar levels to the EU. Eurostat (2003) show that the percentage of people aged twenty-five to sixty-four who either work in S&T occupations or have a tertiary-level education is similar in the EU and the CEE countries (26.6 per cent in EU15 versus 21 per cent in Accession Countries in 2002). Moreover, although the proportion of people with tertiary education was lower in the Accession Countries, these people were subsequently more likely to work in S&T occupations than their EU counterparts (61 per cent versus 55 per cent). A likely explanation of this apparent paradox is the relative stability of these jobs which were largely funded by public resources.

The Research Function of the University

In many CEE countries in the 1990s a strengthening of the academic research function started on the basis of new legislative frameworks for higher education and research that gave back to universities research prerogatives that had been lost in the Soviet model of dividing basic and applied research capabilities between the academies of sciences and public research institutes, respectively. For example, the Higher Education Law introduced in 1993 in Hungary stated the importance of university research and gave back to the university the right to award doctoral degrees and appropriated new resources for university research. This led to a significant change in the collaboration between universities, the Hungarian Academy of Sciences and public research institutes modelled after the German Fraunhofer Foundation (OECD 1998). Similarly, in 1990 Slovakia adopted a very liberal higher education law that granted autonomy and full academic liberties to universities (Navrat and Molnar 1998).

The research potential of universities has been strengthened since the 1990s through a variety of policies and programs supported by the government and, more recently, by the EU, such as[9]:

- Slovenia's *Research Group Funding Program*, introduced in 1999, provides support for long-term basic research (three to six years funding) in natural sciences, engineering, medical sciences, biotechnology sciences, social sciences and humanities. Since its inception, this has been the largest source of public funding for research, being open to universities as well as public research institutes and firms.
- Latvia's *Support Scheme for the Implementation of Doctoral Programs and Postdoctoral Research*, introduced in 2004, promotes the renewal of academic and research staff in higher education establishments and research institutes, by supporting new researchers, PhD students and PhD graduates in natural sciences, engineering sciences and technologies, health, agriculture and forestry sciences. The program is currently implemented in five

Latvian universities and is 75 per cent co-financed by the European Social Fund.

- Hungary's *Support Program to Basic Research for Innovation* funds mission-oriented basic research for economic or social benefits, in life sciences, technical and natural sciences. The program was launched in 2007 in cooperation between the National Scientific Research Fund (OTKA) and the National Office for Research and Technology (NKTH), the two main national organizations funding basic and applied R&D, respectively.

The strengthening of the university research function is also visible in the high numbers of R&D personnel in the higher education sector in most CEE countries (e.g. Estonia, Latvia, Lithuania, Hungary, Poland, Slovakia), in contrast to EU15, where the business sector comes ahead with 0.74 per cent R&D personnel in the labour force. By field of science, Romania, Bulgaria, Czech Republic and Slovenia stood out with the highest shares of R&D personnel in engineering and technical sciences in 2002, whereas natural sciences concentrated the largest share of higher education researchers in Estonia and Latvia and humanities was the most important field of science for Hungary and Poland (Figure 14.3).

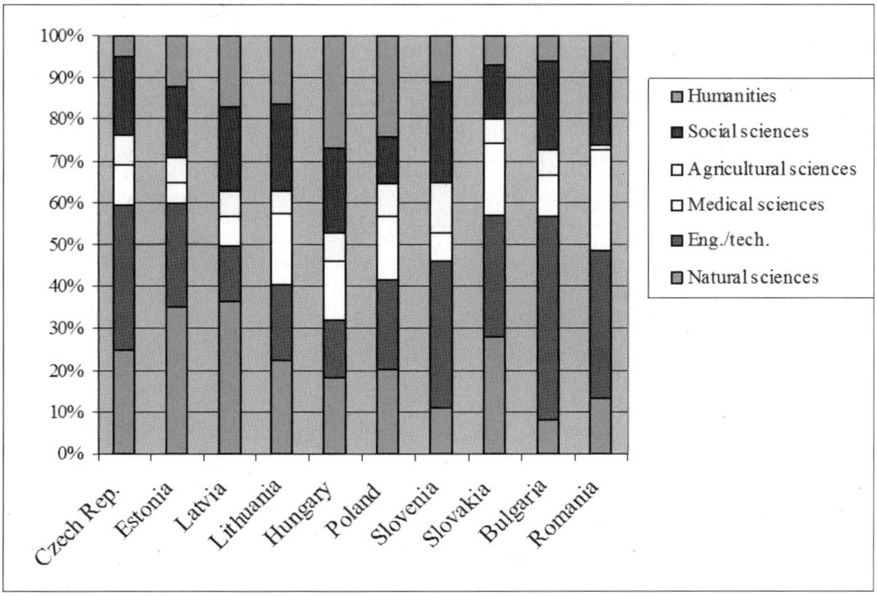

Figure 14.3 Percentage of researchers in the higher education sector by fields of science, 2002.

Source: Eurostat (2005)), Science, technology and innovation in Europe. Available at: http://epp.eurostat.ec.europa.eu/cache/ITY_OFFPUB/KS-76-06-203/EN/KS-76-06-203-EN.PDF. Last accessed: 8 September 2010..

The strengthening of the university research function has also created new opportunities for collaborative links with industry and paved the way for the introduction of entrepreneurial activities[10] and the creation of 'hybrid' institutions at the science–industry interface, like technology transfer offices, science parks, business incubators, etc., which will be discussed in detail in the following, as elements of a further stage of a triple helix–based innovation strategy.

Industry

All CEE countries experienced a sharp economic decline in the early days of transition. Recovery started only very hesitantly in 1993 in a few Central Europe countries and accelerated by the end of the decade in the whole region, although for some of the countries the economic rebound and return to pre-1990 levels of the GDP was a slow and lengthy process that took more than a decade. Economic growth and recovery emerged as the result of several factors, including financial stabilization and structural reforms, market liberalization and opening to EU and international trade, privatization and market-friendly government rules and institutions and production improvements stimulated by FDI.

Technological change in industry was only marginally related to R&D (Havrylyshyn 2001; Dyker and Radosevic 1999). Some of the key factors influencing this process will be briefly discussed in the following:

- FDI flows
- the innovative capacity of industry
- small and medium-sized enterprises (SMEs)

FDI Flows in the CEE Countries

FDI flows have been driven by several factors, including: increasing institutional and political stability; relatively low labour costs in the region and proximity to consumers (Lankes and Venables 1996); FDI-supportive government policies (Pavlinek 2002); enforcement of intellectual protection rights laws, especially in the high-tech sector (Smarzynska 2004); traditional market size and cost factors; and the expectation of EU membership for the more advanced countries in the region (Clausing and Dorobantu 2005). The EU membership prospect, in particular, has been a key driver for anticipatory investment in the region as a way to gain a foothold in potentially profitable future markets. Also, social relations such as political, migration, trade and cultural relations between investors and hosts had a significant influence on FDI flows in the region (Bandelj 2002). Under the cumulated influence of these factors, FDI intensity by country varied significantly over 1997–2004 (Table 14.3).

Table 14.3 FDI Intensity in CEE Countries, 1997–2004

	2004	*2003*	*2002*	*2001*	*2000*	*1999*	*1998*	*1997*
EU25	0.9	1.3	1.3	2.4	:	:	:	:
EU15	1	1.3	1.4	2.4	3.5	2.8	2	1.1
Czech Rep.	2.3	1.3	5.9	4.8	4.6	5.5	3.1	1.2
Latvia	2.9	1.5	1.4	0.9	2.7	2.6	3.1	4.3
Lithuania	2.3	0.6	2.6	1.9	1.7	2.3	4.2	1.9
Estonia	5.8	5.9	3.0	6.2	4.1	3.5	5.2	4.1
Hungary	:	2.3	2.5	4.1	2.3	2.3	2.7	2.9
Poland	2.6	1.2	1.1	1.5	2.7	2.2	1.9	1.6
Slovenia	1.7	2.4	2.4	1.4	:	:	:	:
Slovakia	1	1.1	7.9	3.7	5.3	-0.1	1.6	0.6
Bulgaria	5.4	5.3	2	2.6	4	3	:	:
Romania	:	:	1.3	1.4	1.4	1.5	2.4	:

Source: Eurostat (2005). Eurostat (2005), Tables, Graphs and Maps Interface (TGM) table. Available at: http://epp.eurostat.ec.europa.eu/tgm/table.do?tab=table&init=1&language=en&pcode=tsier130&plugin=1. Last accessed: 8 September 2010.

CEE countries have become an increasingly attractive destination for FDI, starting from a modest base in the early 1990s and reaching $174.1bilion of FDI between 2001 and 2006.[11]

The amount of actual investment in 2004 was $1.1 billion in Slovakia, $4.2 billion in Hungary, $4.5 billion in the Czech Republic and $6.2 billion in Poland (Allen and Overy 2006). Most FDI inflows came from the EU, in particular, Germany, Austria, the Netherlands, France and UK,[12] and to a smaller extent from non-EU investors[13] and international donors.[14]

Most FDI was concentrated in low-technology sectors (both manufacturing and services), and much less in high-technology manufacturing, due to the lack of marketing and technological innovation in this sector, but also to relative instability and uncertainty at macro- and microeconomic levels in the host countries (Resmini 2000). Two major trends in the FDI flows have been observed: (a) a move from traditional manufacturing to service industries (banking, IT, telecoms, etc.) in the late 1990s and early 2000s, and more recently, a return to manufacturing (for example, the recent major investments by Western car makers in automobile plants in the Czech Republic and particularly Slovakia); and (b) a shift from repatriating to reinvesting profits (Allen and Overy 2006). Initially perceived by Western investors as an amorphous group, the CEE countries developed a distinct profile over time (Disdier and Mayer 2004).

FDI flows remain, however, a 'mixed blessing'. On the one hand, there are some undeniable positive aspects, such as access to investment capital, sale and distribution networks of parent companies, technology transfer (Pavlinek

2002), technology upgrading, skill improvements, training and qualification, new supplier networks and improvement of management methods. In some cases, influences on the education and business systems, with some visible effects at the regional level were observed, as well as improved access to finance, investments and better quality of the small local suppliers through vertical and horizontal spillovers in the supply chain (Dries and Swinnen 2004). FDIs were also credited with strengthening the political and financial stability of the CEE countries, creating anti-corruption pressures and introducing environmentally friendly technologies (Sakova and Hansenova 2002).

On the other hand, FDI tended to reinforce existing disparities by crowding out less competitive firms, as spillovers were confined to domestic companies with sufficient technological absorptive capacity (Srholec 2003). However, technology-intensive firms were rarely able to improve their technology base through venturing with multinationals (Sadowski 2001). FDI flows in the Czech Republic, Hungary, Poland and Slovakia in the 1990s increased regional development disparities, led to a dual economy, failed to develop linkages with local and regional economies, concentrated in unstable industries, depended upon cheap labour and increased the trade deficit and regional economic instability (Pavlinek 2004; Sakova and Hansenova 2002). The uneven effects are mostly related to underlying regional differences in capability, skills, human resources base, etc. Also, the learning effects of FDI remain a controversial issue (Konings 2001). Large FDI inflows were accompanied by slow GDP growth, with over 70 per cent of the total inflows coming through privatization and having only short-term positive effects (Ferencikova and Dudas 2005).

The consolidation of production linkages through FDI flows and subcontracting increased the dependence on specialized subsidiaries in low-value-added activities (e.g. clothing and furniture in Romania and Poland). If FDIs have been relatively successful in facilitating the EU integration in terms of markets and production locations, they did not promote a similarly successful technological integration, which depends to a large extent on the local knowledge base that remains weak in most CEE countries (Radosevic 2002). For instance, networked companies are heavily dependent on networks as sources of information, learning and innovation. However, close cooperation was not found to always mean deep cooperation, and high levels of integration into international networks did not necessarily mean high levels of dependence (Dyker et al. 2003).

As a new form of FDI, investment of private equity funds in emerging markets (also known as financial foreign direct investment [FFDI]) is at an early stage. Although the first fund-raisings and investments were completed in the early 1990s, the bulk of activities did not start until the mid-1990s and late 1990s. Hence, many of the private equity funds have not yet completed a full cycle of investments and exits, and the overall returns of the industry to date may not be fully meaningful.

In 2005, private equity investments across the CEE region totalled EUR 508 million, some 7 per cent below the 2004 level. In 2005, over 85 per cent of the region's investment activity was concentrated in Hungary, the Czech Republic, Poland and Romania. Hungary, in particular, received the largest investment pool, totalling EUR 147 million, up by 21 per cent from 2004. The total investment as a percentage of the GDP in 2005 was by far the highest in Hungary (0.167 per cent) followed by Estonia and Czech Republic (0.12 per cent and 0.11 per cent, respectively). About half of the total investment was used for buyouts, whereas start-up capital represented only about 2 per cent (European Private Equity and Venture Capital Association [EVCA] 2006).

The positive effects of private equity funds in the CEE countries are scarcely documented, but the available evidence suggests that they brought strong advantages for early entrants. For instance, the venture capital industry in Poland developed during 1990–2003 through three distinct phases (development, expansion and correction) that broadly followed a normal Western-type venture capital cycle. A strong local presence of venture capital funds seemed to facilitate higher returns and an increasing specialization of venture capital firms once the industry matured (Klonowski 2005).

The Innovative Capacity of Industry in the CEE Countries

Much of the situation described in the preceding originates in the weak innovative capacity of industry in the CEE countries, which is much lower than the EU27 average, except for Estonia (Figure 14.4).

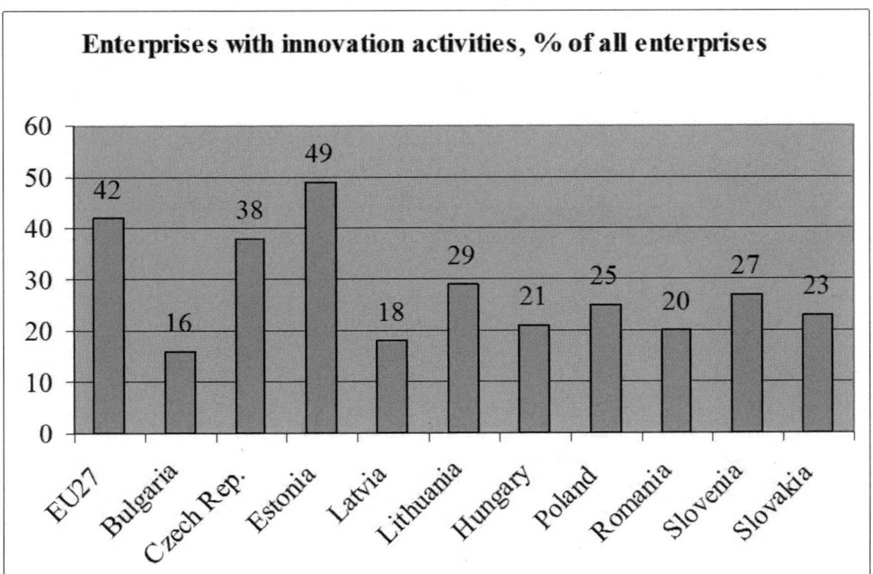

Figure 14.4 Enterprises with innovation activity in CEE countries, 2002–2004.

Note: Data for Czech Republic correspond to the reference period 2003–2005.

Source: Eurostat. 2007. Fourth Community Survey. Available at: http://europa.eu/rapid/pressReleasesAction.do?reference=STAT/07/27&format=HTML&aged=1&language=EN&guiLanguage=en.

Small and Medium-Sized Enterprises (SMEs) in the CEE Countries

SMEs account for about 99 per cent of the firm population in the CEE, but they perform limited production and innovation activities. Such activities remain concentrated in a few leading large firms that are most often favourite recipients of FDIs. The links between large firms and SMEs are underdeveloped, and the poor innovative capacity of SMEs remains one of the weakest parts of the CEE countries' innovation systems.

SME development has been a key economic reform issue addressed by the post-communist governments in the transition years. SME numbers have increased significantly, primarily as a result of the break-up of larger inefficient state enterprises and, to a much smaller extent, of mass privatization of smaller units, such as retail outlets. In 2001, SMEs accounted for 72 per cent of total employment in the then Candidate Countries, versus 66 per cent for EU19. The difference was concentrated in micro enterprises, which provided 40 per cent of total employment in the Candidate Countries, versus only 34 per cent in EU19. A convergence to the EU in terms of decreasing average enterprise size was observed, particularly in the Baltic countries and Hungary, as a result of changes in the economic structure, such as restructuring of former state-owned enterprises, improvements in the institutional environment for starting new businesses, and the rise of the services sector (European Commission 2002). However, in terms of innovative capacity, SMEs in the CEE generally remain below the EU average, with the exception of Estonia, which scores above the EU25 average in most innovation and entrepreneurship indicators (Table 14.4).

Table 14.4 Innovation and Entrepreneurship Indicators for CEE Countries (% of EU25 Average)

	SMEs innovating in-house	*% of all SMEs collaborating on innovation*	*Early stage venture capital*	*Innovation expenditure*	*Non-technical change*	*ICT expenditure*
Bulgaria	36	25	0	38	20	137
Czech Rep.	91	46	4	61	94	113
Estonia	115	123	0	79	124	137
Latvia	59	53	0	93	84	113
Lithuania	85	134	0	96	72	121
Hungary	52	285	6	20	69	92
Poland	48	71	28	150	-	114
Romania	48	29	10	67	182	24
Slovenia	59	76	0	61	120	95
Slovakia	62	33	6	160	24	83

Source: European Trend Chart on Innovation (2005). Comparative Analysis of Innovation Performance. Available at: www.proinno-europe.eu/extranet/admin/uploaded.../EIS_2005.pdf. Last accessed: 7 September 2010.

Very few SMEs are seriously engaged in manufacturing involving innovative capital investment and new technology, and they are much smaller and less experienced than those in the EU. In general, SMEs are very small and are typically family firms with only a few employees.

Government

Government's role in restructuring national R&D and innovation systems has gradually moved from a relatively passive stance in the first half of the 1990s, marked by a mix of shock therapy and gradualist strategies (with Poland and the Czech Republic coming closest to using shock therapies, and Romania at the opposite end, of gradualist strategies aimed to ensure the stabilization of the economy, inflation and unemployment), to a more active involvement in the late 1990s and the 2000s, in a shift from a 'hands-off' to an 'assisted linear' model of policy intervention (Etzkowitz 2006).

In the R&D and innovation area, the key challenge for the CEE governments was to stop the dissolution of national R&D systems that posed a major threat especially in the first half of the 1990s. This arose primarily from the massive loss of researchers, either through internal migration to other, more profitable economic areas or through external migration (Brain Drain) to the West. Other major challenges included: scarce public resources for R&D, which made the survival of many R&D institutions difficult, so that many of them disappeared or split into smaller units that have been partially or totally privatized; outdated equipment and research infrastructure; weak science–industry linkages; and poor absorptive capacity of public and private R&D institutions. In addition, R&D institutes—a specific structural characteristic of R&D systems in the CEE—continued to play an important role, in their contradictory evolution between reinforcement and transformation, or as Radosevic (1996) put it, between 'shock' and 'gradualism' reform strategies.

Some R&D institutes have had their traditional functions strengthened through successful appeals to autonomy, political connections and gradually increased funding in reviving economies. For instance, in Estonia, they have undergone favourable restructuring and established new types of interactions with business firms and international partners (Paasi 1998). In Romania R&D institutes account for the highest share of national R&D-performing institutions (European Commission 2007) but their reform has been a slow process. In Hungary, some small institutes have been incorporated into universities as research centres, and many others have disappeared within 'an awkward structural change' (Mosoni Fried 1995), being transformed into SMEs or wound up by the state. The main policy focus of R&D institutes is placed on strengthening their linkages with other innovation actors and better integration into industrial and regional development strategies.

The distinction between R&D and innovation policies became more apparent in the 1990s, with some notable differences among countries. For instance, Estonia and Slovenia are much more advanced in this process, whereas Romania and Bulgaria have only recently moved towards more differentiated policies. New institutions have been created to address the challenges of restructuring the R&D systems and developing full-fledged, functional innovation systems at national and regional levels. In addition to the ministries in charge of science and technology, education or economic affairs, a wide range of new government agencies, councils, bureaus, etc., have been created, but this institutional renewal often had the adverse effect of weakening the process of decision-making and horizontal coordination. A certain improvement in the horizontal coordination between R&D and innovation government policy-making structures emerged after the EU accession (2004 and 2007 enlargements) in the implementation of the operational programs funded by the Structural Funds and to some extent by national funds, as their management usually involves several government agencies. R&D and innovation policies have gained higher visibility in such programs and are related to broader development strategies, especially industrial, human resources and regional policies.

An important stimulus for all CEE governments in the effort to increase public R&D and innovation funding came from their commitment to meet the EU's Lisbon Strategy objectives and the so-called 'Barcelona target', i.e. raising the R&D and innovation funding to 3 per cent of the GDP, of which 1 per cent would be from public funds and 2 per cent from the private sector. In this context, one of the most significant funding increases emerged in Romania, where public expenditure for R&D and innovation increased from 0.15 per cent of the GDP in 2004 to 0.26 per cent in 2005, 0.38 per cent in 2006 and 0.56 per cent in 2007, with prospects to go up gradually to 1 per cent by 2010 (European Commission 2007). This is a significant improvement, meeting multiple needs within the R&D and innovation system that has been severely under-funded for more than a decade.

INFLUENCE OF ONE HELIX UPON ANOTHER AND THE EMERGENCE OF 'HYBRID' ORGANIZATIONS AT THE INTERFACES BETWEEN INSTITUTIONAL SPHERES

The changes in the triple helix institutional spheres discussed earlier created new opportunities for industry–science collaboration and stimulated the emergence of interface 'hybrid' organizations, such as science parks, business incubators, technology transfer centres and spin-offs in universities. The collaborative links of innovative enterprises in the CEE countries with universities and other higher education institutions, government and public research organizations often outrun the EU27 average (Table 14.5).

Table 14.5 Innovation Activity and Cooperation during 2002–2004

	Enterprises with innovation activities, % of all enterprises	*All types of cooperation with other enterprises or other institutions*	*Cooperation partners*			
			Suppliers	*Clients or customers*	*Universities or other higher education institutions*	*Government or public research institutes*
			% of all innovative enterprises			
EU27	**42**	**26**	**17**	**14**	**9**	**6**
Bulgaria	16	22	16	13	6	4
Czech Rep.	38	38	31	26	13	7
Estonia	49	35	23	23	9	6
Latvia	18	39	33	29	14	12
Lithuania	29	56	45	35	12	10
Hungary	21	37	26	20	14	5
Poland	25	42	28	16	6	9
Romania	20	17	14	10	4	4
Slovenia	27	47	38	33	19	13
Slovakia	23	38	32	30	15	11

Note: Data for Czech Republic corresponds to the reference period 2003–2005.
Source: Fourth Community Innovation Survey.

Industry–science linkages have been supported by a wide range of government policy measures, some initiated in preparation for EU accession, others adopted after accession, in the framework of the operational programs. Such measures aim to enhance knowledge transfer to industry, stimulate the mobility of researchers from the public to the private sector, encourage the creation of consortia, excellence poles, technology platforms, clusters, etc., and promote a stronger involvement of universities and firms in regional development (Table 14.6).

The effectiveness of such programs is, however, difficult to assess, as formal evaluations are not systematically available and evidence from the literature is scarce. For instance, Inzelt (2004) shows that interaction is still limited in the move towards a knowledge economy. Bartlett and Cuckovic (2006) argue that in spite of a successful development of policies to develop technology networks and business incubators in Slovenia, such entities have not generated much spin-off activity. Similarly, a Romanian study (Group of Applied Economics [GEA] 2004) concludes that investments in industrial parks were disappointing, because of important structural deficiencies (e.g. inadequate location, design and structure, an investment-based rather than a market-based development, lack of managerial experience, fuzzy legislation, etc.) and lack of synergy or network effects.

The Innovation Relay Centres (IRC)—an EU-funded initiative that supports innovation and transnational technological cooperation in Europe[15]—had a notable contribution to the development of science–industry linkages.

Table 14.6 Policy Measures Supporting Science—Industry Links in CEE Countries

	Program/objectives
Czech Rep.	*Park II Program (2002–2003)*—Support to S&T Parks, business incubators, innovative SMEs.
	TANDEM (2004–2010)—Improve cooperation between industrial and R&D organizations, technology support to SMEs, transfer of basic research results towards industrial R&D.
	Prosperity (2004–2006)—Development of industrial R&D infrastructure for start-ups, business incubators or business innovation centres. Technology development and innovation focused on new technologies, competitive products and services.
	Cooperation (2007–2013)—Development of clusters, poles of excellence, technology platforms and cooperation projects at regional, supra-regional and international level. Creation of a friendly business environment and improvement of links between research, universities and business.
Estonia	*Joint spin-off program of Tallinn Technical University and Tallinn Technical University Innovation Centre Foundation (1999–2004)*—Provides basic training for start-ups and individual mentoring and consultations (engaging Finnish experts) leading to the formulation of a business plan. Aimed primarily at university staff but in the future will also include students.
	SPINNO program (2001–2006)—Strengthen universities' entrepreneurial capacity and ability to manage spin-off processes. Development of a regulatory framework, patent and license policy, promotion of spin-offs, access to capital markets and cooperation networks with enterprises.
	Tartu University Innovation Office spin-off program (1998–2003)—The majority of shareholders are university researchers and students, spin-offs seen as an outlet for the potential of university researchers and as a training basis for students.
Latvia	*Development of innovation centres and business incubators (2007–2013)*—Supports HEIs and national research institutes to establish new innovation centres, helps regional or local governments to develop new business incubators. Provision of services to start-ups by innovation centres and business incubators, development of new innovative firms.
	Support to technology transfer (2005–2007)—Support to cooperation of scientists and entrepreneurs, set up of technology transfer offices to promote.
	Support to market-oriented research (1992–present)—Funding for applied research projects for integrating science and industry, develop technology-oriented sectors and create new jobs, encourage researchers from universities, research institutes and SMEs to develop new competitive products and facilitate the development of new start-ups.

(*continued*)

Table 14.6 (continued)

	Program/objectives
Lithuania	*Development of S&T parks, technology centres (2004–2006)* *Participation of educational/academic establishments and businesses in EU Technology Platforms (2006–2008)*
Hungary	*'Pazmany Peter' Program (Regional Knowledge Centres at Universities) (2004–2008)* *S&T cooperation between business and public research units (2004–2006)* *Cooperative Research Centres between companies and higher education institutes (2005–present)*
Poland	*Strengthening cooperation between public R&D units and business (2004–2006)*
Slovenia	*Development of business incubators at universities (2002–present)* *Co-financing employment of researchers in enterprises (2006–2008)*—Stimulates the mobility of researchers from public research institutions to the business sector. In the long run, improve science–industry linkages and minimize cost of transfer of innovations.
Slovakia	*Business Incubators, Technology Parks and R&D Centres (2004–2006)*—Set up of business incubators, technology parks for various industries and R&D centres for applied research, which should be based by Slovak municipalities and cooperate with universities.
Romania	*TRANSINO Program (2002–2006)*—Promotes technology transfer and innovation between R&D units, firms and universities. *INFRATECH Program (2004–2008)*—Support for the development of innovation infrastructure (S&T parks, business incubators, technology transfer offices, etc.).

Source: Selected from the PRO-INNO Europe Policy Measures Database, http://www.proinno-europe.eu.

CREATION OF A NEW OVERLAY OF LATERAL NETWORKS AND ORGANIZATIONS AMONG THE THREE HELICES: CLUSTER POLICIES AND REGIONAL DEVELOPMENT

Cluster policies, especially as a regional development instrument, are a new concept in the CEE countries; the geographic concentrations of interconnected firms and associated institutions leading to localized collective learning and regional integration were practically unknown in the socialist regime. Regional development policies became a relevant issue in the preparations for EU accession and the economic changes after accession, promoted in all the EU Member States primarily by the regional operational programs.

In the 1990s, several clusters emerged in the CEE countries, with inter-firm linkages promoted mainly through traditional business intermediaries (federations, associations, etc.), inter-firm mobility of skilled people and regular exchange of market information and partnerships around specific projects, rather than through technology production chains and horizontal or vertical integration. At present, in spite of some remarkable developments, there still are striking differences between clusters in the EU15 and the New Member States (NMS; Innobarometer 2006):

- Only 9 per cent of companies in the NMS work in a cluster environment, versus 28 per cent in EU15.
- Only 41 per cent of the top-level managers from the NMS claimed to be familiar with the relatively new concept of clusters, compared to 62 per cent in EU25.
- NMS clusters are generally much less involved in networks, e.g. only 51 per cent of clusters in the Czech Republic were found to have an active participation in business networks, 48 per cent in Hungary, 41 per cent in Slovakia and 39 per cent in Slovenia.

At the country level, there is a variety of national approaches in the coordination and design of cluster policies and tools. In the Czech Republic the national cluster policy has been coordinated since 2001 by Czech Invest, in Hungary and Slovenia by the Ministry of Economy, whereas in Latvia, a cluster program was promoted with PHARE support in 2000, but was interrupted (Ketels and Sölvell 2006).

Cluster policy design also presents a variety of features (OECD 2005):

- *Slovenia* adopted a 'dynamic concentric circles' approach of clustering SMEs around a lead (large) company. In 2003 there were eleven clusters encompassing seven hundred firms, working on over 150 joint projects in marketing, production, R&D and internationalization. Slovenian cluster policy emerged in the late 1990s to enhance industry productivity and is systematically integrated into long-term economic policy goals (see Box 14.1).
- *Slovakia* supported SME development at all levels of governance since the early 1990s, but has no specific cluster policy. However, a certain trend to cluster support is present in the automotive industry and industrial parks policies, mainly driven by FDI inflows and employment policy.
- *Poland* has both traditional and high-tech clusters with strong regional elements and spontaneous bottom-up networking (see Box 14.1), but no specific cluster support was adopted, in spite of a large number of policies and institutions supporting SMEs. Many of the emerging regional innovation systems present cluster features, especially in high-tech sectors; regional innovation strategies are seen as a key ingredient of regional development.
- *Hungary* has several clusters in the automotive industry (e.g. the Pannon Automotive Cluster PANAC, representing 10 per cent of GDP, see Box 14.1), logistics, construction and tourism. The investment-based, export-oriented machinery and automotive industry

Box 14.1 Country Experiences in Cluster Development

The Hungarian Automotive Cluster

'PANAC, founded in 2000 and supported by the Ministry of Economics, is a pilot cluster project located in the Györ region in Hungary, now having 67 members in a cooperative network. Most of the automotive companies in Hungary participate in it: Audi Hungaria, Opel Hungary, Hungarian Suzuki, LuK Savaria Ltd. The cooperative activities include assessment of automotive requirements for suppliers, exchange of know-how and expertise, and partnerships with universities. PANAC is developing a coordination role in the automotive industry, which is one of the most important sectors in the Hungarian economy. This experience suggests that through the cluster, foreign investors diversify their local activity, while on the other hand also attracting increasing numbers of firms with related businesses. The cluster emergence was made possible by the use of advanced IT, which contributed considerably to decreasing the costs of communication, monitoring and transaction. In addition, Hungarian firms could become suppliers of large multi-national companies not only in the automotive industry but also in IT-related services.

The Slovenian Transport Logistics Cluster

In the late 1990s, the Slovenian government started an effort of cluster building and support, and in 2001 the first generation of clusters was launched, including the Transport Logistics Cluster (TLC). It has the mission to provide joint education programmes, market and competition analyses, market presence, equipment procurement and integral development of services. The cluster members seek to offer integrates transport-logistics services along the Slovenian corridor as the south gateway for markets in Central and South-Eastern Europe. The TLC cluster is constituted by a mix of companies and academic institutions, including transport companies, shipping agents, port reloading companies, companies dealing with environmental problems, institutions of higher learning and research (altogether employing more than 13,000 people). It is backed by strong ICT support and applications including e-work applications (e-Portal and e-Stock-Exchange of logistic services) and integrated information support systems.

Two Polish Clusters in the Making

Poland does not have programme of support to clusters at the national level but has 16 regional programmes that encourage cluster creation. So far, two clusters are emerging as a result of bottom up initiatives. The Polish aviation cluster in the Southeastern region of Podkarpackie associates 40 companies and the Rzeszow University of Technology. The Podkarpackie region has 100 years of aviation industry history, and thanks to that attracted the US United Technologies Corporation to invest in the local aircraft engine producer WSK PZL Rzeszów, taking the lead in building the Aviation Valley cluster. Cluster members are now producing parts for the aviation industry worldwide, including Airbus in Europe, and the cluster now numbers 50 companies. The second cluster is the furniture cluster in region Wielkopolska which covers all aspects of the supply chain, from specialised research facilities like the Institute for Wood Technology in Poznan and the Institute of Natural Fibres, through subcontractors such as paint and chemical suppliers, to the furniture manufacturers themselves – in all covering 3,700 businesses. Furniture manufacturers took the initiative to form the cluster in co-operation with Poznan University of Economy. An evaluation of this cluster (Gorynia et al. 2007) indicated that the companies are still reluctant to enter into co-operative relationships, especially with competitors, and they also do not see the benefits accruing from co-operation with rivals. This experience highlights the importance of the social and cultural context, and the need for dedicated efforts to overcome some of the related difficulties.

(*continued*)

Box 14.1 (continued)

Clusters in Slovakia

No explicit cluster policy measures have been applied in Slovakia so far. Rather, national-wide policies were adopted in particular policy areas (e.g. entrepreneurship and SME policy, sectoral industry policy, S&T and regional policies etc.). These are non-discriminative and do not target specific regions and/or industries (except for the car industry). Cluster development was added to 'wish lists' of some innovation policies. These, however, supported establishment of industrial parks rather than creation of specific clusters. The development of clusters in Slovakia rather followed the logic of market forces, economic geography and flows of the FDI, than any formal cluster policies.

Source: UN-ECE (2007)

acted as front-runners in cluster development, which was fuelled by FDI inflows, especially in the West of the country. However, as the regional disparities in the FDI distribution created a widening West–East economic divide, support to domestic clusters is now much needed to counteract the overdependence on FDI and to reduce regional gaps. In that respect, the Pannon Economic Initiative (2001) supports alternative strategies of economic and regional development through a cooperative partnership between regional and economic development organizations, with private funds and assistance from international financial institutions in regional development.

- The *Czech Republic* has firm concentrations in old industrial areas (e.g. metallurgy, engineering in Moravia), as well as countrywide supplier networks for large international firms (e.g. Volkswagen/Skoda). Here, a bottom-up SMEs clustering emerged to withstand asymmetric power relations in supplier networks, in transnational supermarket chains or the automotive industry. A policy priority for the country is the increased support to cross-border cooperation, in view of participation in an emerging supranational automotive multicluster in Central Europe with its core in the north-eastern part of the country.
- In *Romania* several studies (e.g. GEA 2004; Ferrari 1999; Ionescu 1999) point out the existence of 'proto-clusters' or 'emergent clusters' of firms in several industrial sectors, such as the software, textile, footwear, ceramic and wood processing industries. They are spread throughout the country, but the highest concentration is in the west and north-west regions, primarily due to the significant Italian investment in the area, as well as the proximity to Western markets (GEA 2004). The software industry cluster in Bucharest region stands out for its high cluster generation potential fostered by horizontal

> collaboration between firms and the network characteristics of the local production system, whereby business relations often based on personal relations and the local specialized workforce generated a strong feeling of civic and social inclusion (Ionescu 1999).

Many of these clusters include academic institutions, e.g. the Slovenian Transport Logistics Cluster, the Polish aviation cluster in the Southeastern region of Podkarpackie, which associates the Rzeszow University of Technology, or the furniture cluster that includes Poznan University of Economy. In Romania, one of the first measures adopted in the context of the Regional Innovation Strategy of the North-West region was the creation of the first Regional Institute for Education, Research and Technology Transfer (IRECTT) in May 2007, by the local regional development agency together with fourteen public authorities and seven universities in the area. Created on the model of the European Institute for Technology, IRECTT aims to develop competences in the education-R&D-innovation triangle and to bridge public authorities, academia and business.[16] All these examples suggest that although at an early stage, the creation of a new overlay of lateral networks and organizations from the interactions among the three helices is already underway in the CEE countries, with effects that are expected to appear in the near future.

THE RECURSIVE EFFECT OF TRILATERAL NETWORKS AMONG INSTITUTIONAL SPHERES ON INNOVATION

This most advanced stage in the development of a triple helix regime is not yet in place in the CEE countries. However, positive results can be expected in the years to come, on the basis of the transformations now taking place within each helix and the broad range of measures to enhance linkages between them described earlier. The process of strengthening each triple helix institutional sphere and increasing interactions and mobility of resources within and among institutional spheres needs to be continued and enhanced to realize this potential. Although several national programs have been instituted to encourage the return and integration of S&T Diaspora, their effect remains marginal because of inadequate resources allocated to these programs. To date, a relatively few expatriates who have completed careers abroad have returned to head up R&D groups in their home countries or take university leadership roles. However, some leading scientific figures, like Janos Kornai, Professor of Economics at Harvard University, lead 'dual lives', simultaneously serving in a high-economic advisory position in his native Hungary on a part-time basis (Kornai 1998). The head of a Romanian biotechnology research institute also held a professorship at Yale University, acquiring grants from the U.S. National Institute of Health, utilizing them to support joint research endeavours in both countries. On the other hand, linkages with the S&T Diaspora abroad have the potential to instigate high-tech industries in the future, as shown by the experiences of China, Taiwan and Ireland. All these linkages provide new ways for strengthening the R&D and innovation capabilities of the CEE countries and enhancing their international visibility and competitiveness.

CONCLUSIONS: BUILDING A TRIPLE HELIX–BASED 'CREATIVE RECONSTRUCTION' STRATEGY IN THE CEE COUNTRIES

We have shown in the previous sections that a triple helix–based 'creative reconstruction' process has been primarily driven by two key factors that exerted distinct influences on the different stages:

1. The FDI flows, on the economic level, with a significant impact on the industry helix, but also on R&D institutes with a stronger applied research profile.
2. National and regional R&D and innovation policies and funding, complemented in recent years by EU programs and funding, with a stronger impact on the government and university, and to a lesser extent, the industry helix.

The latter is a longer-term growth strategy than the former, but it is ultimately more efficacious as it creates an endogenous innovation capacity that is essential for the transition to a knowledge-based society.

The balance between the two factors in this process is a dynamic one, shifting from a stronger influence of FDI flows in the 1990s decade to an increasing role of national, regional and EU policies in the 2000s decade. Much of the FDI flow is likely to be lost to lower-cost venues, but higher levels may be retained if infused with advanced technology from indigenous sources, as the best chance of creating 'stickiness' to restrain footloose multinationals. Facilitating collaboration among start-ups in product development and marketing, networking incubators and other support mechanisms, allows SMEs to attain some of the distributive capabilities internalized within multinational firms that allow them to transcend national boundaries and industrial sectors (Papagiannidis et al. 2009).

In this chapter, we attempted to assess the feasibility of a triple helix–based innovation strategy in the CEE countries, by analysing the changes occurring within and between the triple helix actors over the last two decades. In the light of the evidence provided in this chapter, we conclude that such a strategy is viable, with a scale-up of nascent trends in the advancement towards the entrepreneurial university, as well as further development of support structures for innovation.

Several areas of intervention can be identified in this respect.

Consolidate the Entrepreneurial Potential of Universities, by Strengthening Their Research Capabilities and Their Commercialization Skills and Infrastructure

This is a major objective for enhancing development, especially in regions lacking strong research universities. This is also the case in most CEE countries, where entrepreneurial universities are still in their infancy, given the weak academic research capacity and the low domestic demand for R&D. Therefore, it is necessary to focus international, national and regional resources for reaching this target.

Foster the Process of Innovative Firm Formation and the Absorptive Capacity of Local Firms

Innovation capacities may be enhanced through creation of hybrid organizations such as science parks, business incubators, etc. Positive developments in this respect have emerged especially over the last decade or so, as a result of various government programs. Also, some EU initiatives played a catalytic role in this process, such as the DG Enterprise's initiative PAXIS (Pilot Action of Excellence on Innovative Start-Ups) that aims to boost the transfer of local and regional excellence in innovation, and to foster the cooperation and the exchange of tacit knowledge and learning among local innovation stakeholders.[17] Also, the Innovating Regions in Europe (IRE) network was created by the European Commission in the mid-1990s to facilitate the development and implementation of regional innovation strategies and schemes among European regions. The majority of the IRE regions in the CEE have undertaken Regional Innovation Strategy (RIS) projects with support from the European Commission. In many regions, the RIS project has been the starting point for successful initiatives boosting innovation.[18]

In addition, the industrial clusters that emerged in the CEE region over the last two decades are expected to foster the technological integration into the EU and generate new clusters over the longer term. This process may take decades. For example, the firm formation activity that was observed in Silicon Valley in the 1960s and 1970s resulted from initiatives dating from the end of the nineteenth century of encouraging students from the Stanford engineering school to form firms. It was realized that a great technological university required a support structure of firms. Otherwise graduates would move elsewhere and the university would remain a small isolated entity without this industrial base. This process is being replicated *in nucleo* in the neighbourhood of the Technical University in Budapest and in the Białystok Science and Technology Park (BPN-T) in Poland. The development of the Skype software by graduates of the Tallinn Technology University into a multibillion dollar firm sold to eBay and then bought back is the most significant start-up success story in the CEE and indeed of all Europe. The Skype case, linking London financial investors to a Tallinn-based start-up, mentored by Turku Science Park in Finland, exemplifies integration of the FDI and indigenous development strategies and has potential for generalization.

Increase the Availability of and Access to Venture Capital

The availability of and access to venture capital remain much lower in the CEE countries than in the rest of the EU, in spite of some measures adopted by national governments. The JEREMIE (Joint European Resources for Micro to Medium Enterprises) initiative of the European Commission, in collaboration with the European Investment Bank and the European Investment Fund initiative, aims to reduce this gap by promoting increased access to finance for micro enterprises and SMEs in the EU regions.

In the CEE context, yet another triple helix idea deserves special attention: who shall take the lead as Innovation Organizer in countries where governments, universities and industry are still relatively isolated? The answer to this question will likely determine the viability of a triple helix innovation system in the CEE. At present, much of this role is performed by the recipient and managing institutions of FDI flows, national, regional and EU regional operational programs supported by structural funds. However, some changes in this configuration can be expected in the future, along with the gradual shift of CEE countries from a factor- and investment-based development model towards an innovation-based model, with a strengthened local innovation and knowledge base. From a triple helix perspective, the recombination and hybridization of existing elements to create new organizational forms and relationships have the potential to break the path dependency and create several possible development scenarios. Not one but several development strategies thus become possible, creating the premises for realignment and reconfigurations in the years ahead.

NOTES

1. The CEE countries referred to in this chapter are: the Czech Republic, Estonia, Hungary, Latvia, Lithuania, Poland, Slovakia, Slovenia, Bulgaria and Romania.

2. See Tanzi (1999).
3. Liberals support anti-trust policies to keep firms apart in order to control tendencies to monopolization and enhance competition, whereas conservatives believe in the efficacy of the 'invisible hand' of the market and tend to object to any intrusion by government on its presumed automatic equilibration.
4. The U.S. innovation system consists of highly interactive dynamics among the triple helix institutional spheres, often hidden by conservative ideological prescriptions and consequently typically instituted by indirect measures. An analytical model in the U.S. crafted to reveal the inner workings of its innovation system has been translated into a normative model.
5. *The Enlarged European Union: A Statistical Handbook* (2008); available at http://www.euromonitor.com/The_Enlarged_European_Union (acessed 16 December 2009).
6. Approximately ten years for Estonia, Czech Republic, Lithuania and Cyprus, fifteen years for Slovenia, twenty years for Poland, Latvia, Bulgaria, Slovakia and Romania.
7. This strategy is very similar to Mexico's *maquiladora* approach of importing, assembling and re-exporting materials and equipment on a duty-free and tariff-free basis, usually back to the originating country, which has been in place over the last forty years or so, ensuring a large share of the country's exports (45 per cent of the country's exports, according to Gruben and Kiser 2001).
8. In 1989, gross enrolment rates (GER) ranged from below 10 per cent (Romania), 10–25 per cent (Slovakia, Hungary, Czech Republic, Poland), 20–25 per cent (Slovenia, Bulgaria, Latvia), up to 28 per cent (Lithuania) and 36 per cent (Estonia). Ten years later, in 1999, GER often exceeded 40 per cent, with Estonia, Latvia and Slovenia topping the list with 45 per cent, 46 per cent and 51 per cent, respectively (Kwiek 2006b).
9. Selected from the PRO INNO TrendChart Database, http://www.proinno-europe.eu/index.cfm?fuseaction=page.display&topicID=262&parentID=52 (accessed 19 December 2009).
10. A similar process of entrepreneurial academic transition has been identified in Brazil (see Maculan and Mello 2009).
11. *Foreign Direct Investment Magazine* (2007).
12. France invested mainly in Poland, whereas the Netherlands and Germany invested in Poland, Hungary and the Czech Republic (Sakova and Hansenova 2005). Austria is a major investor in Slovakia and Slovenia.
13. Non-EU investors, like Japan and Republic of Korea showed relatively weaker interest in the region and are involved in a small number of large projects. U.S. investors preferred larger investments, especially in Poland. Turkey and Greece invested mostly in Bulgaria and Romania due to neighbourly ties. Russia is not a significant direct investor in the region, except in Poland (World Bank 2001).
14. For instance, World Bank's Multilateral Investment Guarantee Agency (MIGA), in which some CEE countries like Romania, Slovenia and Czech Republic were involved as host and investor country.
15. The first IRCs were established in 1995 with the aim to create a pan-European platform for transnational technology transfer and innovation services. Today, there are seventy-one regional IRCs involving almost 220 partner organizations in the twenty-seven EU Member States, and also Iceland, Israel, Norway, Switzerland, Turkey and Chile. Most IRCs are operated by consortia of regional organizations, such as Chambers of Commerce, regional

development agencies and university technology centres, and have facilitated over one thousand transnational transfers of technology—agreements for sale, licensing, distribution or joint development of new technologies. On 1 April 2008, the IRC network and the Euro Info Centre merged into 'Enterprise Europe Network'.

16. IRECTT aims to maximize the North-West region's competitive advantage. It operates on three main dimensions: education, research and technology transfer, along the following activities: accreditation of new occupations, development of curricula for the conversion and training of adults for jobs demanded on the labour market, joint development with higher academic institutions of new service packages for postgraduate education, technology transfer and support to innovation through new products, materials and services (see further details at http://www.nord-vest.ro/genpage.aspx?pc=SPI-IRECTT).
17. The project was launched in 1999 and works through Thematic Networks (including twenty-two European cities and regions with an outstanding track record in supporting the creation and growth of innovative start-ups); Projects, which aim to develop, validate and disseminate concepts to encourage innovative start-up creation; and a set of Accompanying Measures, which are designed to analyse and disseminate PAXIS results and good practices to all interested parties across Europe (see http://cordis.europa.eu/paxis/src/about_paxis.htm).
18. See http://ec.europa.eu/enterprise/ire/Innovating-regions/www.innovating-regions.org/index.html. Last accessed: 7 September 2010.

REFERENCES

Allen and Overy. 2006. *Foreign direct investment in Central and Eastern Europe.* http://www.visegradgroup.eu/download.php?ctag=download&docID=93 (accessed 19 December 2009).

Bartlett, W., and N. Cuckovic. 2006. 'Knowledge transfer, institutions, and innovation in Croatia and Slovenia'. *Drustvena Istrazivanja* 15 (3): 371–399.

Bandelj, N. 2002. Embedded economies: Social relations as determinants of foreign direct investment in Central and Eastern Europe. *Social Forces* 81(2): 411–444.

Clausing, K. A., and C. L. Dorobantu. 2005. 'Re-entering Europe: Does European Union candidacy boost foreign direct investment?' *Economics of Transition* 13 (1): 77–103.

Disdier, A. C., and T. Mayer. 2004. 'How different is Eastern Europe? Structure and determinants of location choices by French firms in Eastern and Western Europe'. *Journal of Comparative Economics* 32 (2): 280–296.

Dries, L., and J. F. M. Swinnen. 2004. 'Foreign direct investment, vertical integration, and local suppliers: Evidence from the Polish dairy sector'. *World Development* 32 (9): 1525–1544.

Dyker D., A. Nagy, P. Stanovnik, J. Turk, H. Usenik and P. Vince. 2003. '"East"–"West" networks and their alignment: Industrial networks in Hungary and Slovenia'. *Technovation* 23 (7): 603–616.

Dyker, D., and S. Radosevic. 1999. 'What quantitative analysis of trends in science and technology tell us about patterns of transformation and growth in the post-socialist countries?' In *Innovation and structural change in post-socialism*, ed.

D. Dyker and S. Radosevic, 1–28. Dordrecht: NATO Science Series, Kluwer Academic Publishers.

Edquist, C. 1997. 'Introduction'. In *Systems of innovation: Technologies, organizations, and institutions*, ed. C. Edquist, 1–35. London: Pinter Publishers/ Casell Academic.

Edquist, C., and L. Hommen. 1999. 'Systems of innovation: Theory and policy for the demand side'. *Technology in Society* 21:63–79.

Eisemon, T. O., I. Mihailescu, L. Vlasceanu, C. Zamfir, J. Sheehan and C. H. Davis. 1995. 'Higher education reform in Romania'. *Higher Education* 30 (2): 135–152.

Etzkowitz, H. 2003a. 'Innovation in innovation: The triple helix of university-industry-government relations'. *Social Science Information* 42 (3): 293–337.

———. 2006. 'Assisted linear model. The new visible hand: An assisted linear model of science and innovation policy'. *Science and Public Policy* 33 (5): 310–320.

———. 2008. *The triple helix: University-industry-government interactions for innovation*. London: Routledge.

European Commission. 2002. *SMEs in Europe, including a first glance at EU Candidate Countries, observatory of European SMEs, No. 2*. European Commission, DG Entreprise: Luxembourg: Officer for Official Publications of the European Communities.

———. 2007. 'INNO-Policy TrendChart—Policy trends and appraisal'. Report Romania 2007. Available at: http://www.proinno-europe.eu/page/innovation-and-innovation-policy-romania (accessed 7 September 2010).

European Private Equity and Venture Capital Association. 2006. 'Central and Eastern Europe statistics 2005—an EVCA special paper'. November.

Eurostat. 2003. 'Statistics on science and technology in Europe'. http://www.insme.org/documenti/Statistic_Report_part1.pdf (accessed 20 December 2009).

———. 2008. 'Statistics in Focus, 29/2008'. http://epp.eurostat.ec.europa.eu/cache/ITY_OFFPUB/KS-SF-08–029/EN/KS-SF-08–029-EN.PDF (accessed 17 December 2009).

Eurostat. 2007. Fourth Community Survey. Avaliable at: http://europa.eu/rapid/pressReleaseAction.do?reference=STAT/07/278format=HTML&aged=^&language=EN&guillanguag-en.

European Trend Chart on Innovation. 2005. European Innovation Scoreboard 2005. Comparative analysis of Innovation Performance. Available at: www.proinno-europe.eu/extranet/admin/uploaded.../EIS_2005.pdf. (accessed: 7 September 2010).

Ferencikova, S., and T. Dudas. 2005. 'The impact of foreign direct investment inflows on the economic growth in the new EU member states from Central and Eastern Europe'. *Ekonomicky Casopis* 53 (3): 261–272.

Ferrari, M. R. 1999. *Small enterprise clusters for local development in transition context: The case of Romania*. Milan: Bocconi University.

Foreign Direct Investment Magazine. 2007. 'Eastern promise'. 6 June. http://www.fdimagazine.com/news/fullstory.php/aid/2061/Eastern_promise.html (accessed 18 December 2009).

Freeman, C. 1987. *Technology policy and economic performance*. London: Pinter.

Gibbons, M., C. Limoges, H. Nowotny, S. Schwartzman, P. Scott and M. Trow. 1994. *The new production of knowledge: The dynamics of science and research in contemporary societies*. Sage Publications: London.

Group of Applied Economics. 2004. 'Towards an industrial policy based on competitive economic agglomerations—clusters (III)'. http://www.gea.org.ro/documente/ro/clustere/instrumentedepoliticadragospislaru.pdf (in Romanian) (accessed 7 September 2010).

Gruben, W. C., and S. L. Kiser. 2001. *The border economy: NAFTA and maquiladoras: Is the growth connected?* Dallas, TX: The Border Economy, Federal Reserve Bank of Dallas.
Havrylyshyn, O. 2001. 'Recovery and growth in transition: A decade of evidence'. *IMF Staff Papers, Special Issue, Washington, IMF* 48:53–87. http://www.imf.org/external/pubs/ft/staffp/2001/04/pdf/havrylys.pdf (accessed 7 September 2010).
Innobarometer. 2006. 'Innobarometer on clusters' role in facilitating innovation in Europe: Analytical report'. July, Flash Eurobarometer 187, The Gallup Organization, European Commission.
Inzelt, A. 2004. 'The evolution of university-industry-government relationships during transition'. *Research Policy* 33 (6–7): 975–995.
Ionescu, V. 1999. 'Supply-side strategy for productivity, competitiveness and convergence between the CEECs and (in) the EU—Romania case study'.
Ketels, C., and O. Sölvell. 2006. 'Clusters in the EU-10 New Member Countries'. Study commissioned by EU DG Enterprise, Europe Innova.
Klonowski, D. 2005. 'The evolution of the venture capital industry in transition economies: The case of Poland'. *Post-Communist Economies* 17 (3): 331–348.
Konings, J. 2001. 'The effects of foreign direct investment on domestic firms—evidence from firm-level panel data in emerging economies'. *Economics of Transition* 9 (3): 619–633.
Kornai, J. 1998. Interview with Henry Etzkowitz, Cambridge, MA.
Kwiek, M. 2006a. 'Academic entrepreneurialism and private higher education (in a comparative perspective', *CHEPS*, 12 October.
———. 2006b. *On accessibility and equity, market forces, and entrepreneurship: Developments in higher education in Central and Eastern Europe.* London: Institute of Education, University of London.
Lankes, H. P., and A. J. Venables. 1996. 'Foreign direct investment in economic transition: The changing pattern of investments'. *Economics of Transition* 4 (2): 331–347.
Lundvall, B. -A. 1992. *National systems of innovation.* London: Pinter.
Lundvall, B. -A., B. Johnson, E. S. Andersen and B. Dalum. 2002. 'National systems of production, innovation and competence building'. *Research Policy* 31 (2): 213–231.
Maculan, A. M., and J. M. C. de Mello. 2009. "University start-ups for breaking lock-ins of the Brazilian economy." *Science and Public Policy* 36 (2): 109–114.
Mosoni Fried, J. 1995. 'Industrial research in Hungary—a victim of structural change'. *Social Studies of Science* 25 (4): 777–803.
Müller, K., and H. Etzkowitz. 2000. 'S&T human resources: The comparative advantage of the post-socialist countries'. *Science and Public Policy* 27 (4): 285–291.
Navrat, P., and L. Molnar. 1998. 'Curricula transformation in the countries in transition: An experience from Slovakia'. *IEEE Transactions on Education* 41 (2): 88–91.
Nelson, R. 1993. *National innovation systems.* New York: Oxford University Press.
Nelson, R., and N. Rosenberg. 1993. 'Technical innovation and national systems'. In *National innovation systems*, ed. R. R. Nelson, 3–21. Oxford: Oxford University Press.
OECD. 1998. *University research in transition.* OECD Publishing: Paris.
———. 2005. *Business clusters: Promoting enterprise in Central and Eastern Europe.* OECD Publishing: Paris.

Paasi, M. 1998. 'The emergence of innovative firms in Estonia'. In *Transforming science and technology systems—the endless transition?*, ed. W. Meske, J. Mosoni-Fried, H. Ezkowitz and G. Nesvetailov, 211–221. Amsterdam: IOS Press/NATO Scientific Affairs Division.

Papagiannidis, S., F. Li, H. Etzkowitz and M. Clouser. 2009. 'Entrepreneurial networks: A triple helix approach for brokering human and social capital'. *Journal of International Entrepreneurship* 7:215–235.

Pavlinek, P. 2002. 'The role of foreign direct investment in the privatisation and restructuring of the Czech motor industry'. *Post-Communist Economies* 14 (3): 359–379.

———. 2004. 'Regional development implications of foreign direct investment in Central Europe'. *European Urban and Regional Studies* 11 (1): 47–70.

Radošević, S. 1996. 'Restructuring of R & D institutes in post-socialist economies: Emerging patterns and issues'. In *Building new bases for innovation: The transformation of the R&D system in post-socialist states*, ed. A. Webster. Cambridge: Anglia Polytechnic University, Cambridge, UK: 8–30.

———. 1998. 'The transformation of national systems of innovation in Eastern Europe: Between restructuring and erosion'. *Industrial and Corporate Change* 7 (1): 77–108.

———. 1999. 'Transformation of science and technology systems into systems of innovation in Central and Eastern Europe: The emerging patterns and determinants'. *Structural Change and Economic Dynamics* 10:277–320.

———. 2002. 'Mobilising the regional foresight potential for an enlarged European Union'. Paper prepared for the STRATA-ETAN Expert Group Action, European Commission, DG Research.

Radošević, S., and Kriaucioniene, M. 2006. 'The role of higher education in national innovation systems in Central and Eastern Europe'. In *Higher education for national development: Universities and societies in transition*, ed. Bridges et al. Routledge, Abingdon, UK: 135–160.

Resmini, L. 2000. 'The determinants of FDI in the CEECs: New evidence from sectoral patterns'. *Economics of Transition* 8 (3): 665–689.

Sadowski, B. M. 2001. 'Towards market repositioning in Central and Eastern Europe: International cooperative ventures in Hungary, Poland and the Czech Republic'. *Research Policy* 30 (5): 711–724.

Sakova, B., and H. Hansenova. 2002. 'The consequences of Slovak accession to the European Union in the field of foreign direct investment'. *Ekonomicky Casopis* 50 (6): 926–938.

Schumpeter, J. 1942. *Capitalism, socialism and democracy*. New York: Harper.

Scott, P. 2002. 'Reflections on the reform of higher education in Central and Eastern Europe'. *Higher Education in Europe* 27 (12): 137–152.

Smarzynska, B. 2004. 'Composition of foreign direct investment and protection of intellectual property rights: Evidence from transition economies'. *European Economic Review* 48 (1): 39–62.

Srholec, M. 2003. 'Foreign direct investment and technology change'. *Politicka Ekonomie* 51 (5): 695–713.

Tanzi, V. 1999. 'Transition and the changing role of government'. *Finance and Development Magazine* 36 (2). See http://www.imf.org/external/pubs/ft/fandd/^999/ob/tanzi.htm (accessed 7 September 2010).

UN-ECE. 2007. 'Creating a conducive environment for higher competitiveness and effective national innovation systems: Lessons learned from the experiences of UNECE countries'. United Nations: New York and Geneva. Available at: http://vi.unctad.org/resources-mainmenu-64/digital_

library?act=browse&by=browse-by-category&cl=1.1.4. (accessed 7 September 2010).

World Bank. 2001. 'Foreign direct investment: New trends in transition countries'. *Transition Newsletter*. http://www.worldbank.org/html/prddr/trans/so99/pgs7-8.htm (accessed 7 September 2010).

15 Not Yet Triple Helix III?

Japanese MOT Policies and the Problem of Technology Exploitation

Ingyu Oh

INTRODUCTION

Leydesdorff and Etzkowitz (1998) present an evolutionary explanation of the triple helix. According to their narrative, in Triple Helix I, 'three spheres are defined institutionally (university, industry, and government)', whose interaction is mediated by liaison offices, technology licensing offices (TLO) and contract offices for a stable technology development, transfer and spin-off. Universities, industries and governments 'enter into a reciprocal relationship' in which 'each attempts to enhance the performance of the other' (Etzkowitz 2008, 8).

If Triple Helix I is the first phase of the entire evolutionary process, Triple Helix II is a more advanced institution of technological innovation and transfer. In this second phase 'the helices are defined as different communication systems' that would develop unique forms and contents of communication to enhance the process of operating the market, introducing technological innovations and governing the entire operation of the triple helix (Leydesforff 1997; Etzkowitz and Leydesdorff 1997; Leydesdorff and Etzkowitz 1998; Etzkowitz 2008). Under the new rubric of interfaced communication between the three components of the triple helix, participants can create new forms of communication between helices for a sustainable technology transfer interface or for patent legislation.

Finally, Triple Helix III is the most advanced, terminal state of the triple helix, where each unit in the helix assumes the role of the others in addition to carrying out their conventional activities. Universities, for example, can create an industrial organizational platform or perform a quasi-governmental role as a regional or local innovation organizer, whereas firms can take educational and research roles (Gulbrandsen 1997; Pires and Castro 1997; Leydesdorff and Etzkowitz 1998; Etzkowitz 2008, 38).

This chapter intends to apply the concept of the triple helix to the case of Japan and see how the Japanese system of national and regional innovation clusters, or what they call *sangakurenkei* (industry–university clusters) or *sangakukan renkei* (industry-university-government clusters), are faring vis-à-vis its North American and European counterparts. This question is significant on two grounds. First, Western scholars were writing articles and books in support of the Japanese national innovation system (NIS) in general or the Japanese corporate innovation strategy in particular during the 1980s (Freeman 1987; Posen 1992; Teece and Mowery 1993; Kindleberger 1996). Their argument held that narratives of innovation had to take into account the structural mechanism of the Japanese style of innovation to shed light on the importance of incremental innovation through improvements, or *kaizen* (Goto 2006, 2).

Second, the Japanese NIS and corporate research and development (R&D) capacities of *kaizen* were suddenly threatened by the long-term recession in the 1990s, raising scepticism about the longevity and sustainability of the Japanese NIS (Taguchi 2003; Goto and Kodama 2006; Sakakibara 2007; Imai 2008). These critics of the Japanese NIS point fingers at the lacklustre performance of the Japanese triple helix, or *sangakurenkei*, for the economic downturn (see also Itô 2008). Therefore, a new theoretical necessity arises to analyse the question of whether Japan can recover the previous level of process innovation (i.e. incremental innovation) and induce genuine breakthroughs in high technologies (i.e. radical innovation), for which Triple Helix III is required. This chapter aims to achieve the following three objectives.

1. To provide a brief history of the Japanese triple helix
2. To sketch the current evolution of the triple helix toward the second stage
3. To provide a critical alternative to the current debate on the potential upgrading of the Japanese NIS to Triple Helix III. The chapter will use cases of local and regional triple helix experiments to highlight both the success and failure of the current Japanese *sangakukan renkei* and will provide future implications.

HISTORY OF JAPANESE TRIPLE HELIX

The Japanese innovation system maintains two fundamental characteristics that its North American or European counterparts don't. First, the whole purpose of technological innovation or corporate R&D was to improve the quality of products that the West had developed through product innovation (i.e. incremental innovation or process innovation; for further elaboration, see Abernathy 1978, 78). Therefore, exploiting existing knowledge through knowhow accumulation, along with the improvement of the extant products through new technology applications, was the common goal of the corporate innovation strategy (Taguchi 2003, 70–73; Nemoto 2006, 30–31; Miyata and Nishimura 2007; Sakakibara 2007, 38–39).

Second, the Japanese government and universities, however, have traditionally emphasized fundamental scientific research projects, not encouraging government R&D laboratories and universities to provide private firms with applicable technological knowledge that is mature enough for immediate product development and/or improvement. The so-called 'Devil's River' and 'Death Valley' phenomena, which prohibit the transfer and commercialization of new technology seeds developed by the fundamental science project, was widely observed in the Japanese case (Taguchi 2003, 3; Degawa 2006, 41; Shimoyama 2006, 49; Sakakibara 2007).

One of the consequences of these two contradictory aspects of the Japanese NIS and the corporate innovation strategy was the belated introduction and development of the triple helix in that country. Compared to the U.S. legislation of the Stevenson-Wydler Technology Innovation Act (1980) and the Bayh-Dole Act (1980), the Japanese legislative introduction of the TLO Law came out only in 1998, about eighteen years after the U.S. precedence (Nagahira 2006, 8).

As shown in Table 15.1, the U.S. has systematically and continuously established and expanded legal and institutional foundations for the triple helix since 1979. The central feature within the American NIS was to establish liaison facilities between universities, small and medium firms and the government agencies. A central feature of this system is to motivate universities and firms to be strategically linked up together to transfer technologies for future commercialization. Technology and R&D outsourcing in the form of A&D (acquire and develop) or C&D (connect and develop) loom large in this stage of NIS evolution, empowering firms and universities to tap into a new value-added market of technology management (Misawa 2007, 47–51).

Up until the early 1990s, the Japanese response to the development in the U.S. and Europe in the area of the triple helix had been either indifference or ignorance, partly because Japanese manufacturing firms achieved arguably the world's highest level of process innovation that substantially lowered production costs with high product quality. However, the economic recession, which had been prolonged for more than fifteen years since the early 1990s, invited suspicions about the Japanese NIS, especially about its incompetence in adapting to the rapidly changing global economic milieu. The initial governmental response to the long spell of economic recession, which was made in a collective fashion with its industrial and academic partners, was to inaugurate the U.S.-style triple helix for the first time in Japan.

As Table 15.1 shows, the Japanese government enacted the TLO Law in 1998 to legalize the opening of a technology licensing organization for both national and private universities. TLO's purpose was none other than to liaise among the Ministry of Education, Culture, Sports, Science, and Technology (MEXT), the Ministry of Economy, Trade, and Industry (METI), national and private universities, and private firms (e.g. small and medium enterprises [SMEs], large firms, venture firms). The Japanese-style Bayh-Dole Act and SBIR were also introduced the next year to complete the post-industrial transition to the triple helix regime, making the government

Table 15.1 U.S. and Japanese Triple Helix History

	U.S.		*Japan*
1979	President Carter calls for strong intellectual property protection	1998	TLO Law to ease transfer of knowledge from universities to firms
1980	Stevenson-Wydler Technology Innovation Act	1999	Japanese SBIR; Japanese Bayh-Dole Act; JABEE
	Bayh-Dole Act	2000	Amendments to TLO Law
	Patent rights to organizations that developed new technologies despite federal grants	2001	Hiranuma Plan for 1,000+ university-based ventures
	Copyrights to computer software	2002	Amendments to TLO Law
1982	Small Business Innovation Research Act (SBIR); CAFC began specializing in patent rights violation	2003	Intellectual Property Law; School Education Law for school accreditation
1984	National Cooperative Research Act; Semiconductor Chip Law	2004	National University Law to allow professors to join TLO; Amendments to Patent Law to lower TLO fees
1986	Federal Technology Transfer Act; GATT Uruguay Round		
1988	Omnibus Trade & Competitiveness Act		
1989	National Competitiveness Technology Transfer Act		
1992	Defence Conversion, Reinvestment & Transition Assistance Act		

Source: Nagahira (2006, 8–10); Taguchi (2003, 11).

properties and R&D facilities, including public funding, more accessible to universities and private firms than before (Oh et al. 2005). The Hiranuma Plan of 2001 was a nationwide effort to motivate universities and firms to create up to one thousand venture firms, especially venture spin-offs from universities, all in three years. The final legal provision for the institutionalization of the Japanese triple helix was the 2004 decision to convert national universities into corporate entities (*hôjinka*) so that professors can take non-academic positions within the TLO in addition to their traditional teaching and research posts within universities (Tamura 2006).

By 2004, therefore, Japan finished the first phase of the triple helix evolution, or Triple Helix I, where three distinctive institutional boundaries were linked up together through new institutions called the TLO and 'incorporated' national universities. This does not mean, however, that private firms, national universities and the government agencies gave up their traditional functions entirely. Firms still relied heavily on their central R&D

facilities, and outsourced R&D projects to foreign universities and R&D firms; national universities still vied for government funding for fundamental and long-term science projects; and the government still maintained and monitored traditional NIS projects allocated to national universities on a long-term basis.

During the stage of Triple Helix I, the three types of the triple helix projects mostly found in Japan were: (a) referred projects (*jutaku kenkyû*); (b) collaborative projects (*kyôdô kenkyû*); and (c) others. Referred projects are usually funded by the government for university-based research activities, and the fund is widely distributed among natural sciences, engineering, agriculture and biomedicine. Some referred projects are also funded by corporations, although they do not dispatch researchers from their central R&D units.

Collaborative projects, on the other hand, require a mutual cooperation between firms and universities for one or multiple research topics, and they are usually funded by corporations in exchange for future technology licensing and/or spin-offs. At the moment, roughly 60 per cent or more of all collaborative projects are awarded to natural science and engineering colleges, whereas only 20 per cent of them are granted to the biomedical field. Unlike in referred projects, corporate researchers are dispatched to universities for collaborative projects.

'Others' refers to individually organized research projects that are funded by corporations for a university professor who has informal ties with the firm. Rewards for this kind of 'other' research projects are typically paid to the university in the form of corporate scholarships (*shôgakukin*). Traditionally, the majority of the Japanese *sangakurenkei* has been this 'other' type of projects. In 2005, 73.3 per cent of all national university *sangakurenkei* projects belonged to the category of this 'other' type, whereas the same figure for private universities was 62.4 per cent (Miyata and Nishimura 2007, 5).

The reason why the scholarship contribution as a form of *sangakurenkei* was predominant during the phase of Triple Helix I was not because private firms wanted university professors to generate research results that were supposed to be used for corporate R&D or technology commercialization. The reason was to maintain university–firm relations for information leading to recent trends in fundamental research projects carried out by national universities and to secure a human resource (HR) pool for the recruitment of future corporate R&D researchers. Most of these new recruits, therefore, will work for the very firm that provided the scholarship fund for their study at a national university. This is why Triple Helix I in Japan was predominantly HR-oriented in an effort to network national universities and firms. On the other hand, private firms heavily relied on either their own central R&D organization for the development of application technologies or global R&D organizations in the U.S. or Western Europe for technology licensing (Itô 2008, 90–99).

As Table 15.2 indicates, the Japanese investment in the triple helix project was much more than what the UK universities received for similar purposes. However, the royalty income UK universities earned in 2002 was more than ten times the amount their Japanese counterparts received in 2003. Furthermore, UK universities launched more university-based venture firms than their counterparts in Japan did. Even though the Japanese level of spending in triple helix projects was more than 61 per cent of the U.S. total, actual project outputs between the two countries differed substantially, as the U.S. royalty income was almost three hundred times the amount Japanese universities gained.

Several reasons may explain why the first phase of the Japanese triple helix (i.e. Triple Helix I) was not as successful as it was first envisaged. As Katô and Enomoto (2006) report, most Japanese corporations do not think the government is doing a good job in governing Triple Helix I. The government is found to still hold on to its conventional norm of concentrating its resources on fundamental scientific projects for national universities without any clear guidelines as to how referred and collaborative projects can be applied to actual corporate technological needs.

Firms also tenaciously uphold the so-called 'not invented here' (NIH) syndrome, systematically derailing any efforts at commercializing technological seeds developed externally, such as the fundamental research projects universities carried out. Overly committed to developing process technologies incrementally to reduce production costs while upgrading quality, Japanese firms with strong R&D infrastructures also tend to slight the contribution universities can make through their fundamental scientific research (Miyata and Nishimura 2007). What makes firms shy away from fundamental studies is the potential of such projects to create radical innovations, which in turn can make core corporate technologies obsolete. Therefore, only spin-off firms from universities and SMEs tend to utilize

Table 15.2 Performance of Triple Helix Systems in Japan, the U.S. and the UK

	Japan	*U.S.*	*UK*
Funding Received by Universities	\$33 billion (2003)	\$54 billion (2002)	\$10 billion (2002)
Patent Applications	1,680 (2003)	6,509 (2002)	967 (2001)
Licensing Activities	531 (2003)	3,739 (2002)	758 (2002)
Patents Granted (%)	32% (2003)	57% (2002)	64% (2001)
Royalty Incomes	\$5.5 million (2003)	\$1.5 billion (2002)	\$85.6 million (2002)
University-Based Venture Firms	179 (2003)	364 (2002)	197 (2002)

Source: Miyata and Nishimura (2007, 6).

university research results. Even if large corporations sometimes rely on the breakthroughs university research can generate, the former will soon develop their own replicate technologies in order not to pay royalties to universities (Miyata and Nishimura 2007).

The majority of Triple Helix I projects in Japan, namely, those that belonged to the category of 'others', also failed to generate desirable performance outcomes. Traditionally, the role of scholarship contribution to universities from corporations was pivotal in the Japanese triple helix, because most of the university–industry relation was done in an informal way between university inventors and their corporate friends. University researchers often sold their patents to corporations in exchange for the scholarship fund contributed to the university. Due to this symbolic exchange of knowledge and money at the university level in the form of scholarship donation, it was unlikely for corporations to think seriously about commercializing patents they bought. The fear of creating conflicts of interests between university researchers and corporate managers was always present, too (Miyata and Nishimura 2007).

In order to upgrade Triple Helix I to II, participants must realize the importance of creating new knowledge and technology bases for further economic development in a region. Therefore, at this stage, centres of knowledge creation, such as universities and other R&D facilities, become important. Thus the facilitating bodies, such as local authorities and central government agencies, have to beef up the functions of these knowledge creation centres with clear goals, means and communication interfaces. Participating organizations must arrange one or multiple collaborative projects to arrive at a terminal state of creating a venture spin-off that actively utilizes innovations made available by researchers. Universities and other knowledge-creating institutions will replace the industry as the 'core spiral', all for the purpose of commencing collaborative research, initiating technology transfers through the TLO and developing management of technology (MOT) professionals within the HR section of the TLO (Tamura 2006; Etzkowitz 2008, 8–9).

However, the Japanese triple helix suffers from the following five main inhibitors.

1. unclear agreements between firms and universities on how to utilize research results
2. abrupt stoppages of research projects without any further thoughts or guidelines on technology commercialization
3. unclear or conflicting research goals held among university and corporate researchers
4. few research topics that can actually motivate firms to develop tools of commercial applications
5. lack of agreements on research speed or lead time (Katô and Enomoto 2006; Nishio 2007)

As the 2004 amendments to the Patent Law indicate, Japanese universities showed an enormous level of uneasiness about sharing research overheads. Firms, however, detested the idea of information sharing between universities and corporate R&D centres because they did not want to see their corporate secrets being disclosed to the participating universities. Universities hesitated paying for the patent application fees, expecting additional subsidies from the firm or the government under the rubric of Triple Helix I. The 2004 amendments reduced patenting fees as long as university TLOs sent in patent applications just to ease this tension between universities and private firms (Miyata and Nishimura 2007).

The fact that Japanese corporations spent $1.83 billion in 2003 for R&D projects with foreign R&D institutes and universities, whereas they spent only $834 million for projects with Japanese universities in the same year, may augur firms' uneasiness with these structural problems of Triple Helix I in Japan (Miyata and Nishimura 2007, 5). However, active networking with foreign R&D institutes and universities actually made these firms change their attitudes toward technology outsourcing or external procurement of technologies, an important element of Triple Helix II, which is investigated in the next section.

TRIPLE HELIX II IN JAPAN

The 2004 amendment to the Patent Law and to the National University Law certainly changed the attitude among Japanese universities toward the triple helix experimentation. They became much more serious about the whole concept than in the past, as the Patent Law amendment lowered costs related to university patent applications, whereas the National University Law amendment also allowed national university professors to take up non-academic jobs within the TLO.

These attempts at ameliorating the unproductive system of Triple Helix I coincided with the second phase of the Science and Technology Basic Plan (2001–2006). The Japanese government pledged to provide $240 billion to science and technology research projects over the course of five years, the second largest sum in the world after the U.S. expenditure (Taguchi 2003, 156). During this period, the Japanese *sangankurenkei* succeeded in institutionalizing Triple Helix II by filling the gap between university-originating technologies and their commercialization by private firms and by introducing systemic governance schemes to reduce uncertainties surrounding the communication interface between the government, universities and private firms.

Filling the gap between research and knowledge exploitation is a pivotal process during the transition to Triple Helix II. The concept of 'development-gap technology' was first introduced by Atkinson (1994), who found out after the review of Harvard medical research projects that had taken place between 1977 and 1980 that some university-originating technologies were not adequate for simple licensing through the TLO. These technologies were so ingenious that firms failed to develop application knowledge for a series of new future products despite the fact that a relatively long period of time had elapsed. Faced with this new problem in the triple helix, Harvard responded by offering a joint research project to achieve commercialization of the development-gap technologies through a newly established venture firm within the university and additionally opening up a venture capital firm in alliance with private firms (see also Kodama and Suzuki 2006).

What is required to determine whether this kind of new joint venture will work smoothly is the combination of the degree of maturity of the development-gap technology and the firm's absorptive capacity (Cohen and Levinthal 1990; Kanô 2002; Kodama and Suzuki 2006, 37–38). Therefore, the more mature the technology is, or the higher the absorptive capacity is, the higher the chance to succeed in this venture.

The main reason why the development gap occurs is the lack of interfaced communication between firms and universities regarding product and process development. Universities do not possess any organizational tools with which they can attack the thorny issues of marketing, commercialization, financing and securing HR resources dedicated to these issues (i.e. technological maturity). Concomitantly, firms do not have organizational tools with which they can increase their ability in unpacking the results of the fundamental research performed by the university and apply the findings to their firm-specific technological know-how (i.e. absorptive capacity). The mismatch between the technological maturity for commercialization (university-side problem) and the absorptive capacity of university technologies (firm-side problem) creates the problem of the development gap (see Figure 15.1).

Corporate R&D Centre Model: Corporations only

Triple Helix I: University + [TLO] + Corporations (TLO was often neglected)

Gap Model: University (maturity) → [Gap] ← Corporations (absorptive capacity)

Triple Helix II: University + Venture Firm [or beefed up TLO] + Corporations

Triple Helix III: University + Univ. Venture Firm + Univ. Regulatory Agency

Figure 15.1 Evolution of the triple helix.
Source: Kodama and Suzuki (2006, 40); revised by the author.

In Triple Helix II the beefed up TLO or the venture firm, which is placed between the university and the corporation, enhances interfaced communication *qua* liaisons. As explained in the following, the liaison organization will neutralize technological uncertainties and/or communicational fuzziness between two innovative actors in the helices by sophisticating communication media and expanding the organizational and communicative boundaries between the two. Many authors have already substantiated the correlation between effective communicative means and successful technology transfers and/or innovations (see, *inter alia*, Ettlie 1976; Fischer 1979; Tushman 1977; Allen, Lee Denis and Tushman 1980; Barley 1990; Weick 1990).

These studies demonstrated how organizational 'gatekeepers' can collect and process external information to augment 'information richness' and provide fine-grained fresh information to innovation specialists within organizations. Furthermore, various communication media that organizations make available in this process of information collection and processing, such as face-to-face meetings, emails and other forms of written and verbal communication, can substantially reduce the so-called 'informational equivocality', which often intervenes in the process and threatens the sustainability of information richness that is essential in the triple helix (Weick 1979; Daft and Wiginton 1979; Ebadi and Utterback 1984; Rice 1992).

Starting from 2001, and specifically after 2004, Japan entered the stage of Triple Helix II, as the TLO, which had grossly been neglected or unutilized in Triple Helix I, beefed up its intended functions, on the one hand, and universities and corporations began opening up their mutual venture firms to ease the communication difficulties in the transfer of university technologies to corporations on the other. A piece of notable evidence of success toward Triple Helix II is the emergence of what Nishio (2007) calls 'inter-organizational collaborative research' (IOCR).

As Table 15.3 shows, the purpose of IOCR is to obtain research funding for universities and to obtain technological 'seeds' for future commercialization by participating firms. Under the IOCR rubric, the inter-organizational relationship during the research is very stable, as it relies on the beefed up TLO for inter-organizational communication during technology transfers. In this version of the TLO, both firms and universities dispatch their R&D professionals in order to maximize the effectiveness of communicative activities toward their collaborative technology transfers. Due to the clear advantage presented to participating firms, the number of IOCR projects jumped to thirty-eight in 2004 from ten in 2002 (Nishio 2007, 66).

Albeit similar in looks to traditional collaborative research, what makes IOCR distinctive is its ability to resolve the problems of communication and governance, which were detrimental to Triple Helix I in Japan during the previous stage, through organization–organization communication, instead of individual–individual communication. Concomitantly, contractual governance with an oversight committee, comprised of university and

Table 15.3 IOCR and Traditional Research Collaboration

	IOCR	*Traditional Research Collaboration*
Content	Multiple research projects	Single research project
	Technology transfer and MOT HR development in addition to research projects	Exchange of information; technological instructions
Channels of communication	Organization-to-organization communication	Corporate R&D researcher and university researcher
Topic selection	Systematic coordination by university	Individual decision-making between corporate and university researchers
	Both parties decide the scope of research; openly invite competition from researchers within university	
	Firms set up booth to familiarize professors with corporate needs; professor suggests research topics; firms and universities decide topic	
Fund and period	$100,000 per year per firm	$10,000 per year per firm
	Also uses national funds	
	Three to five years	One year
Governance	University–Industry Committee will evaluate research results based on the initial agreements	Report to corporate and university supervisors only
	Amendments or cancellation of research	No amendments or cancellation
	Frequent progress reports	Few progress reports
	Contractual governance	Scholarship contribution agreements only
	High transaction costs due to contractual governance	Impossible to control opportunist behaviours, except governance by trust

Source: Nishio (2007).

industry representatives, is permanently established within the university. Although room for opportunism is high under the contractual governance, the trust-based Japanese organizational culture seems to reduce such threats, especially when institutional complementarities are high between contracting organizations (Nishio 2007).

Katô and Enomoto (2006), however, find problems with the TLO-based governance structure within the IOCR. First and foremost, Japanese universities are still only awkwardly, and often only if threatened, assuming a leading entrepreneurial roles with private firms. Private firms are also reluctant to take an initiative in the network. In the U.S., for example, private firms can directly ask professors to develop a particular technology for an immediate transfer. Also, in Korea, the situation is not much different from that in the U.S., as there are several cases of new technology transfers from the university to the firm at the request of the latter (Oh 2003a, 2003b). The situation in Japan is still very conservative, as professors insist on choosing their own research topics, although in theory they should respond to corporate requests. Their primary goal of research participation is journal publications, not technology commercialization. It will certainly take more time in Japan for each participant to adapt to this new Triple Helix II regime because of their radically different organizational culture (Katô and Enomoto 2006; Itô 2008).

Consequently, the absence of clear leadership, just like in Triple Helix I, augurs additional problems in the area of coordination, scheduling, budgeting and other critical administrative procedures. Participants in the Japanese triple helix deal often do not seem interested in putting effort into understanding the whole new process or the significance of joint research. We can think of several reasons for this observation, although the cultural difference between participating organizations is certainly the most important one.

The triple helix governance is also fraught with difficulties. Committees can be a very flexible form of a controlling mechanism, sitting on the comfort of contractual forms of control and the inter-organizational relationship bounded by social trust that is also backed up by the MEXT and the METI. However, the problems are the infrequency of the committee meetings and the absence of a clear definition of the distribution of authority and responsibility within the committee. It is found that university-based committees seldom convene meetings, nor do they delegate authority and responsibilities to an administrative leader who can manage project teams on a full-time basis. Whenever decisions are not made or the project is halted, it is not always clear who should be held responsible in the current framework of the Japanese IOCR (Katô and Enomoto 2006).

Finally, the current IOCR discriminates against SMEs and private universities. According to the 2001 Science and Technology Basic Plan, which is the backbone of Triple Helix II, policymakers clarify that the government will assist SMEs for their managerial innovation, not for setting up IOCR facilities with national universities (Taguchi 2003, 159). According to the MEXT, the government will process requests from SMEs for technology transfers in association with the SME Revitalization Project Team, although there is no comment whatsoever about whether the project team will match SMEs with local universities (Taguchi 2003, 170). The most important case is the Technology Advanced Metropolitan Area (TAMA) project, where SMEs elect to purchase TAMA membership in exchange for assistance they can get from the TAMA. However, an empirical study found that the effectiveness of technology transfer to SMEs from the TAMA universities for new product development was unclear at best (Kodama 2006, 120). The technological capacity in the form of absorptive capacity must be high within SMEs if they want to be successful in the TAMA (Motohashi 2006, 164). Disadvantages to private universities are also very obvious, as the number of TLOs that were established within private universities between 1998 and 2002 was only three (four, if we include Ritsumeikan's collaboration with the TLO at Kyôtô National University) out of twenty-seven (Taguchi 2003, 180).

The TLO system in Japan has certainly increased more university patents than during Triple Helix I. However, given the problems mentioned earlier, and other financial problems that Nakahara (2006, 22) reports, such as the negligible profitability of the TLO and TLO budget shortages, the current TLO-based triple helix looks unsustainable. But what about university ventures?

The 2001 Hiranuma Plan of creating more than one thousand venture firms in three years, all of which are spun out of university research projects, has been an important motivator for Triple Helix II in Japan (Ogura 2008). By 2007 Japanese universities, mostly national, have created 1,574 firms (see Table 15.4). By 2004, the total number of university ventures surpassed the target number of one thousand. However, since 2005, the number has been in decline. Out of these 1,574 firms, sixty of them went belly-up, whereas fifteen firms were sold in leveraged buyout deals. Twelve firms have successfully listed their company in the stock market through initial stock offerings (IPOs). Therefore, 1,499 firms are still functioning in Triple Helix II, whereas there are 1,487 firms that are still developing or being incubated.

Like the TLOs, of all the university ventures, 63.4 per cent were created by national universities, whereas the number for private universities was only 29.5 per cent (Ogura 2008, 133). Moreover, the majority of these start-ups

Table 15.4 Situations of University Venture Firms (March 2007)

Total Established (1)	*1,574 Firms*
Bankrupt, Shut Down, Disassembled, Hibernating (2)	60 Firms (3.8%)
IPOs (3)	12 Firms (0.8%)
Buyouts (4)	15 Firms (1.0%)
Current Number [1—(2 + 4)]	1,499 Firms (95.2%)
Still Developing [1—(2 + 3 + 4)]	1,487 Firms (94.5%)

Source: Ogura (2008, 133).

are concentrated in the areas of life sciences and information technologies (53.7 per cent). Only 10 per cent of these firms carry out the practice of technology commercialization in the area of manufacturing, such as automobiles (Ogura 2008, 135). This is clearly a Japanese phenomenon, because the manufacturing sector has traditionally relied on their own R&D centres or C&D with U.S. and EU research firms and/or universities for technology sharing. On the other hand, Japanese national universities are known for their strength in life science research, along with robotics.

Policies and venture creation governance mechanisms are much more systematic and specialized than the TLO or Triple Helix I. For example, in the case of Tokyo University, which currently has the largest number of venture firms among all the universities in Japan (107 firms by March 2007), university-wide policies of intellectual property rights protection, conflicts of interests among participants and multiple occupation holding were established all in 2004 to resolve governance problems the TLO and Triple Helix I had experienced until 2003 (Ogura 2008, 142).

The central communication window for inter-organizational communication toward the creation and management of new university ventures is located within Tokyo University under the name of *Sangakureneki Honbu* (Triple Helix Headquarters). In addition to this, all patenting and other administrative tasks are taken care of by the extant TLO with clearly beefed up functions. The venture capital functions, which were not addressed in the previous period, are now under the control of Tokyo University Edge Capital (UTEC). From 2006 all participants can also acquire stocks of new venture firms in lieu

Table 15.5 Top Ten Universities with Most Venture Start-Ups (March 2007)

Rank	*University*	*Type*	*No. of Ventures*	*Concentration Ratio*
1	Tokyo	National	107	6.40%
2	Waseda	Private	96	5.70%
3	Osaka	National	68	4.00%
4	Tsukuba	National	61	3.60%
5	Keio	Private	52	3.10%
6	Kyushu	National	48	2.90%
7	Kyoto	National	45	2.70%
8	Kobe	National	42	2.50%
9	TIT	National	40	2.40%
10	KIT	National	39	2.30%
Total			598	35.60%

Source: Ogura (2008, 136).

of traditional scholarship donation or licensing fees, as long as such transactions are reported to the MEXT. Finally, Tokyo University also runs venture incubation facilities, which can be rented by any interested community group for nominal fees. Incubation rooms in Sangakurenkei Plaza at Tokyo University currently charge \$2 per m^2, whereas Tokyo University Kashiwa Venture Plaza charges \$900 for 300 m^2 (Ogura 2008, 142). Figure 15.2 roughly captures the whole governance structure of Triple Helix II at Tokyo University.

The process at Tokyo University was a success given its rapid and continuous venture creation on campus. However, that many university venture start-ups are still being incubated or growing (94.5 per cent) augurs that these university start-ups have faced various problems. Nakahara (2006) attributes the stagnated growth of the university venture start-ups to three important factors. First, 53 per cent of all university ventures (*daigaku-hatsu bencha-*) are suffering from revenue deficits. Second, these start-ups do not possess professional HR resources for venture management. Third, university culture is often not complementary to venture culture. Therefore, Nakahara (2006, 22) reports that only 30 per cent of the university venture incubators or Venture Business Laboratories (VBLs), which were installed in forty-five universities in 2001, are still in business. Other related figures are also desolate, as 68 per cent of these VBLs are not receiving any researchers dispatched from corporations, and 92 per cent of these are not sending any students for practical internship to corporations.

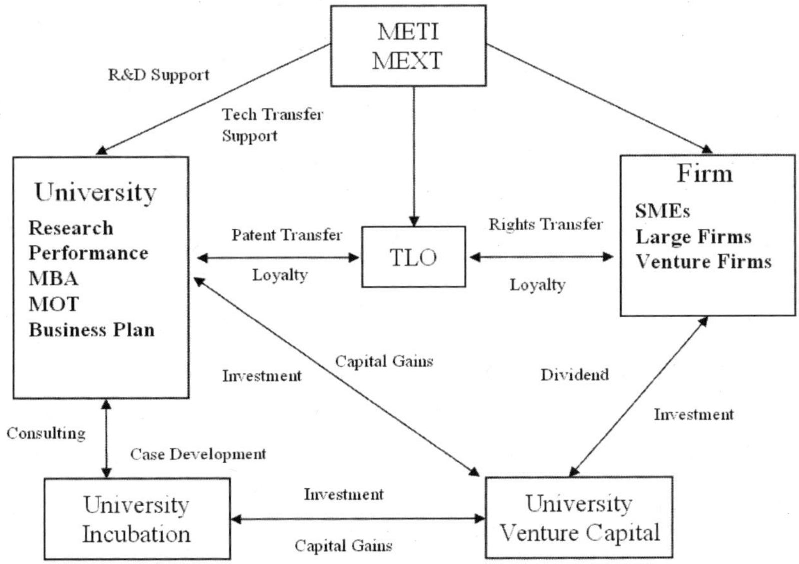

Figure 15.2 Triple Helix II toward venture creation.
Source: Tamura (2006)

What are the causes of the problems Triple Helix II is facing now in Japan? Would Japan be able to countervail the difficulties and upgrade the current system to Triple Helix III, where the regulatory roles of the university can be matured, in addition to the TLO (patenting regulation), venture and venture capital roles Triple Helix II currently has?

DISCOURSES ABOUT FURTHER EVOLUTION

AIL, a Tokyo- and Kobe-based semiconductor designing company, is a spin-off firm that came out of Kobe University. The founder, Taki Kazuo, was a former professor of Kobe University, and he was the first Japanese national university professor who used the new university-wide system of taking a year off to set up a venture firm by himself. However, even in this success case, Dr. Taki could not continue to hold on to his university post after a year, because Kobe University did not want to run the venture using professors as the firm's directors. Therefore, Dr. Taki had to quit his professorial position and became a full-time executive for his own company.

The AIL case reveals some of the fundamental problems of the Japanese triple helix. According to Kitachi (2006, 203), university ventures face long-term and short-term tasks they have to complete in one way or another. Long-term problems include cultural conflicts, social environmental problems and lack of experiences and know-how among the triple helix participants, whereas short-term problems are the conflict of interests among participants, patent-related issues and tax systems. According to Kitachi (2006), the real cause of the lacklustre performance of the university venture firms in Japan are the long-term problems, which require the wholesale transformation of Japanese society.

The speed of institutionalization of the Japanese triple helix, as it has been sketched out so far, was very fast, not giving enough time for university and corporate participants to adapt to new institutions and new culture, mostly deriving from the U.S. model. At the same time, minute details in the execution of new rules and policies, especially regarding short-term issues (e.g. taxation, intellectual property protection, conflict of interests) were discussed over a long period of time among policymakers, which in turn failed to give venture firms timely opportunities to utilize their technological know-how (Kitachi 2006, 204).

Three facets of the long-term problems loom large. First, universities don't seem to develop rapport with the spirit of entrepreneurialism. If professors are encouraged to move between different boundaries freely, and if universities are encouraged to create venture firms, all these new activities unnecessarily subject universities and professors to new uncertainties and risks. However, neither universities nor professors have means with which to fight these risks.

The traditional pyramid of university R&D structure with one professor on top and others scattered around different layers of the pyramid makes it difficult for young researchers to bypass the whole hierarchy to create their venture firms. It is not easy for professors to presume all the risk by themselves either. Delegating power and responsibility to external representatives from corporations or the government can be a solution to this bottleneck, although the culture of not trusting outsiders in Japan can be a detriment to this alternative.

What is worse, many professors do not want to participate in venture businesses because their performance evaluations are usually based on professional publications instead of their entrepreneurial skills. Furthermore, university faculty members, who are used to the notion of stability and tenure, are not willing to take unstable and non-tenured positions within venture firms (Kitachi 2006, 205–206).

Second, the Japanese triple helix constantly suffers from the shortage of specialists in the area of corporate tax laws, accounting and patent registration. Although Japan has many qualified lawyers, chartered accountants, tax consultants and patent specialists, when it comes to venture activities and *sangakurenkei* it is not easy to locate real specialists with ample experiences in each specialized field. This is mainly because venture firm activities are fairly recent developments in Japan. Furthermore, because the triple helix cannot function very well without the presence and support of these professionals on campus or in the vicinity of the university, it is difficult to invite these professional to the campus, not to mention to locate them in the market (Kitachi 2006, 207).

Third, social values and institutions do not support venture spirits in Japan. Japanese culture does not usually tolerate failures and therefore discourages people from participating in university-based ventures for fear of causing failures. Internal factional politics, character assassination and jealousy toward leaders are typical cultural hindrances to venture development in the university setting as well. In the cultural tradition where sticking out is considered bad (*yokonarabi*) and where cultures of shame (*haji no bunka*) and indifference (*kotonakare shugi*) are taken as the best strategy in organizational politics, a radical social reform toward Triple Helix III is unlikely (Kitachi 2006, 209). If these social values are coupled with the pyramidal hierarchy and the lack of professional support, it is highly recommended for universities and firms to fight these three undesirable cultural traits.

These three features of the long-term problems of the Japanese triple helix will take a long time to fix. The distance between what is intended by policy means and what has actually occurred as both intended and unintended consequences is huge and contradictory, making most of the discernible solutions to the discrepancy unfeasible or unmanageable. Therefore, the triple helix development policy must address issues of filling the gap between technology development and commercialization, HR development, support facilities development, time management for the institutional reform in favour of venture firms and radical cultural changes within the participating organizations, including universities.

FUTURE IMPLICATIONS

Japan is the first country in Asia that succeeded in the modernization of the nation through radical reforms of its feudalistic institutions during the Meiji era (1868–1912). Previous successes of modernization and economic development even during the post-war period made the Japanese proud of their innovation and technology management. However, many people are suspicious of Japan's ability to organize rapid and recurrent reforms at the time of cyclical crises (Imai 2008).

During the fieldwork for this research, it was also noticed that the higher education reform toward educational globalization that the university carried out by expending billions of dollars was only superficially successful, as the traditional culture (e.g. bureaucratic governance, *nemawshi* or decision by consensus, job rotation vis-à-vis specialization, payment by seniority, *amakudari* or taking posts through political connections) of the university had rarely been reformed. The current situation with the Japanese triple helix is facing a similar fate, as the institutional and cultural core of the Japanese MOT will not budge, whereas organizational and legal parameters are rapidly changing.

Unless Japan is willing to develop new skills of introducing recurrent reforms at the fast pace of global change, an upgrade to Triple Helix III is most likely futile or unfeasible in the current Japanese context. One policy implication for the current deadlock is to introduce an alternative leadership pattern within the triple helix, the one that does not have any roots in the Japanese traditional organizational heritage, as some of the success cases in the Japanese private sector indicate (e.g. Carols Ghosn at Nissan-Renault).

CONCLUSION

This chapter intended a theoretical discussion of why the Japanese national innovation system, or the triple helix, maintains the world's second highest level of public expenditures to bring about the rapid development and success of university-based venture firms (Triple Helix II), although the actual profitability of the venture through technology transfers and commercialization is relatively in its infantile stage compared to North American and Western European standards (Triple Helix III).

The chapter argued that various factors intervene in this evolutionary transition to Triple Helix III, and one of the critical aspects of it is the financial and the regulatory role universities must play in the creation of university-based ventures. However, the current Japanese triple helix is fraught with difficulties because of both long-term and short-term problems. Long-term problems derive from cultural differences between Japan and the West, where Triple Helix III is fully blossoming. Short-term problems are organizational in nature, as they refer to the problems of the conflict of interests, patent registration and taxation. Among these, the chapter suggest that the long-term problems are the serious hindrance to this transition to Triple Helix III.

Future studies can be designed to address the possibility of locating and introducing the triple helix leadership to Japan from foreign countries, where leaders' cultural heritage is not rooted in the Japanese tradition, making it easier for organizational actors to challenge bureaucratic inertia.

ACKNOWLEDGEMENTS

This study was partically funded by the Departmental Research Grant, which Bristol Business School generously provided to the author of this chapter. The author would also like to thank Mohammed Saad, Judith Jordan, Will Brett and Emiko Hatayama for their invaluable support and assistance.

REFERENCES

Abernathy, W. J. 1978. *The productivity dilemma.* Baltimore, MD: The Johns Hopkins University Press.

Allen, T. J., M. S. Lee Denis and M. Tushman. 1980. 'R&D performance as a function of internal communication, project management, and the nature of the work'. *IEEE Transactions on Engineering Management* 27 (1): 2–11.

Atkinson, S. H. 1994. ' University affiliated venture capital funds.' *Health Affairs* (Summer): 159–175.

Barley, S. R. 1990. 'The alignment of technology and structure through roles and networks'. *Administrative Science Quarterly* 35 (1): 61–103.

Cohen, W. M., and D. A. Levinthal. 1990. 'Absorptive capacity: A new perspective on learning and innovation'. *Administrative Science Quarterly* 35:128–152.

Daft, R. L., and J. Wiginton. 1979. 'Language and organization'. *Academy of Management Review* 4 (2): 179–191.

Degawa, T. 2006. 'Sangyôkai no sangakukan renkei katsudô'. In *Kyôsôryokukyôkani Muketa Sangakukan Renkei Manejimento*, ed. A. Nagahira and K. Nishio, 33–67. Tokyo: Chuo Keizaisha.

Ebadi, Yar M., and J. Utterback. 1984. 'The effects of communication on technological innovation'. *Management Science* 30 (5): 572–585.

Ettlie, J. E. 1976. 'The timing and sources of information for the adoption and implementation of product innovations'. *IEEE Transactions on Engineering Management* 23 (1): 62–68.

Etzkowitz, H. 2008. *The triple helix: University-industry-government innovation in action.* London: Routledge.

Etzkowitz, H., and L. Leydesdorff, eds. 1997. *Universities in the global economy: A triple helix of university-industry-government relations*. London: Pinter.

Fischer, W. 1979. 'The acquisition of technical information by R&D managers for problem solving in non-routine contingency situations'. *IEEE Transactions on Engineering Management* 26 (1): 8–14.

Freeman, C. 1987. *Technology policy and economic performance: Lessons from Japan*. London: Pinter.

Goto, A. 2006. 'Nihon no inobe-shon sisutemu: tsuyomi to yowami'. In *Nihon no Inobe-shon Sisutemi: Nihonkeizai Fukkatsu no Kibankôchiku ni Mukete*, ed. A. Goto and T. Kodama, 1–17. Tokyo: Tôkyôdaigaku Shuppankai.

Goto, A., and T. Kodama, eds. 2006. *Nihon no Inobe-shon Sisutemi: Nihonkeizai Fukkatsu no Kibankôchiku ni Mukete*. Tokyo: Tôkyôdaigaku Shuppankai.

Gulbrandsen, M. 1997. 'Universities and industrial competitive advantage'. In *Universities in the global economy: A triple helix of university-industry-government relations*, ed. H. Etzkowitz and L. Leydesdorff, 121–131. London: Pinter.

Imai, K. 2008. *Sôzôteki Hakai toha Nanika?: Nihonsangyô no Saichôsen*. Tokyo: Tôyôkeizai Shinposha.

Itô, K. 2008. *Nihon ni No-berushôga Kuru Riyû*. Tokyo: Asahishinbun Shuppan.

Kanô, S. 2002. 'Sangyôrenkei no gyappu chôsei mekanizumu ni kansuru kenkyû'. PhD thesis, Tokyo University.

Katô, K., and S. Enomoto. 2006. 'Sangakurenkei kenkyûkaihatsu purojekuto niokeru kadai to sono kaizenjirei'. *Purojekuto Manejimento Gakkaishi* 8 (1): 17–22.

Kindleberger, C. P. 1996. *World economic primacy: 1500–1990*. Oxford: Oxford University Press.

Kitachi, T. 2006. 'Nihon niokeru daigakutohatsu bencha- kigyo no kadai'. In *Kyôsôryoku Kyôka ni Muketa Sangakukan Renkei Manejimento*, ed. A. Nagahira and K. Nishi, 202–217. Tokyo: Chûôkeizaisha.

Kodama, T. 2006. 'Sangyô kurasuta- keisei niokeru seishin kaihatsugata chûshôkigyô no yakuwari: TAMA (gijutsu senshin shutoken chiiki) ni kansuru jisshô bunseki ni motozuite'. I n *Nihon no Inobe-shon Sisutemi: Nihonkeizai Fukkatsu no Kibankôchiku ni Mukete*, ed. A. Goto and T. Kodama, 79–136. Tokyo: Tôkyôdaigaku Shuppankai.

Kodama, T., and J. Suzuki. 2006. 'Sangakurenkei no bunseki wakugumi: genjitsu o chokushishita gijutsu iten moderu no kaihatsu'. In *Nihon no Inobe-shon Sisutemi: Nihonkeizai Fukkatsu no Kibankôchiku ni Mukete*, ed. A. Goto and T. Kodama, 35–51. Tokyo: Tôkyôdaigaku Shuppankai.

Leydesdorff, L. 1997. 'The new communication regime of university-industry-government relations'. In *Universities in the global economy: A triple helix of university-industry-government relations*, ed. H. Etzkowitz and L. Leydesdorff, 106–117. London: Pinter.

Leydesdorff, L., and H. Etzkowitz. 1998. 'The triple helix as a model for innovation studies'. *Science and Public Policy* 24 (3): 195–203.

Misawa, K. 2007. *Gijutsu Manejimento Nyûmon*. Tokyo: Nihonkeizaishibun Shuppansha.

Miyata, Y., and Y. Nishimura. 2007. 'Nihongata inobe-shon to sangakurenkei no jittai'. *Ôsaka Furitsudaigaku Keizai Kenkyû* 52 (4): 1–26.

Motohashi, K. 2006. 'Chushokigyo no sangakurenkei to kenkyukaihatsu nettowa-ku: henkakukini aru nihon no inobe-shon sisutemu niokeru ichizuke'. In *Nihon no Inobe-shon Sisutemi: Nihonkeizai Fukkatsu no Kibankôchiku ni Mukete*, ed. A. Goto and T. Kodama, 137–167. Tokyo: Tôkyôdaigaku Shuppankai.

Nagahira, A. 2006. 'Sangakureneki sokushin seisaku: chitekizaisan kanrenn seisaku no gaiyô'. In *Kyôsôryoku Kyôka ni Muketa Sangakukan Renkei Manejimento*, ed. A. Nagahira and K. Nishio, 3–13. Tokyo: Chûôkeizaisha.

Nakahara, H. 2006. 'Inobe-shon suishin no tameno sangaku renkei keitai no yûki-teki tôgô'. *Chibadaigaku Keizai Kenkyû* 21 (3): 1–43.
Nemoto, T. 2006. 'Gijutsu keiei to inobe-shon'. In *Gijutsu Keiei: Gijutsu Senryaku to MOT*, ed. T. Nemoto and Y. Utashiro, 30–44. Tokyo: Gakubunsha.
Nishio, K. 2007. 'Nihon niokeru sangakukan no soshikiteki kenkyû kyôryoku ni kansuru kenkyû'. *Kenkyû Gijutsu Keikaku* 22 (1): 65–81.
Ogura, M. 2008. 'Nihon no daigakutôhatsu bencha- to seifukei kenkyûhatsu bencha-'. In *Akademiku Inobe-shon: Sangakurenkei to Suta-to Appusu Sôshutsu*, ed. T. Watanabe, 129–151. Tokyo: Hakutô Shobô.
Oh, I. 2003a. *The development of cluster tool controllers (CTC) for semiconductor manufacturing: The case of KAIST and Jusung Engineering, Co.* Beppu, Japan: Ritsumeikan Asia Pacific University, Case Development Centre.
———. 2003b. *Finding a niche market in times of recession—Lucid Korea*. Beppu, Japan: Ritsumeikan Asia Pacific University, Case Development Centre.
Oh, I., H. Park, S. Yoneyama and H. Kim. 2005. *Mad technology: How East Asian companies are defending their technological advantages*. New York: Palgrave.
Pires, A., and E. Anselmo de Castro. 1997. 'Can a strategic project for a university be strategic to regional development?' *Science and Public Policy* 24 (1): 15–20.
Posen, A. 2002. 'Japan'. In *Technology innovation and economic performance*, ed. B. Steil, D. Victor and R. Nelson, 74–110. Princeton, NJ: Princeton University Press.
Rice, R. 1992. 'Task analyzability, use of new media, and effectiveness: A multi-site exploration of media richness'. *Organization Science* 3 (4): 475–500.
Sakakibara, K. 200 7 *Inobe-shon no Shûekika: Gijutsu Keiei no Kadai to Bunseki*. Tokyo: Yûhikaku.
Shimoyama, K. 2006. 'Gijutsu teikei to kyûdôkenkyû'. In *Gijutsu Keiei: Gijutsu Senryaku to MOT*, ed. T. Nemoto and Y. Utashiro, 47–58. Tokyo: Gakubunsha.
Taguchi, T. 2003. *Sangakukyôdô to Kenkyûkaihatsu Senryaku*. Tokyo: Hakutô Shobô.
Tamura, Y. 2006. 'MOT to sangakurenkei manejimento'. *Ofisu O-tome-shon* 26 (4): 35–44.
Teece, D. J., and D. C. Mowery. 1993. 'Japan's growing capabilities in industrial technology: Implications for U.S. managers and policymakers'. *California Management Review* 35 (2): 9–34.
Tushman, M. 1977. 'Communication across organizational boundaries: Special boundary roles in the innovation process'. *Administrative Science Quarterly* 22:587–650.
Weick, K. E. 1979. *The social psychology of organizing*. Reading: Addison-Wesley.
———. 1990. 'Technology as equivoque: Sense making in new technologies'. In *Technology in organizations*, ed. P. Goodman and L. Sproull, 1–43. San Francisco: Jossey-Bass.

Contributors

Azley Abd Razak is a Lecturer at the Bristol Business School, Department of Strategy and Operations Management, at the University of West of England in Bristol. Previously, he was a Research and Teaching Associate at the Bristol Business School. Currently, Azley is pursuing a doctoral degree at the Bristol Business School. His research interests lie in the area of the triple helix model of innovation and science and technology policy, specifically in developing countries. Bristol Business School, School of Operations and Information Management, University of West of England, Bristol, BS16 1QY UK.

Dr. Mariza Almeida is Assistant Professor at the Master Course of Local Development, Augusto Motta University Center, Brazil. Her research and scholarly interests include sciences and technology, social network analysis and developmental policy.

Amira Bouhamed is a PhD student at the Faculty of Economics and Management of Sfax in Tunisia. She is a member of a Research Unit on Management Studies. Her research interests include triple helix, entrepreneurship, public–private partnerships, learning and strategy. Faculté des Sciences Economiques et de Gestion de Sfax, BP 1088, 3018 Sfax, University of Sfax, Tunisie.

Najeh Bouraoui is a PhD student at the Faculty of Economics and Management of Sfax in Tunisia. She is a member of a Research Unit on Management Studies. Her research interests include alliances and networks, learning, strategy and triple helix. Faculté des Sciences Economiques et de Gestion de Sfax, BP 1088, 3018 Sfax, University of Sfax, Tunisie.

Jamil Chaabouni is Professor in Management Studies at the Faculty of Economics and Management of Sfax in Tunisia. He leads a Research Unit on Management Studies. His research interests include governance, strategy and organizational structure, organizational change and information management. Faculté des Sciences Economiques et de Gestion de Sfax, BP 1088, 3018 Sfax, University of Sfax, Tunisie.

Norman Clark is Professor of Innovation Systems and Development at the Open University, UK, and Director of Research and Technical Advisor to the Director at the African Centre for Technology Studies, Nairobi. Previously he was Vice Chancellor of Kabarak University, Nakuru, Kenya, and before that, Professor of Environmental Studies and Director of the Graduate School of Environmental Studies at the University of Strathclyde, where he is now an Emeritus Professor. He also held academic posts at Sussex University, where he acted as the Founding Director of Graduate Studies at the Science Policy Research Unit (SPRU). He is a development economist specializing in science, technology and innovation policy issues with particular relevance to Third World problems. He has also acted as Founding Director of the Technology Planning and Development Unit, University of Ife, Nigeria; Visiting Professor, Institute for Advanced Studies, University of Sao Paulo, Brazil; and Director of the Capacity Development Program at the African Centre for Technology Studies.

Surja Datta received his doctorate from Bristol Business School, University of the West of England. Surja also holds a MBA from Strathclyde University, and a MA in Applied Social Research from the University of the West of England. Dr. Datta is currently a senior lecturer of strategy in the Business School, Oxford Brookes University. His research interests include innovation systems and outsourcing. Department of Economics and International Business, Oxford Brookes University.

Dr. James Dzisah is Assistant Professor at the Department of Sociology, Nipissing University, Ontario, Canada. His research and scholarly interests revolve around science, technology and society; knowledge-based development; and the global South.

Professor Henry Etzkowitz is Faculty Research Fellow at Stanford University and Visiting Professor at Edinburgh University Business School. Henry is President of the Triple Helix Association and General Advisor to the International Triple Helix Institute, LaSalle University, Madrid. His previous work on Central and Eastern Europe includes co-editorship of *Transforming Science and Technology Systems—The Endless Transition?* (NATO Science Series 4, 1998) and articles in *Science and Public Policy*. His most recent book is *The Triple Helix: University-Industry-Government Innovation in Action* (London: Routledge, 2008). His next work is *Gender Dynamics in Science and Technology: The Vanish Box Phenomenon* (co-authored with Marina Ranga) for Cambridge University Press.

Maria Del Pilar Montserrat Perez Hernandez (MSc) is a Researcher at National Polytechnic Institute, Mexico, and her main research interests include incubators and entrepreneurship.

Dr. Dessy Irawati is a researcher based at KITE/Knowledge Innovation Technology Enterprise in Newcastle University Business School. Her research interests include knowledge, innovation, network and technology. She continues to develop her research interests on innovation and regional development in sustainable innovation and green manufacturing.

Dr. Leigh W. Jerome is the President and CEO of the Institute for Triple Helix Innovation, in Honolulu, Hawaii. The Institute is a federally founded non-profit organization whose mission is 'Accelerating the transfer of knowledge for innovation, profit and the social good'. Dr. Jerome is a clinical psychologist with roots in both research and practice. She has provided leadership and consulted with the American Psychological Association, the World Health Organization and the National Institutes of Health regarding technology, telehealth and using information and communication technologies to facilitate collaboration and global eHealth. Dr. Jerome's areas of expertise include boundary spanning; social networks; knowledge clusters; multimedia technologies; communities of practice; transformative learning; empathy; mind–body integration; biosensors; virtual environments; and the synthesis of healthcare, technology, science and the arts. The Institute for Triple Helix Innovation, Honolulu, HI 96822.

Dr. John Kirkland is Deputy Secretary-General of the Association of Commonwealth Universities, where he also directs the ACU Research Management program. Prior to arriving at the ACU in 1999, he spent five years as Company Secretary at the UK National Institute of Economic and Social Research, and was, from 1988 to 1994, Director of Research Services at Brunel University. Amongst his other roles, Dr. John Kirkland is Executive Secretary to the Commonwealth Scholarship Commission in the United Kingdom and the Marshall Aid Commemoration Commission. He is also a Visiting Professor at the Bristol Business School (University of the West of England).

Anne-Marie Maculan is Associate Professor at the Production Engineering Program of COPPE-Federal University of Rio de Janeiro, Brazil. She received a PhD in socio-economics at the University of Quebec at Montreal. She has been working as a consultant for the Ministry of Science and Technology, World Bank and various Brazilian federal or state agencies for research and innovation. Her main academic areas of interest include technology transfer, academic entrepreneurship and the role of the scientific research institutions in the local innovation system.

Dr. Chandra Malairaja has been a civil servant in the Malaysian government with responsibilities in policy matters relating to the development of business enterprises through joint ventures and technology transfer and technology development initiatives. He had a long stint at the Ministry of Science, Technology and Innovation (MPSTI) as a principal assistant director after which he took study leave to undertake PhD studies at the University of Strathclyde in Glasgow, UK. He has researched and published on the effectiveness of science parks as a strategy for technological progress and industrialization in emerging economies like Malaysia. He is currently engaged as a senior staff in the Ministry of Natural Resources and Environment of Malaysia, based in Kuala Lumpur.

Professor José Manoel Carvalho de Mello holds a PhD in Production Engineering from the University of Birmingham (England). He is a Visiting Professor at the Production Engineering Graduate Program and Advisor at the Innovation Agency at the Fluminense Federal University, Brazil. Dr. Mello is the Vice President of the Triple Helix Association. He was a Visiting Fellow at the Management Centre of the University of Bradford, England (1973); at the Centre Science, Société et Technologie in Paris (1984–1985) and at the State University of New York (1997–1998). Dr. Mello is the author of numerous papers on science, technology and innovation policy, local systems of innovation, triple helix models of innovation and on entrepreneurial university.

Dr. Ingyu Oh is Senior Lecturer at Bristol Business School, University of the West of England. He formerly taught at Waikato University, Ritsumeikan University and UC Berkeley. He has published numerous books and articles on the national innovation system, Asian business and technology transfer. Bristol Business School, School of Operations and Information Management, University of West of England, Bristol.

Dr. Marina Ranga is Research Fellow at Stanford University, Visiting Assistant Professor at Stony Brook University New York and Visiting Fellow at SPRU, Sussex University, UK. Her research interests include national and regional innovation systems and the European Research Area, triple helix and the entrepreneurial university, gender in science, technology and innovation. She is a member of the UN Economic Commission for Europe's Expert Group on Innovation and Competitiveness Policies and of the Gender Advisory Board of the UN Commission for S&T for Development. She is Country Correspondent for Romania in the European Union's PRO INNO TrendChart, ERAWATCH Research Inventory, ERANET, METRIS, European Observatory of Research Universities.

Professor Mohammed Saad (License Sciences Economiques, DMS, MSc, PhD) is a Professor of Innovation and Technology Management at the Bristol Business School, Department of Strategy and Operations Management (the University of West of England in Bristol). He is a visiting professor at the University of Sfax (Tunisia) and the Besançon Business School (France). He also worked as a Lecturer of Management in Algeria for ten years. His research activity spans several fields—mainstream operations management, innovation and technology management and more global policy-related issues of knowledge development, transfer and institutional collaborative learning in the context of developing countries. He is co-founder and co-editor of the *International Journal of Technology Management and Sustainable Development*, which is published in collaboration with the Association of Commonwealth Universities. Bristol Business School, University of the West of England, Bristol.

Dr. Matthew A. Shapiro is Assistant Professor of Political Science in the Department of Social Sciences, Illinois Institute of Technology (IIT), U.S. He received his PhD in Political Economy and Public Policy from the University of Southern California, U.S. His recent publications include *Korea's Environmental Sustainability Leadership in East Asia and Beyond* (2009) and *Public–Private R&D Collaboration in Korea: A Survey of Public Sector Institutes* (2007). He is a reviewer for several journals, including *Research Policy, Contemporary Economic Policy* and *Review of Policy Research.*

Julie Stackhouse is Senior Program Officer (Research) at the Association of Commonwealth Universities (ACU) and amongst other roles works on the ACU's Research Management program. Julie Stackhouse has worked at the ACU since 1998 and has an MA in Sociology and Women's Studies from Lancaster University and a BA (Hons) in Sociology from the University of Leeds.

Dr. Branca Terra is Adjunct Professor at the School of Administration and Finance at the Rio de Janeiro State University and Castelo Branco University, Brazil. Her research and teaching activities include entrepreneurial university, innovation and technology.

Professor María Isabel Rivera Vargas holds a PhD in Sociology from the State University of New York (2000). She is currently a full-time professor and researcher at the Department of Regional Culture Studies at the University of Guadalajara, Mexico, where she acts as a research coordinator. Since 2001 Professor Rivera Vargas has been leading the doctoral program of Business and Economic Studies at the same university. She has been a member of the National System of Researchers in Mexico since 2001, and founder member of the Jalisco's Science Academy. She is author of several national and international publications.

Dr. Girma Zawdie (BA, MSc, PhD) is a lecturer in the David Livingstone Centre for Sustainability at the University of Strathclyde, Glasgow, UK. He is an economist with over thirty years of teaching and research experience gained from work in the UK and overseas. Areas of his research interest cover technology and innovation management; policy issues relating to environmental management; world poverty; and sustainable development. He has published papers widely in peer-reviewed journals, conference proceedings, books and monographs. He is co-founder and co-editor of the *International Journal of Technology Management and Sustainable Development*, which is published in collaboration with the Association of Commonwealth Universities. He also had a stint as managing editor of *Science, Technology and Development*, journal of the Third World Science, Technology and Development Forum.

Index

A

accountability 88
acquis communautaire 250
active internationalization 27
Africa: dysfunctionality of triple helix 4; research managers 94–5; *see also individual countries*
agricultural science 98–115; agenda for 100; Andhra Pradesh Netherlands Biotechnology Programme 105–8; fodder for livestock 104–5; high-yielding hybrid crops 98; livestock diseases 102–4; R&D 100
alliance specialists 82
alliances 67–83; capabilities of 74–5; international 73–4; knowledge acquisition 72–3; learning within 73–4; management of 82–3; *see also* joint ventures
ALPHA companies 67, 75–83
analytical framework 26–9
analytical ideology 112
Andhra Pradesh Netherlands Biotechnology Programme 105–8
Anglo-Gold 157
anticipatory systems 111–12
Apache 212
applications engineering 108
applied research linkages 38
Asian Development Bank 171
Association of Commonwealth Universities 90
Association for Overseas Technical Scholarship (AOTS) 168
Association of Research Managers and Administrators 90
Association of University Industry Research Links 90
Astra Automotive 167–8
Astra Manufacturing Polytechnic 168
Australasian Research Management Association 90

B

Biggs, Stephen 113
Black Book of Outsourcing 224
blue sky research 231
bootstrapping 226
brain drain 215–18
Brazil 52–64; *Ação da Cidadania contra a Fome, a Miséria e pela Vida* 183; Anprotec 182, 184; *Comitê de Entidades Públicas no Combate à Fome e pela Vida* 183; cooperative incubators 183; economic crisis 179–80; First Company Program 62–3; 'Good Act' 61–2; growth of incubators 184; import substitution industrialization 56–7; *Incubadora Tecnológicade Cooperativas Populares* 183; incubators 176–87; industrialization 55; innovation capacity 57–8; Innovation Law 53, 61; Inovar Project 186; National Advanced Technology Enterprise Promoter Entity (Anprotec) 176; National Association of Organizations for the Promotion of Entrepreneurships Innovators 59; National Council for Scientific and Technological Research 180; National Institute of Industrial Property 59; National Scientific and Technological Development Fund 55; patents 59–60; political, scientific and social context

179–81; *Programa de Implantação de Parques de Tecnologia* 180; *Programa Nacional de Apoio a Incubadoras de Empresas do Ministério da Ciência e Tecnologia* 176; regulatory environment 52–3, 54, 61–3; Science, Technology and Innovation (ST&I) policy 53; science and technology policy 55, 57; teaching and research 54–7; Technological Innovation Nuclei 62; traditional incubators 182–3; triple helix 53; universities 52, 58–63; University Reform Law 55
Bulgaria: cooperation 267; FDI 261; innovation activities 263; innovation performance indicators 257; innovation/entrepreneurship indicators 264; university graduates 257; university-based researchers 259
business incubators 59

C
CAD/CAM 32
CEE *see* Central and Eastern Europe
Central and Eastern Europe 249–78; clusters and regional development 269–73; FDI 260–3; GDP 250; government 265–6; hybrid organizations 266–9; industry 260–5; innovation activities 263; innovation performance indicators 257; R&D 251, 265–6; Regional Innovation Strategy 275; science-industry links 268–9; SMEs 264–5; transition 255–66; triple helix effects 273–4; universities 255–60; university graduates 257; university-based researchers 259; venture capital 276; *see also individual countries*
chaebols 25, 35, 123
China, GDP 250
class 210–11
clusters 17–18; advantages for SMEs 171–3; CEE 269–73; Indonesia 163–5, 168–71; innovation 17; Japan 283–4; knowledge 17–18
collaboration 12–13, 149; public-private 125, 132; R&D 123–4; university-industry 230–44
collaborative innovation 13–15, 20
collaborative output, effects of new/pre-existing capital 131, 132, 133
collaborative research 13
commercialization 201–2
common perspectives 74
communities 15
Communities of Practice 18
community service 167–8
community-based organizations 100
Compac 36
competition 68
complementarity 28
contract equipment manufacturers (CEMs) 36
cooperation: CEE 267; North-North 69; North-South 69; South-South 75–83
cooperative incubators 183
Council for Scientific and Industrial Research (CSIR) 152
country characteristics 26
creative destruction 249
creative imitation 31–2
creative reconstruction 249–78
cultural capital 210–11
cultural differences 73–4
Czech Republic: clusters and regional development 272; cooperation 267; FDI 261; innovation activities 263; innovation performance indicators 257; innovation/entrepreneurship indicators 264; science-industry links 268; university graduates 257; university-based researchers 259

D
Death Valley phenomenon 285
developing countries: Ghana 146–60; India 207–27; Indonesia 161–74; lack of interaction 1–2; Malaysia 190–205; micro/macro-innovation 215–16; research managers 94–5; technology transfer to 14; triple helix system in 2, 3; welfare considerations 15
development planning 111; Ghana 152–4; success of 112
development thinking/practice 147–8
development-gap technology 290–1
Devil's River phenomenon 285
Diaspora networks 217–18

distributed networks 16–17; knowledge clusters 17–18; requirements of 18
domains of responsibility 2
duplicative imitation 30–1

E
ecological creativity model 110
economic capital 209–10
education 147–8
Egypt, joint venture with Tunisia (STEFM) 76, 77–80
employees: distribution 77; skilled labour 215–18, 221; training 43; turnover 43
endogenous technology 56
entrepreneurial universities 191, 254; academic revolutions 254
entrepreneurs 2; 'founder' 225
entrepreneurship, CEE 264
Estonia: cooperation 267; FDI 261; innovation activities 263; innovation performance indicators 257; innovation/entrepreneurship indicators 264; science-industry links 268; university graduates 257; university-based researchers 259
etatistic model 178
European Association of Research Managers and Administrators 90
European Commission, JEREMIE initiative 276
European Union: Innovation Relay Centres 268; Structural Funds 251; TEMPUS program 256
evolution of triple helix 193–5, 200–4, 291; Japan 290–300; *see also various triple helix phases*
expertise 89

F
FDI 27, 250, 251; Central and Eastern Europe 260–3; financial 262; Malaysia 234; Mexico 36–7
Federation of Malaysian Manufacturers 238–9
financial FDI 262
focus groups 104
fodder for livestock 104–5
Ford, partnership with Hyundai 30
Fordism 56
foreign direct investment *see* FDI
foreign technology, acquisition of 31
forms of capital 208–11; cultural capital 210–11; economic 209–10; human capital 210; relation to innovation 208–9, 211–13; social capital 2, 210
'founder' entrepreneurs 225
Freescale 41
fuzziness 111

G
gatekeeper role 88–9, 292
GDP: Ghana 150; India 222; Indonesia 161
Ghana 146–60; development planning 152–4; development thinking/practice 147–8; economic growth and decline 150–2; GDP 150; Kwame Nkrumah University of Science and Technology (KNUST) 155; Ministry of Environment, Science and Technology (MEST) 153; National Council on Tertiary Education 152; R&D 154–5; R&D funding 156–7; science and technology 152–4, 156–7; Sir Gordon Guggisberg Plan 152; socio-economic development mix 157–8; Structural Adjustment Program 150–1; Teaching and Learning Innovation Fund 156; triple helix model 148–50; universities 148–9, 155–6; University of Development Studies 155
Ghana Academy of Arts and Science 152
Ghana Atomic Energy Commission 152
Ghana Education Trust Fund (GETFund) 156
Ghana Standards Boards 152
global connectivity 11
global innovation networks 15–17
global opportunity 10
globalization, role of language in 219–20
government policies 1, 15, 68; CEE 265–6; Indonesia 168; Malaysia 203–4
government research institutes 119, 120
Green Revolution 98, 99, 100
gross domestic product *see* GDP
groundnut fodder 104–5
Guinness Ghana Limited 157

H

Hewlett-Packard 36, 38, 41
high-yielding hybrid crops 98
highly indebted low-income countries 151; *see also* developing countries
HIPC *see* highly indebted low-income countries
hôjinka 286
human capital 210; skilled labour 215–16
human resources 201
humanitarian ventures 15–16
Hungarian Automotive Cluster 271
Hungary: clusters and regional development 270, 272; cooperation 267; FDI 261; innovation activities 263; innovation performance indicators 257; innovation/entrepreneurship indicators 264; science-industry links 269; university graduates 257; university-based researchers 259
hybrid organizations 266–9
hybrid triple helix model 194–5, 255
Hyundai 29, 30–5; assimilation of assembly technology 30–1; deepening of design technology 31–2; initial design technology 31; international R&D network 33; own design technology 32–4; R&D investment 34; synthesis and analysis 34–5

I

IBM 36, 38, 41
incubators 176–87; cooperative 183; growth of 184; Malaysia 237; Mexico 176–87; models of 184–5; traditional 182–3; types of 182–4; *see also* science parks
India 207–27; constitution 220; emigration to USA 221–2; forms of capital 208–11; GDP 222; innovation 207–9; language 219–20; migration of skilled labour 221; outsourcing 217, 223; self-sufficiency 220–1; service exports 218–19; service industry evolution 219–23; US affiliations 225–6
Indian Institute of Technologies 221
indigenous suppliers, training of 44–5
Indonesia 161–74; active clusters 170; analysis for improvement 171–3; artisan clusters 170; Bandung Institution of Technology 171; Broad Outlines of Government Policy 168; clusters 163–5, 168–71; dynamic clusters 170; GDP 161; Institut Teknologi Bandung 171; *pribumi* 169; Pusat Inkubator Bisnis 171; research/study 166–7; small and medium enterprises 163–5, 168–71; TRI DHARMA PERGURUAN TINGGI program 166; triple elements of higher education 165–8; triple helix model 162–3; universities 171–3; University of Gajah Mada 171; University of Indonesia 171
industrial internationalization 28–9; passive 27
industrial parks 232; *see also* science parks
industry 1; CEE 260–5; *see also individual countries and companies*
industry-university links 230–1
information 112; brokering 217; equivocality 292; sharing 15–16
information and communications technologies 17
infrastructure 14
innovating firms: fostering of 275–6; ownership of 196
innovation 10, 19; capital mix 211–13; CEE 263; changing landscape 10–12; collaborative 13–15, 20; as driver of transition 250–2; global networks 15–17; India 207–8; links with production 28; open model 12; system perspective 213–15; and technological advance 10–11; triple helix relationships 12–13, 67–9
innovation clusters 17
innovation indicators, CEE 264
Innovation Relay Centres 268
innovation systems 99, 109; Mexico 37–8; science 99–102
innovators 2
Intel 36, 41
intellectual property 59–60, 120; decentralization of 162–3; Malaysia 203

inter-organizational collaborative research 292–3
inter-organizational learning 67–83
interactions 149
interactive bottom-up approach 106
interactive model 253
interdisciplinarity 12–13; barriers to 13
International Association of Outsourcing Professionals (IAOP) 224
International Associations of Science Parks 231
International Development Association 156
International Laboratory for Research on Animal Diseases 103
International Monetary Fund 150, 151
internationalization: active 27; passive 27
Internet 17, 212
interoperability 16
intra-organizational learning 70–1
Invest Penang 199

J
Jabil 41
Jalisco electronics cluster 36
Japan 283–301; clusters 283–4; collaborative projects 287; *haji no bunka* 299; Hiranuma Plan 286, 295; *hôjinka* 286; *jutaku kenkyû* 287; *kaizen* 284; *kotonakare shugi* 299; *kyôdô kenkyû* 287; national innovation system 284; National University Law 290; 'not invented here' syndrome 288; Patent Law 290; *sangakukan renkei* 283; *Sangakureneki Honbu* 296; *sangakurenkei* 283, 284, 287, 290; Science and Technology Basic Plan 294; *shôgakukin* 287; small and medium enterprises 294–5; Technology Advanced Metropolitan Area project 295; TLO Law 285, 290–1, 295; triple helix evolution 290–300; Triple Helix II 290–8; triple helix model 284–90; triple helix performance 288; triple helix problems 298–300; university venture firms 295–7; *yokonarabi* 299
Japan International Co-operation Agency 171
Japan Vocational Ability Development Association (JAVADA) 168
JEREMIE initiative 276
joint ventures 67; humanitarian 15–16; Tunisia-Egypt 76, 77–80; Tunisia-Libya 76, 80–4; *see also* strategic alliances
jutaku kenkyû 287

K
kaizen 284
knowledge acquisition 72, 74; Japan 284
knowledge clusters 17–18
knowledge flow 101
knowledge infrastructure 122
knowledge sharing 2
knowledge transfer 89
knowledge-based economy 146; features of 232; Malaysia 234–5
Kodak 36
Korea 29–35, 119–43; catch-up process 29; *chaebols* 25, 35, 123; collaborative tendencies 123, 124; Hyundai 30–5; patents 127, 129; personal ties 136; repartnering 134; triple helix presence 35
Korean National Innovation System 29
KORTAI R&D data set 120, 121, 123, 125
Kulim Hi-Tech Park 237, 238
Kwame Nkrumah University of Science and Technology (KNUST) 155
kyôdô kenkyû 287

L
laissez-faire model 178, 194, 253
language, role in globalization 219–20
large enterprises 161
Latvia: cooperation 267; FDI 261; GDP 250; innovation activities 263; innovation performance indicators 257; innovation/entrepreneurship indicators 264; science-industry links 268; university graduates 257; university-based researchers 259
learning: acquisition of 72; with alliance partners 73–4; inter-organizational 67–83; interdependence 83; intra-organizational 70–1; organizational 69–71; strategic alliances 71–2
learning by doing 74–5

Lewis, Sir Arthur 150
Libya, joint venture with Tunisia (STL) 76, 80–3
licensed technology: assimilation of 31–2; technology licensing officers 283
linkages 149
Linux 212
literature review 121–2
Lithuania: cooperation 267; FDI 261; innovation activities 263; innovation performance indicators 257; innovation/entrepreneurship indicators 264; science-industry links 269; university graduates 257; university-based researchers 259
livestock: diseases 102–4; fodder for 104–5
local authorities 161
local value added 45
Lucent Technologies 36, 38
Lundvall, Bengt-Åke, *National Systems of Innovation* 214

M

macro-innovation 215–16
Malaysia 190–205; Action Plan for Industrial Technology Development 235; commercialization issues 201–2; data analysis 198–9; Eighth Malaysia Plan 190, 195; Engineering Innovation and Technology Development 235–6; evolution of triple helix 193–5, 200–4; FDI 234; government policies 203–4; incubators 237; intellectual property 203; intensification of research in priority areas (IRPA) 195; Invest Penang 199; as knowledge-based economy 234–5; main stakeholders 202; Medical Innovation and Technology Development Unit 236; Ministry of Science, Technology and Innovation (MOSTI) 198, 241; Multimedia Super Corridor 237; national innovation system 234; Ninth Malaysia Plan 196; ownership of innovating firms 196; policies, procedures and processes 201; science parks 237–8; technological and human resources 201; tertiary education 236; TPM Academy 242; triple helix model 193, 195–7, 199–200; universities 191, 197, 235–7; university-industry collaboration 230–44; USains Holdings 196, 235; work cultures 203
Malaysian Science and Technology Information Centre (MASTIC) 195, 201, 238
Malaysian Technology Development Corporation 238
market failure 112
mental models 73
Mexican National Innovation System 35
Mexican-United States Foundation for Science (FUMEC) 177
Mexico 35–45; Advanced Program in Electronics Design and Software (PADTS) 40–1; CINVESTAV 38–9, 40; FDI 36–7; Fund for the Small and Medium Business 182; growth of incubators 184; Incubator of Base Technology Enterprise Program 177; incubators 176–87; innovation systems 37–8; Integral Program for the Development of Higher Education (PROIEDES) 37; Jalisco electronics cluster 36, 39; local value added 45; MNCs in 39; National Association of Universities and Higher Education Institutions (ANUIES) 37; National Chamber of Electronics and Telecommunications Industry (CANIETI) 42; National Program for Educational Development 37; political, scientific and social context 181–2; Productive Chain of Electronics Industry Suppliers (CADELEC) 42; Program for the Creation of Technological Entrepreneurship 41; Program for the Promotion of the Computer Industry 36; Semiconductors Research Centre 40; State Council of Science and Technology (COECYTJAL) 40, 42; technological learning process 41–2; technology spillovers 42–5; technology transfer 37–8, 41; training of employees 43; triple helix presence 39–40

micro-innovation 215–16
Millennium Development Goals 9, 14; contribution of science/ technology 15
Mitsubishi 31–2
MNCs *see* multinational corporations
Motorola 36
Multimedia Development Corporation (MDC) 238
multinational corporations 27, 198; employee training in host countries 43

N

national innovation systems 215; Japan 284; Korea 29; Malaysia 234; Mexico 35
NEC 36
needs assessment survey 106
networking 138
new capital: and collaborative output 131, 133; and non-collaborative output 132, 133
newly industrializing countries 101
non-collaborative output, effects of new/pre-existing capital 132
non-governmental organizations 15, 100, 107; *see also individual organizations*
North American Free Trade Agreement 36, 181
North-North cooperation 69
North-South cooperation 69
'not invented here' syndrome 288

O

open innovation model 12
organizational learning 69–71
original equipment manufacturers (OEMs) 36
outsourcing in India 217, 223; social capital of 223–7
own design technology 32–4

P

participatory rural appraisal 104
participatory technology development 106
passive internationalization 27
patents 75, 126; Brazil 59–60; Japan 290; Korea 127, 129; Taiwan 128, 129
PAXIS initiative 275
personal ties 126; Korea 136; Taiwan 136
PINTEC innovation surveys 57–8
Poland: clusters and regional development 270, 271; cooperation 267; FDI 261; innovation activities 263; innovation performance indicators 257; innovation/ entrepreneurship indicators 264; science-industry links 269; university graduates 257; university-based researchers 259
policy research 112
Polman Astra 168
pre-existing capital: and collaborative output 131, 133; and non-collaborative output 132, 133
predatory-prey relationship 110–11
pribumi 169
private education providers 256
private sector 15, 120
production, links with innovation 28
public-private collaborations 125, 132
Pusat Inkubator Bisnis 171

R

R&D 11, 33; agricultural science 100; CEE 251, 265–6; collaboration 123–4; cross-sectoral 11; Ghana 154–5
rapid rural appraisal 104
regional development, CEE 269–73
regulatory environment, Brazil 52–3, 54, 61–3
relational capital 3
repartnering 126; Korea 134; reasons for 134, 135; Taiwan 135
research: blue sky 231; Brazil 54–7; learning by 32–4; university-based 258–60
research and development *see* R&D
research managers 87–97; challenges faced by 93–4; developing countries 94–5; expertise 89; gatekeeper role 88–9; job description 92–3; origins of 89–91; skills 92; and triple helix model 95–7
reverse engineering 43–4, 122, 251
Ricardo Engineering Company 33
Romania: clusters and regional development 272–3; cooperation 267; FDI 261; GDP 250; innovation activities 263; innovation performance indicators 257;

innovation/entrepreneurship indicators 264; science-industry links 269; university graduates 257; university-based researchers 259
rules of the game 214

S
Samsung 29
sangakukan renkei 283
Sangakureneki Honbu 296
sangakurenkei 283, 284, 287, 290
Schumpeterian thinking 2, 211; creative destruction 249
Science Council of the Consultative Group for International Agricultural Research 98
science and innovation systems 99–102
science parks 59, 230, 231–4; Brazil 180; Malaysia 237–8; off-park firms 240; as resource network 233; university links 240; vs industrial parks 232; *see also* incubators
science and technology 119; Brazil 55, 57; Ghana 152–4, 156–7
scientific discovery: mode 1 paradigm 99, 109, 113; mode 2 paradigm 99–100, 109, 113
Selangor Science Park 237
semi-knock-down system 30
shôgakukin 287
Siemens 36, 41
Silicon Valley 17
Sir Gordon Guggisberg Plan 152
skilled labour 215–16; emigration of 215–18; migration from India 221
Slovakia: clusters and regional development 270; cooperation 267; FDI 261; innovation activities 263; innovation performance indicators 257; innovation/entrepreneurship indicators 264; science-industry links 269; university graduates 257; university-based researchers 259
Slovenia: clusters and regional development 270, 272; cooperation 267; FDI 261; innovation activities 263; innovation performance indicators 257; innovation/entrepreneurship indicators 264; science-industry links 269; university graduates 257; university-based researchers 259
Slovenian Transport Logistics Cluster 271
small and medium enterprises 123, 161; advantages of cluster approach 171–3; CEE 264–5; features of 232; Indonesia 163–5, 168–71; Japan 294–5; perception of universities 202; productivity and efficiency 165
Small and Medium Industries Development Corporation (SMIDEC) 239
SMEs *see* small and medium enterprises
Smith, Adam 209
social capital 2, 210; and outsourcing 223–7
Society for Research Administrators International 89
socio-economic development 157–8
South Korea *see* Korea
South-South cooperation 75–83; STEFM 76, 77–80; STL 76, 80–3
Southern and West African Research and Innovation Management Associations 90
specialized suppliers 36
stakeholders 202
statist model 253
status 210–11
STEFM 76, 77–80; distribution of employees 77; financial turnover 78
STL 76, 80–3; employees 81
strategic alliances *see* alliances
structural holes 217
sub-Saharan Africa *see* Africa
substitution of technologies 28
suppliers, training of 44–5
symbiosis 110
system perspective of innovation 213–15

T
Taiwan 119–43; collaborative tendencies 124, 125; patents 128, 129; personal ties 136; repartnering in 135
teaching function of universities 256–8
technological change 10–11
Technological Innovation Nuclei 186
technological learning process 41–2
technology: development-gap 290–1; endogenous 56; foreign,

acquisition of 31; licensed, assimilation of 31–2; own design 32–4; *see also* science and technology
technology licensing officers 283
technology linkages 101
Technology Park Malaysia 237, 238, 242
technology parks *see* science parks
technology spillovers 42–5
technology transfer 2, 13–15, 59, 231, 242; Brazil 59; developing countries 14; Mexico 37–8, 41
Technovation Park 237
Telmex 36
Toyota Indonesia 167–8
traditional incubators 182–3
traditional research collaboration 293
training 19, 33–4, 43; of indigenous suppliers 44–5; *see also* learning
transformative networks 19–20; potential of 19–20
transition 255–66; innovation as driver of 250–2
transnational social circulation 216–18
TRI DHARMA PERGURUAN TINGGI program: community service-research/study 167–8; education-community service 167; education-research/study 166–7
trilateral partnerships 16–17, 19; transformational capacity 19
triple helix model 178–9, 191–3, 252–5; advantages of 9–10; definitions 1, 68, 214–15; and development 148–50; evolution of 193–5, 200–4, 291; Japan 290–300; Ghana 148–50; hybrid 194–5, 255; Indonesia 162–3; innovation under 12–13, 67–9; interactive 253; Japan 284–90; *laissez-faire* 178, 194, 253; Malaysia 193, 195–7, 199–200; statist 253; USA 286
Triple Helix I 283, 286–7
Triple Helix II 283; Japan 290–8; upgrading to 289
Triple Helix III 283–301
Trypanosoma congolese 102
Trypanosoma vivax 102
trypanosomiasis 102–4; control of 102–3; impact on livestock 102; vaccine research 103
Tunisia: ALPHA companies 67, 75–83; joint venture with Egypt (STEFM) 76, 77–80; joint venture with Libya (STL) 76, 80–3

U

UK, triple helix performance 288
Unilever 157
United Kingdom Science Parks Association 231
United Nations: Brahimi Report 15; Millennium Development Goals *see* Millennium Development goals
United States-Mexico Foundation for Science (FUMEC) 42
universities 1, 68, 192; Brazil 52, 54–7, 58–63; Central and Eastern Europe 255–60; commercialization issues 201–2; entrepreneurial 191, 254, 275; Ghana 148–9, 155–6; Indonesia 165–8, 171–3; initiatives from 58–61; Malaysia 191, 197, 235–7; perceptions of 202; policies, procedures and processes 201; research function 258–60; research managers 87–97; science park links 240; teaching function 256–8; technology transfer *see* technology transfer; work cultures 203
university-industry links 230–1
UPM-MTDC Incubation Centre 237
USA: Bayh-Dole Act 285; immigration from India 221–2; Indian affiliations 225–6; Stevenson-Wydler Technology Innovation Act 285; triple helix model 286; triple helix performance 288
USains Holdings 196, 235

V

venture capital 276
vicious circle phenomenon 46
virtual spaces 12

W

Weber, Max, *The Protestant Ethic and the Spirit of Capitalism* 209
welfare, targeting of 15
Wikipedia 212
work cultures 203
World Bank 147–8, 150, 171; HIPC initiative 151–2